Developing Mathematics through Applications: Intermediate

Instructor Resources

Preliminary Edition

Developed by
COMAP, Inc.

Project Leadership

Solomon Garfunkel
COMAP, Inc., Lexington, MA
Susan Forman
Bronx Community College, CUNY, Bronx, NY

Authors

Nancy Crisler
COMAP, Inc., Lexington, MA
Gary Simundza
Wentworth Institute of Technology, Boston, MA

Key College Publishing
Innovators in Higher Education

The Consortium for Mathematics and Its Applications (COMAP)
Lexington, MA

This book was developed with the support of NSF Grant DUE 9950036. However, any opinions, findings, conclusions, and/or recommendations herein are those of the authors and do not necessarily reflect the views of NSF.

Key College Publishing was founded in 1999 as a division of Key Curriculum Press® in cooperation with Springer-Verlag New York, Inc. We publish innovative texts and courseware for the undergraduate curriculum in mathematics and statistics as well as mathematics and statistics education. For more information, visit us at www.keycollege.com.

Key College Publishing
1150 65th Street
Emeryville, CA 94608
(510) 595-7000
info@keycollege.com
www.keycollege.com
© 2003 by Key College Publishing. All rights reserved.
Published by Key College Publishing, an imprint of Key Curriculum Press.

Key College Publishing and
Key Curriculum Press:
DEVELOPMENT EDITORS: Jacqueline Meijer-Irons, Allyndreth Cassidy
PRODUCTION PROJECT MANAGER: Michele Julien
EDITORIAL PROOFREADER: Linda Ward
PRODUCTION DIRECTOR: Diana Jean Parks
COVER DESIGNER: Design Deluxe
COVER ILLUSTRATOR: Jeff Brice
PRINTER: Data Reproductions Corporation
EXECUTIVE EDITOR: Richard Bonacci
GENERAL MANAGER: Mike Simpson
PUBLISHER: Steven Rasmussen

COMAP:
PROJECT EDITOR: Pauline Wright
PRODUCTION EDITOR: Tim McLean
PRODUCTION DIRECTOR: George Ward
PRODUCTION: Daiva Kiliulis
PHOTO RESEARCH: Lynn Aro

Printed in the United States of America
10 9 8 7 6 5 4 3 2 1   06 05 04 03 02
ISBN 1-931914-06-0

Project Leadership

Solomon Garfunkel, COMAP, Inc., Lexington, MA
Susan Forman, Bronx Community College, CUNY, Bronx, NY

Advisory Board Members

Philip Cheifetz, Nassau Community College, Garden City, NY
William Haver, Virginia Commonwealth University, Richmond, VA
Robert L. Kimball, Jr., Wake Technical College, Raleigh, NC
Karen Larsen, National Alliance of Business, Washington, DC
David R. Mandel, MPR Center for Curriculum and Professional Development, Washington, DC
Pamela Matthews, National Council of Teachers of Mathematics, Reston, VA
Marilyn Mays, North Lake College, Irving, TX
Henry Pollak, Teachers College, Columbia University, New York, NY
William Thomas, University of Toledo, Toledo, OH

Field Testers

Susan Forman
Bronx Community College, CUNY, Bronx, NY

Pei Taverner
Frederick Community College

Jeffrey Morford, Larry Smyrski, Deborah Zopf
Henry Ford Community College, Dearborn, MI

Julane Crabtree
Johnson County Community College

Susan Wood
J.S. Reynolds Community College

Jan Roy
Montcalm Community College

Tim Chappell
Penn Valley Community College

Janet Ray
Seattle Central Community College

Reuben Farley, William Haver,
Gwen Turbeville, Kathryn Wallo
Virginia Commonwealth University, Richmond, VA

Focus Group Participants

Susan Enyart
Otterbien College

Nancy Johnson
Manatee Community College

Michelle Merriweather
Southern Connecticut State University

Charles Patterson
Louisiana Tech University

Don Ransford
Edison Community College

Rochelle Robert
Nassau Community College

Sharon Siegel
Sam Houston State University

About the Authors

Nancy Crisler, COMAP, Inc.

Ms. Crisler was a high school mathematics teacher and for ten years served as the K–12 Mathematics Supervisor for the Pattonville School District in St. Louis County, Missouri. She has taught at Washington University in St. Louis, Missouri. She is the co-author of the COMAP texts *Discrete Mathematics Through Applications* 1st and 2nd editions, a member of the writing team for *Mathematics: Modeling Our World Courses 1–4*, 1st and 2nd editions, and a co-author of *Precalculus: Modeling Our World* (2002) and *College Algebra: Modeling Our World* (2002), all published by W.H. Freeman.

Gary Simundza, Wentworth Institute of Technology

Mr. Simundza is Professor of Mathematics at Wentworth Institute of Technology in Boston, Massachusetts, where he has taught mathematics and physics for over 25 years. As director of the NSF-supported "Mathematics in Technology" project, he led an interdisciplinary team of faculty in creating a set of extensive laboratory activities for college mathematics. These were published by Prentice-Hall as *Precalculus Investigations: A Laboratory Manual* (1999).

Contents

Preface   vii

To the Instructor   ix

To the Student   xxii

Introduction to Book 2   xiv

Instructor Notes

    Chapter 5   I1

    Chapter 6   I17

    Chapter 7   I30

    Chapter 8   I45

Chapter 5   Quadratic Functions and Radicals   1

    5.1 Parabolas   4

    5.2 Quadratic Functions and Their Graphs   10

    5.3 Modeling Data with Quadratic Functions   49

    5.4 Roots and Radicals   89

    5.5 Fractional Exponents and Radical Equations   113

    5.6 Distance in the Plane; Circles   139

    Chapter 5 Review   162

Chapter 6   Rational Expressions and Systems of Equations   171

    6.1 Introduction to Rational Functions   173

    6.2 Modeling with Rational Functions   178

    6.3 Multiplying and Dividing Rational Expressions   213

    6.4 Solving Rational Equations   234

    6.5 Systems of Equations   261

    Chapter 6 Review   294

Chapter 7   Probability   299

   7.1 The Meaning of Probability   302

   7.2 Probabilities for Compound Events   308

   7.3 Finding Probabilities from Data   332

   7.4 Combining Probabilities   345

   7.5 Probability Distributions   367

   7.6 Drawing Conclusions from Data   399

   7.7 Modeling Through Simulations   425

   Chapter 7 Review   442

Chapter 8   Trigonometry   449

   8.1 Similar Triangles   451

   8.2 Properties of Triangles   461

   8.3 Trigonometric Ratios   485

   8.4 Modeling with Right Triangles   508

   8.5 Trigonometric Ratios for Non-Acute Angles   529

   Chapter 8 Review   548

Handouts   556

Blackline Master 8.1   572

Appendix   573

Index to the Student Edition   577

Preface

Developing Mathematics through Applications: Intermediate continues to bring you the application-driven, context-rich, activity-based curriculum that started with *Developing Mathematics through Applications: Elementary*. This second volume completes a one year sequence in developmental mathematics.

Since the 1980s the "typical" college student profile has been changing as more non-traditional students have started their higher education later in life. Some students are entering their post-secondary education after a period of time in the work force, others after starting a family. Often students who have been away from school for extended lengths of time need to refresh and rebuild their mathematical skills and their confidence in their ability to solve challenging problems. Likewise, traditional students need to hone their mathematical skills to enter today's competitive job market. Developmental mathematics courses are offered as the first step in student success in mathematics-related programs.

Students in developmental mathematics courses are particularly in need of course materials that are written at an appropriate reading level and capture adults' interests. It is important to keep in mind that these students often bring with them life experiences, particularly from the workplace, that can be used to connect the mathematics taught in the classroom to their worlds. Yet many developmental mathematics programs simply present concepts and ask students to replicate the results while working through voluminous end-of-section exercises.

We now have an opportunity for real and positive change. Both *Crossroads in Mathematics: Standards for Introductory College Mathematics Before Calculus*, published by the American Mathematical Association of Two-Year Colleges (AMATYC) in 1995, and the National Council of Teachers of Mathematics (NCTM) *Principles and Standards for School Mathematics*, revised in 2000, advocate an integrated approach to mathematics content. They call for an increased emphasis on data analysis, the development of meaningful contemporary applications, and the use of appropriate technologies, as well as activity-based and collaborative learning.

Given students' needs in developmental mathematics, the vision of the AMATYC and NCTM Standards, and COMAP's experience, we have created a one-year sequence, *Developing Mathematics through Applications: Elementary* and *Intermediate*, which takes a different approach from traditional developmental mathematics textbooks.

- The rich applications are equally appropriate for students coming straight from high school with plans to enter technical fields, and those with work experience.

- The approach integrates topics from probability, data analysis, and geometry with the traditional developmental mathematics curriculum, making it a suitable foundation for students in diverse career paths, including engineering and the sciences as well as non-technical occupations.

- The context-rich, activity-based curriculum motivates critical thinking about fundamental concepts while building the analytical and graphical skills needed for further study of mathematics.

- Solving the problems posed in these books calls for integrating technology in a more natural manner then the traditional "drill-and-practice" use of technology commonly employed.

Many people and organizations have been helpful during the creation of this book. We would like to extend a special thank you to Professors James Sandefur of Georgetown University and Bobby Righi of Seattle Central Community College for their ideas and insights during the early development of this book. Thanks, too, to Larry Smyrski for his suggestions for refinement of the manuscript.

We would also like to acknowledge and thank the following:

Professors Peter Rourke, Larry Decker, Marty Kemen, and Philip Comeau, all of Wentworth Institute of Technology; David Joliat, Project Manager with Thoughtforms Corporation; and the people at Zymark Corporation and the USDA Forest Service.

Our sincere gratitude goes to Sue Martin of COMAP. She has always been there when we needed her. And, as always, we wish to thank the members of the COMAP staff. Without their help, this book would not have been possible.

Sol Garfunkel
CO-PRINCIPAL INVESTIGATOR

Susan Forman
CO-PRINCIPAL INVESTIGATOR

Nancy Crisler
AUTHOR

Gary Simundza
AUTHOR

To the Instructor

This book is a different kind of developmental mathematics text. We believe that any mathematics worth learning is best learned in the context of its use by real people in real jobs. Without neglecting necessary practice with fundamental skills, we emphasize connections between mathematics and the workplace, both in the body of the text and in the exercises. Interesting applications motivate students to learn, and guided discovery helps them succeed in their learning by developing habits of persistence necessary for achievement.

The AMATYC Standards have been thoroughly embraced by this book: Active investigation of mathematics is emphasized throughout with the expectation that students will conduct much of this investigation in small, collaborative groups. The content integrates data analysis, measurement, and geometry with traditional topics in elementary and intermediate algebra. A modeling theme runs throughout the text, as students learn to construct a variety of models, using equations, graphs, tables, arrow diagrams, narrative descriptions, geometry, and statistics. Access to computing technology is assumed, although the text can be used successfully without it.

In addition to competence with mathematical skills and concepts, this text intends to provide familiarity with the "new basic skills" demanded by the modern workplace, as reflected in such documents as the SCANS Report and various Industry Skills Standards. These include critical thinking, team-building, interdisciplinary connections, and communications skills.

Unfortunately reading is frequently not a strength of developmental mathematics students. Our philosophy is that students should be encouraged to improve their reading skills as they develop mathematical facility. Many of the contextual situations in the text require students to read for mathematical content. Indeed, an essential part of acquiring the ability to model the world with mathematics is learning to see how mathematics is embedded in various situations. We hope students will be encouraged to increase their reading ability as they see that careful reading helps them achieve mathematics competence. To this end, vocabulary is kept accessible and unfamiliar concepts are clearly explained in terms that students are able to understand.

Some topics that are not always considered part of developmental mathematics are introduced as they arise naturally in context. For example, asymptotes are introduced in Chapter 6 in connection with the graphs of rational functions. Confidence intervals are examined in Chapter 7, based on empirical data and without use of the standard deviation. Basic trigonometry of non-acute angles and an introduction to oblique triangles are included in Chapter 8. Topics such as these provide the underpinnings for more advanced mathematical study and help lay the groundwork for more in-depth work in precalculus and other courses.

Organization

Each chapter is divided into four to six sections sharing similar pedagogical elements.

- **Goals of the chapter:** The first element of every chapter is a list of goals to be accomplished by learning the material presented.

- **Preparation Reading:** Each chapter begins with a brief introduction that sets the stage for the mathematics to come, and may include contextual connections or discussions of modeling aspects.

- **Chapter Review:** This section summarizes the key concepts of the chapter and contains a sampling of exercises from each section.

- **Glossary:** Each chapter concludes with a summary of the new terms introduced in the chapter.

The various sections of each chapter serve different purposes. The sections build on each other, incorporating the following tools.

- **What You Need to Know** and **What You Will Learn:** Each section starts with two lists. The first states the necessary prerequisite knowledge and the second defines the mathematical objectives to be learned.

- **Activities:** The first section of each chapter contains an *Activity* that requires students to work in groups and perform an experiment or otherwise collect data. They are then guided through the exploration of a core mathematical concept that will recur throughout the chapter.

- **Discoveries:** Later sections frequently contain explorations that are similar to *Activities* in that they are intended to be done in groups. A *Discovery* is usually a more focused examination of a single mathematical topic and does not necessarily involve active experimentation. If data are needed, they are often provided rather than collected.

- **Examples:** Numerous examples with detailed solutions are provided in the body of each section.

- **Exercises:** The exercise sets consist of three different types of problems. *Investigations* should be considered as adjuncts to the body of each section and may present new material. Some are in-depth examinations of particular subtopics or applications, and some are guided explorations similar to *Discoveries*. *Investigations* provide an opportunity for instructors to customize the text for their own purposes and students' needs. *Projects* and *Group Activities* are similar to the *Activities* that begin each chapter. They require students to work cooperatively and often involve research or experimentation of some sort. *Additional Practice* problems provide necessary skill practice, including some contextual problems. They are not necessarily in order from easy to difficult, partly so that students will not quit after trying a few easy problems.

Detailed solutions and answers to all questions asked in the *Activities, Discoveries,* and *Exercises* are provided embedded within the text in this Instructor Resources book. At the end of the book, an index is provided. The page numbers on the index correspond to pages in the Student Edition of the text.

Instructor Notes

Detailed instructional suggestions are provided for each chapter. These are intended to help instructors make choices regarding pacing of the course and exercises to assign. Many of the features in the Instructor Notes section are provided for ease of use and convenience. For example, each Instructor Notes section starts by restating the goals of the chapter. Next, the list of objectives to be learned and the list of prerequisites are provided. The bulk of the Instructor Notes is comprehensive material about the teaching of each section, such as amount of time required, and hints and suggestions for presenting the material.

In order to help instructors choose *Investigations* that are most appropriate for their students, the symbol ⇒ is used to indicate any exercise that contains instruction on a topic not otherwise appearing in the body of a section.

Exercises that require, or explain the use of, computing technology for the solution of mathematical problems are indicated with a □ symbol. No particular form of technology is required. Since many students own, or have access to, graphing calculators, calculator screen shots appear throughout the text. (TI-83 screens are shown only because this type of calculator is common, although the particular uses discussed are meant to be generic.) Computer graphing programs such as *Fathom Dynamic Statistics*™, which also provides data analysis capability, can be used, and spreadsheets and geometric drawing utilities such as *The Geometer's Sketchpad*® may also be helpful.

Nancy Crisler   Gary Simundza
AUTHOR   AUTHOR

To the Student

(This material appears in the Student Edition and is repeated for your convenience.)

This book is different from mathematics textbooks you have used before. We believe that mathematics should be learned by seeing how it is used day to day. The reason for learning the mathematics presented here is not simply to prepare you for the next math course but to prepare you for your next job and for using mathematics to solve problems that arise in other areas of your life. Yes, the mathematics that you learn will help prepare you for further schoolwork. But it is also important as a life skill, something you will use and value.

This book encourages active exploration of mathematics. You will learn to construct a variety of mathematical explanations of real problems. To do this, you will use equations, graphs, tables, verbal descriptions, geometry, and statistics. Here are a few tips that will help you use this book successfully:

- Much of the mathematical exploration should occur in small groups. Every chapter of the book includes *Activities* or *Discoveries*, which will most likely be done during class time. Discussions with other students will help you grasp new mathematical concepts.

- You will have to read carefully in order to understand the connections between mathematics and the workplace. This is especially important in some of the exercises.

- Many of the *Activities*, *Discoveries*, and *Exercises* require written responses. Try to write clearly and in complete sentences in order to show your understanding of the mathematics.

- When appropriate, use a graphing calculator or computer.

- Each section begins with *What You Need to Know*. This is a list of prerequisite skills that should be familiar from earlier mathematics courses.

- *What You Will Learn* lists the new mathematics you will study in the section.

- There are three types of exercises at the end of most sections:
    - *Investigations* are "thinking" exercises that will test your understanding and challenge you to apply the mathematics you have just learned. Some investigations will present new mathematics topics that build on the work of the section.
    - *Projects* and *Group Activities* involve experiments or research, many of which are intended to be done in small groups.
    - *Additional Practice* exercises allow you to master the computational skills introduced in the section.

In order to use mathematics to describe the world around you, you need to see how mathematics is part of that world. Our goal is that, by the end of the text, you will be able to recognize certain mathematical principles when you are outside the classroom. We hope that you enjoy using this text to develop your mathematical skills and understanding as much as we enjoyed writing it.

Nancy Crisler
AUTHOR

Gary Simundza
AUTHOR

Introduction to Book 2

Modeling

In an attempt to explain why things happen the way they do or to make predictions about the future, people sometimes create **mathematical models.** A mathematical model can take the form of an equation, a table, a graph, or sometimes combinations of these.

When building models, people often use accumulated knowledge to help them find a relationship between variables. After creating the model, they gather data to check the accuracy of their conjecture.

There are times when it is not possible for the modeler to explain a certain behavior, but predictions are needed anyway. For example, we may not be able to know the exact number of students who will be attending a particular college, but the college needs to make estimates each year in order to plan for faculty, facilities, and materials such as books. In such cases, data are collected and examined. If a pattern in the data is observed, an attempt is made to capture the trend of the pattern with a model. In turn, the resulting model may give "hints" as to why the variables are related as they are.

Thus, data are collected and examined for at least two distinct purposes. One is to check the accuracy of a proposed model. The other is to look for a pattern in collected data in hopes of creating a model based on the pattern. In the first instance, the model is **"theory-driven"** and in the second, the model is **"data-driven."**

$$\text{Theory} \rightarrow \text{Model} \rightarrow \text{Data}$$

$$\text{Data} \quad \rightarrow \quad \text{Model} \rightarrow \text{Theory}$$

Both theory-driven and data-driven models are useful. Each has its advantages and disadvantages.

Taking a closer look at the modeling process itself will help us as we continue in our quest for good mathematical models.

The Modeling Process

Whether a model is driven by data or theory, the process of modeling can be summarized in the following steps:

Step 1. *Identify the Problem:* What is it you would like to do or find out? Make some general observations. Pose a well-defined question asking exactly what you wish to know.

Step 2. *Simplify and Make Assumptions:* Identify the factors that will be used in building a model. Generally you must simplify to get a manageable set of factors.

Step 3. *Build the Model:* Interpret in mathematical terms the features and relationships you have chosen. Your resulting model may be a graph, a table, or an equation. Analyze the model to find answers to the questions originally posed.

Step 4. *Evaluate, Revise, and Interpret:* Your conclusions at this point apply to your mathematical model. Verify your conclusions by collecting data. Does your model yield results that are accurate enough? If not, refine the model by reexamining your assumptions. Based upon the accuracy of your model, relate the mathematical conclusions to the real world.

The following diagram is useful in describing the modeling process:

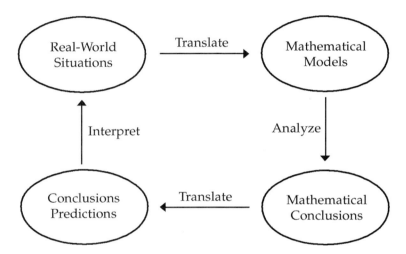

The process begins with a real-world situation. But after a model is built, analyzed, and tested, it is often necessary to revise the model in order to better explain the problem. Therefore, more than one trip around the modeling diagram may be needed in order to find a model that adequately describes the situation. The real-world situation can provide feedback in the modeling process to help us refine our model.

In this text, we will examine a variety of models, both theory-driven and data-driven, for describing real problem situations that arise in the course of people's lives and work. We will sometimes find that a problem has no single "right" answer, but that one model works better than others at describing the problem. A willingness to explore a variety of possibilities is often one of the most important qualities of a good mathematical modeler.

Instructor Notes

Chapter 5: Quadratic Functions and Radicals

Goals of the Chapter

- To solve quadratic equations using analytical and graphical methods
- To model real phenomena with quadratic functions
- To simplify radicals and solve radical equations
- To use fractional exponents to represent radicals

What You Will Learn

Section 5.1

- To describe parabolic shapes with quadratic functions

Section 5.2

- To identify the properties of a quadratic functions
- To sketch the graph of a quadratic function
- To find the vertex of the graph of a quadratic function
- To find the equation of a parabola when the axis of symmetry is the y-axis
- To find the real solutions to quadratic equations using factoring, graphing, and the quadratic formula
- To interpret the meaning of vertices and horizontal intercepts of real-world quadratic models

Section 5.3

- To use quadratic regression to construct models of data
- To choose between linear and quadratic models for data

Section 5.4

- To evaluate square roots and nth roots of numbers
- To simplify radical expressions
- To apply the product and quotient property of radicals
- To add and subtract radical expressions

Section 5.5

- To express radicals using exponent notation
- To simplify expressions containing fractional exponents
- To solve equations that contain radicals

Section 5.6

- To find the distance between two points in a plane
- To find the equation of a circle
- To identify a circle from its equation

What You Need to Know

Section 5.1

- How to identify coordinates of points on a rectangular coordinate system
- How to factor polynomials

Section 5.2

- How to factor polynomials
- How to find the square root of a number

Section 5.3

- How to use linear regression to find a line of best fit

Section 5.4

- How to use the quadratic formula to solve quadratic equations
- How to find the square root of a number

Section 5.5

- How to combine fractions
- How to use the properties of exponents
- How to simplify radicals

Section 5.6

- How to identify coordinates of points on a rectangular coordinate system
- How to use the Pythagorean theorem to find a hypotenuse
- How to simplify radicals

Instructor Notes for Chapter 5

Introduction to Chapter 5

This chapter deals with several topics in Intermediate Algebra. Quadratic equations, their solutions by factoring, and the quadratic formula are introduced in the context of applications that are modeled by quadratic functions. The process of modeling data with quadratic functions using quadratic regression is then explored and contrasted with linear modeling.

As the need arises, radicals, fractional exponents, and radical equations are introduced.

Equations of circles are included because of their quadratic nature and are introduced as an outgrowth of the distance formula. The focus is on circles with centers at the origin, although more general circles are examined in an exercise.

There is an emphasis on applied contexts throughout the chapter. Activity 5.1, the Discoveries, and many of the Investigations and Projects are intended to be thought provoking and to develop model-building and critical-thinking skills. The Additional Practice exercises are primarily intended to reinforce specific skills, but many are contextual and a few involve moderate modeling practice.

It is assumed that students can factor polynomials, specifically common monomials, the difference of two squares, perfect square trinomials, and quadratic trinomials for which the coefficient of the square term is 1.

When appropriate, specific instructional suggestions are included in annotations within activities, discoveries, and some of the longer exercises.

Preparation Reading

This short introductory section introduces the two main topics of the chapter: modeling with quadratic functions and expressions and equations that contain radicals. After a brief reflection on linear modeling, the necessity for nonlinear models is raised.

The "Reflect and Discuss" question is intended to stimulate students' thinking about why we might want to create a data-driven model.

Section 5.1

Activity 5.1 requires one class period.

Activity 5.1

The purpose of the activity is to introduce a quadratic function that models a real object, a simulated bridge cable. The simple materials are inexpensive and available at any hardware store. Hanging the washers (or any other small weights such as fishing sinkers) from the chain simulates hanging a road using vertical cables, but this is primarily intended to give the exercise more of a realistic feel for students. There will probably be insignificant differences in chain shape after the weights are hung because on this scale a parabola and catenary are almost identical. (The mathematical form of a catenary is beyond the scope of this text and is not discussed, because it involves the hyperbolic cosine function made up of exponential functions.)

If graph paper larger than 8.5" x 11" is available, you may want to use it here and in Section 5.2, Exercise 10.

Several new terms, such as *parabola, quadratic function,* and *vertex,* are introduced here along with a review of other terms, such as *axis of symmetry, independent variable,* and *dependent variable.* A review of these terms during a discussion following the Activity could prove very helpful to students.

Extend the Activity

Item 8 poses a "what if" problem that allows students to practice what they have just completed in the Activity by finding an equation for a given parabola.

Items 9 and 10 ask students to consider a different form of their original model by choosing different positionings of the coordinate plane. In item 9, they are asked to place the origin at the vertex of the parabola, and in item 10, they are asked to place the origin 5 units directly below the vertex.

Section 5.2

At least two days are suggested for this section. Discovery 5.1 and reading material up to Quadratic Equations can be introduced on the first day, with a second day devoted to the methods used to solve quadratic equations.

Exercises 1–5, 10, and 12–20 are suitable for use after the Discovery. The remaining exercises require the ability to solve quadratic equations.

Discovery 5.1

This investigation is intended to present quadratic functions as a family of functions whose graphs have common characteristics. It begins by telling students how to find the x-coordinate of the vertex of a parabola. Then by creating a table of values, students are asked to plot points and sketch a graph of a given quadratic.

Students are then asked to investigate the effects of the constants a, b, and c on the graphs of quadratic functions of the form $y = ax^2 + bx + c$. After these investigations, they are asked to apply their observations and to make conjectures about graphs of quadratic functions prior to graphing them.

This investigation is much more powerful when technology is used, but it is possible to complete it by having students graph by hand. If no technology is used, you may wish to consider having students "share the load" and report their findings to the entire class.

Quadratic Equations

The need for solving quadratic equations is introduced as an outgrowth of searching for x-intercepts of quadratic function graphs. Students have already examined x-intercepts of parabolic curves in Section 5.1, so this should make sense as a reasonable thing to do.

The standard form of a quadratic equation is introduced here and the remainder of Section 5.2 deals with methods of solving this type of equation.

Note that the quadratic formula is presented without being derived, the derivation being done as a discovery activity in Exercise 11, in which completing the square is also introduced.

Example 2

This is an example of an approach used by highway engineers and surveyors to plan construction of highways that pass over hills. For the instructor's information, the reason a parabola is most suitable for such use is that its second derivative is constant. Thus, the rate at which the slope changes is constant, and it is easier for drivers to adjust to such a change in their line of sight. A parabola must be chosen so that the value of its derivative matches the slopes of the ascending and descending roads on either side of the hill (at the x-intercepts of the curve in this example).

Example 6

If students need instruction in the use of a graphing calculator, the instructor may wish to use a Texas Instruments View screen for demonstration. Students may need instruction on methods of zooming with a calculator. The Zbox command (used to create Figure 5.9) provides an efficient way to locate intercepts accurately, although you may prefer to use a Zoom In command. (If so, it is probably worth also discussing the SetFactors command on the Zoom Memory menu.) The automatic calculation of zeros through use of the CALC key is purposely avoided here, although some instructors may wish to mention it. Students may find it on their own in any case.

Example 7

This example is intended to give students guidance in creating algebraic models of geometrical problem situations. Many students have considerable difficulty in formulating equations from problem statements. The example suggests that "word equations" be used as intermediate steps in the process.

Special Notes on Exercises

Exercise 1 asks students to practice what they learned about parabolas and their graphs in the Discovery.

Exercises 2 and 3 involve modeling and choice of placement of the coordinate plane.

⇒ Exercise 4 is intended, in part, to counter an assumption made by many students that all U-shaped curves are parabolas. These graphs can be created by hand or through the use of a graphing calculator or computer.

⇒ Exercise 5 examines a fundamental characteristic of quadratic functions: Second differences are constant (which is equivalent to saying that the second derivative is constant, raised here in a constant acceleration context).

⇒ Exercise 6 provides practice with modeling and critical thinking as it includes assessing the reasonableness of solutions. It also introduces the use of the discriminant of a quadratic equation to determine whether one or more real solutions exist.

⇒ Exercise 7 introduces using square roots as an alternative to using factoring to solve some simple quadratic equations. Students should be encouraged to use factoring when possible to avoid overlooking negative roots.

Exercises 8 and 9 continue the emphasis on modeling. Exercise 9 may be particularly difficult for some students, but it is an example of an important business application.

Exercise 10 demonstrates another common parabolic curve, a trajectory, but without the need for a multiple-flash photograph (which is the basis for Exercise 6 of Section 5.3). It can be a bit messy though. A wooden ball is suggested because it will hold ink or paint better than metal balls, although a small rubber ball could be used.

⇒ Exercise 11 asks students to derive the quadratic formula. Some instructors may choose to do this as an instructor-led activity. It requires that students manipulate geometric figures (from Handout 5.1) as part of an area model approach. Each student should receive a copy of the handout.

Exercises 12–16 provide practice writing equations of parabolas when given certain conditions.

Exercises 17–20 provide practice graphing.

Exercise 23 asks students to apply the Zero Product Property to more than two factors and to solve an equation of degree three by factoring.

Exercises 24, 25, and 28, provide practice solving equations (in context) that can be written in a factorable form. In Exercises 26, 27, and 29, the contextual problem must be solved using the quadratic formula.

Students are asked in Exercises 30–33 to determine whether an equation is quadratic. Exercises 34–37 provide additional practice graphing.

In Exercise 38–45 students are asked to solve quadratic equations either by factoring or by using the quadratic formula. In Exercises 46 and 47, they are asked to solve the equation by graphing.

Section 5.3

This section is the most nontraditional of the chapter, as it focuses exclusively on using calculator/computer regression methods to model data. However, even for classes where such technology is not available, Discovery 5.2 is worth doing for its discovery approach to quadratic data. Handout 5.2 is provided for students who need an introduction to the use of calculators for linear regression and the use of residual plots.

One day should be devoted to Discovery 5.2, which introduces quadratic modeling of data, and a second can be used for Discovery 5.3 and Example 8 for a discussion of evaluating models.

Exercises 1, 17, and 18 can be assigned after Discovery 5.2.

All other exercises require the use of residual plots, and so require completion of Discovery 5.3.

Exercise 5 involves choosing among several polynomial regression models and should follow Example 8.

Discovery 5.2

In this activity, students experiment with cutting up "virtual logs" drawn on paper, to simulate cutting lumber. In so doing, they derive a version of a well-known (at least in the lumber industry) formula for determining the amount of lumber that can be obtained from a log. The formula as presented in many textbooks is $L = 0.22d^2 - 0.71d$ and gives the number of board feet of lumber (L) that can be ideally cut from a 4-foot-long log of diameter d (in inches).

Students can draw "logs" of any diameter and should use at least four or five to allow a good pattern to emerge. They may need help with choice of scale. There also may be decisions on whether to saw one board as wide as possible in the center of the log, as in Figure 5.24, or two boards of equal width on either side of the center line. It shouldn't make a significant difference which they choose.

Students' regression equations will not perfectly match the formula given here. For one thing, they are asked to find a formula for a 1-foot-long log. And another factor is that what we call, for example, a 2" x 4" piece of lumber has actual dimensions of 1-1/2" x 3-1/2" when we buy it at a lumber yard. The 2" x 4" dimensions refer to rough-cut lumber before it is smoothly finished.

Discovery 5.3

This activity introduces the use of residual plots as a technique for deciding on the appropriateness of a model for describing data. Residual plots also provide one tool for choosing between competing models. Some students may be familiar with the correlation coefficient r for assessing the strength of a linear model. There is no equivalent for a quadratic model, but the coefficient of determination R^2 (available on graphing calculators) can provide additional information for comparison purposes. R^2 is equal to the fraction of the variation in the values of the response variable that is accounted for by the regression model. A good discussion of the coefficient of determination can be found in an article by Gloria Barrett in *The Mathematics Teacher,* vol. 93, page 230 ff. (March 2000).

Example 8

This example is intended to place quadratic regression in context as one of many data-driven models that can be investigated. The names "Cubic" and "Quartic" regression are used here because they are found on the TI-83 Stat Calc menu, although some instructors may prefer "third-degree" and "fourth-degree" terminology. The examination of the quartic model serves as a caution to those students who are likely to accept uncritically the results of regression computations. The quartic model perfectly fits the five data points but is not a reasonable model for the problem.

Special Notes on Exercises

❑ All exercises in this section involve the use of technology.

Exercises 1 and 2 continue the examination of problem-solving situations that were introduced at the beginning of the chapter.

Exercises 3 and 8–10 involve choosing between linear and quadratic models based on residual plots. Exercise 3 involves additional critical thinking. Exercise 9 in particular illustrates the difficulty of choosing between linear and quadratic models for data that are not obviously nonlinear. Quadratic regression, by minimizing the square term, can produce a function that models essentially linear data very well. In such cases, it is important to look for theoretical grounds to favor one model over the other. In the case of the stress-strain data presented here, a linear model is known to be appropriate.

Exercise 4 is an exploration of several polynomial models for a data set.

Exercise 5 involves a three-point data set, the minimum for determining a unique parabola. An interesting feature of the problem is that all residuals are zero.

Exercise 6 requires use of a CBL (Calculator Based Laboratory) instrument. The TI-83 program entitled FALLING.83p is provided to control data collection. A sample interval of 0.02 seconds works well. Because this produces 50 samples per second, somewhere between 50 and 100 samples should be specified. (Some experimentation is necessary for determining an appropriate number of samples.) The default units for distance are feet, so if meters are desired (as was done in the answers to the text), this must be indicated as an input when prompted. Note that before performing quadratic regression on the collected data, some of the raw data must be discarded. Any data representing times before the ball is released or after it is caught should be deleted from the calculator lists.

Exercise 7 involves measurement on a photograph of a parabolic trajectory.

Exercises 11–16 are essentially contextual drill on constructing quadratic models and their accompanying residual plots.

Exercise 11 asks for a forecast, which must always be treated with caution. And Exercise 12 is adapted from an environmental health study that, as in many real research situations, presents somewhat ambiguous data. The investigators focused on linear regression, but the data are included here because the quadratic model results in $R^2 = 0.85$, compared with 0.80 for a linear model.

Exercises 10, 11, 14, 15, and 16 include calendar year data, and regression equations are given based on retaining the calendar year as explanatory variable. Instructors may prefer to recommend to students that the first year listed be renamed year 0. In such cases, the regression equations given in the solutions will have different forms.

Section 5.4

The material in this section is designed for a single class meeting.

The approach used in this section to introduce radicals is more traditional than approaches taken in other sections. Students are given procedures and shown examples. These skills are then applied to contexts that involve radicals.

Evaluating Roots of Numbers

Several terms such as *radical form, radical sign, radicand, index,* and *principal nth root* are introduced here. These terms are probably not new to students, but it's worth taking time to review them.

Example 10

In this example, when the problem asks for "all" the roots, it is only asking for real roots. At this point, no mention has been made of roots that are imaginary numbers.

Simplifying and Combining Radicals

The simplified form of a radical expression is sometimes referred to as "standard radical form." You may wish to use this terminology in class when asking students to simplify radicals. It simply means that the radical is written in the form $a\sqrt{b}$, where b is an integer with no factors that are perfect squares.

With the easy availability of calculators, it is no longer necessary to simplify radical expressions with numbers as radicands, citing ease of evaluation as the reason for simplification. If your curriculum does not specify "simplifying radicals" as a necessary skill for this course, feel free to skip parts of this section. If that is the case, you will need to refer students to the Product and Quotient Properties so that they will be able to use multiplication and division to combine radicals.

If you do not require students to simplify radical expressions, you will want to skip Exercises 21–31 and make note to students that their answers for other problems may differ from those in the solutions manual.

You may also wish to mention to students the sidebar in Example 12. This points out to students that we will be assuming that the variables in noncontextual radical expressions are nonnegative unless otherwise stated. This avoids the necessity of writing the absolute value symbol when taking the square root of x^2.

Special Notes on Exercises

Exercises 1, 3, and 5 provide contexts that involve radical expressions.

$\Rightarrow$ Exercise 1 is geometric in nature and introduces students to a method of calculating the area of a triangle other than taking "one-half the base time the height." $\Rightarrow$ Exercise 3 discusses arithmetic and the geometric means. Exercise 5 examines a spiral and has students look for patterns.

$\Rightarrow$ Exercise 2 introduces the skill of multiplying two binomials where each is of the form $a + b\sqrt{c}$.

In Exercise 4, if students say that $\sqrt[n]{a}$ when n is even and $a > 0$ can be either positive or negative, remind them that the notation $\sqrt[n]{a}$ refers to the principal root.

$\Rightarrow$ Exercise 6 introduces simplifying radical expressions with roots other than square roots.

The project in Exercise 7 is well worth having students try. In most cases, it will be surprising to students that only the length of the pendulum has any effect on the period of the swing. When students are asked to change the angle of release, try to have them pull the pendulum back no more than about 30 degrees.

Exercises 8–15, 19, and 20 provide practice using radical expressions in context.

Exercises 16–18 provide practice simplifying the results after using the quadratic formula.

Exercises 21–31 provide practice simplifying noncontextual radical expressions.

Exercises 32–45 provide practice adding, subtracting, and multiplying radical expressions.

Section 5.5

This section requires two days to complete. One day should be devoted to fractional exponents and another to solving radical equations.

Exercises 1–4, 6, and 8–33 can be assigned after fractional exponents are studied, but 5, 7, and 34–47 involve solving radical equations.

Discovery 5.4

Provided students are familiar with properties of positive and negative integral exponents, they should be able to discover the meaning of a fractional exponent, especially if working in groups. (Note: For the most part, the term *fractional exponent* is used instead of *rational exponent* because it has more meaning for most students.)

Example 18

In this section, it is expected that evaluation of numerical expressions containing fractional exponents will be done without a calculator so that students will gain familiarity with the meaning of a fractional exponent. Exceptions occur in exercises that involve decimal exponents that are not easily converted to simple fractions, and also where calculator use is specifically discussed (as in Exercises 2 and 3).

Example 19

The statement regarding nonnegative variables (here and in the exercises) eliminates the need to consider whether expressions like $\sqrt{x^2}$ should be simplified to $|x|$.

Example 22

This is most students' first encounter with the concept of an extraneous solution. They should be encouraged to check all solutions to radical equations in order to verify that they are not extraneous. This will also be an issue in Chapter 6, where rational equations are discussed.

Special Notes on Exercises

⇒ Exercise 1 examines the two possible orders of evaluation for a fractional exponent expression of the form $a^{m/n}$. Of course, this is only important when expressions are evaluated without a calculator. It may be worth noting that although $\sqrt[n]{a^m}$ and $\left(\sqrt[n]{a}\right)^m$ are equivalent, the former notation is usually used in formulas, mainly to avoid the need for parentheses.

❏ Exercise 2 discusses calculator use and emphasizes the need for caution in approximating exponents such as 2/3, which have no exact decimal equivalent.

⇒ ❏ Exercise 3 introduces the meaning of more general decimal exponents and provides practice in evaluating them. Exercises 6, 8, 9, 19, and 25 require understanding of decimal exponents, so this exercise must precede them. There are many applied areas, especially in technical fields of study, that make extensive use of formulas involving decimal exponents.

⇒ Exercise 4 examines power functions with fractional exponents and highlights the limited domain of functions of the type $y = ax^{m/n}$ where n is even.

⇒ Exercise 5 examines the solution of more involved radical equations. Many instructors may choose to omit this exercise.

❏ Exercise 6, which examines the equally tempered scale in music, involves practice in evaluation of fractional exponent expressions using a calculator. While frequency ratios for such intervals as a perfect fifth, major third, and minor third are exactly 3/2, 4/3, and 5/4 for stringed instruments such as violins, pianos are tuned with equal temperament and will produce only approximations of these ratios.

❑ Exercises 8 and 9 are experimental activities that involve data modeled by rational power functions. They can only be done in their entirety if students have studied data analysis in Section 5.3. The dependence of the period of a spring-mass system (Exercise 8) on mass is well-known and is given exactly by the function $T = \dfrac{1}{2\pi}\sqrt{\dfrac{m}{k}}$, where m is mass in grams and k is the spring constant in dynes per centimeter. The terminology "weight in grams" is used here even though a gram is correctly a unit of mass. This is done because students may be unfamiliar with the concept of mass unless they have studied physics, and because the (admittedly incorrect) terminology is common in some professions. Exercise 8 is especially important in continuing our emphasis on modeling, although the examination of different models requires use of technology. This experiment is remarkably reliable in producing an exponent in power regression that is very close to 1/2 so that students can be expected to recognize the $\sqrt{W}$ variation. Note the caution about eliminating the ordered pair (0, 0) from a data set before attempting a power regression. This is due to the algorithm for determining the parameters in power regression, which involves logarithms. Because log(0) is undefined, the ordered pair (0, 0) cannot be used. If it is desired to include such a pair in the analysis, replacing (0, 0) with an approximation like (0.01, 0.01) may sometimes be justified.

Exercises 10–33 involve practice with fractional exponent notation.

Exercises 34–47 involve solving radical equations.

Section 5.6

One day can be devoted to Discovery 5.5 and the distance formula, and another to circle equations.

Exercises 1, 2, 5, 9, 10, 11, 12, and 22–30 can be assigned after Discovery 5.5. The remaining exercises involve equations of circles.

Discovery 5.5

This is a contextual development of the distance formula. It is intended to end with an open-ended exploration of different possible configurations (item 6). Students may not discover all possibilities, nor find the optimum system as discussed in the solution. The important thing is that they engage in the search and practice using the distance formula. This exploration can be rather time-consuming and is not essential to the development of the distance formula, so it is possible to end the discovery with item 5. Item 6 could then be assigned as an open-ended homework problem or project.

Special Notes on Exercises

Exercise 1 involves contextual practice with the distance formula.

⇒ Exercise 2 introduces the midpoint formula.

Exercises 3 and 4 involve modeling with circles. Exercise 3 refers to the Global Positioning System but uses a simplified two-dimensional version analogous to the actual three-dimensional GPS. The resulting system will be solved in Chapter 6. Exercise 4 is an example of a problem involving a circular model that reduces to a linear equation and may be difficult for some students.

❑ Exercise 5 examines the peculiarities of graphing circles with a graphing calculator that stem from screen shape and resolution. Because the screen is rectangular, standard windows usually have different scales on the horizontal and vertical axes, distorting geometrical shapes. And the limited number of pixels can cause difficulties in creating complete graphs. Knowledge of the number of pixels in each direction for any particular calculator can make graphing easier.

⇒ Exercise 6 generalizes the standard equation of a circle to circles with center (h, k).

❑ Exercise 7 is a discovery-type problem that uses the distance formula to explore the definition of a parabola as the locus of points equidistant from a point and a line. Here the post is the focus of the parabola and the wall is the directrix. This exercise is a good opportunity to have students use List or Table features of a calculator. If Lists are used, students can be shown the "SEQUENCE" command on the List Ops menu for generating the list of x-values.

Exercises 8–10 involve modeling with circle equations, and 11–12 involve modeling with the distance formula.

In Exercise 11, the assumption is made that distances between points on Earth's surface are measured along straight line segments, when in reality these distances are along great circle arcs. However, for the distances involved in this problem, the differences are insignificant.

Exercises 13–26 are drill problems on circle equations.

Exercises 27–38 are drill problems on the distance formula, with 36–38 placed in a context of triangle identification.

Chapter Review

The breakdown of exercises by section is:

Chapter Summary:	Exercise 1
Sections 5.1 and 5.2:	Exercises 2–16
Section 5.3:	Exercise 17
Section 5.4:	Exercises 18–22
Section 5.5:	Exercises 23–28
Section 5.6:	Exercises 29–33

Instructor Notes

Chapter 6: Rational Expressions and Systems of Equations

Goals of the Chapter

- To explore basic properties of rational functions
- To perform operations with rational expressions
- To solve rational equations
- To solve systems of equations

What You Will Learn

Section 6.1

- To explore the behavior of a non-polynomial function

Section 6.2

- To examine real-world situations that can be modeled with rational functions
- To evaluate a rational function for given values of the independent variable
- To identify asymptotes on the graph of a rational function

Section 6.3

- To reduce rational expressions to lowest terms
- To multiply and divide rational expressions
- To find values for which a rational expression is undefined
- To use rational expressions to model real-world situations

Section 6.4

- To solve rational equations algebraically
- To find common denominators for two or more rational expressions
- To add and subtract rational expressions
- To use rational expressions to model real-world situations

Section 6.5

- To solve linear systems of equations by substitution
- To solve linear systems of equations graphically
- To solve systems containing both a linear and a nonlinear equation by substitution
- To solve systems containing both a linear and a nonlinear equation graphically

What You Need to Know

Section 6.1

- How to evaluate expressions

Section 6.2

- How to use a table of values to construct the graph of a function
- How to find the problem domain for an applied function

Section 6.3

- How to evaluate algebraic expressions
- How to reduce fractions to lowest terms
- How to factor quadratic expressions
- How to find the volumes and surface areas of spheres and rectangular solids

Section 6.4

- How to evaluate rational expressions
- How to simplify rational expressions
- How to multiply rational expressions
- How to add and subtract fractions
- How to find the volume and surface area of a cylinder

Section 6.5

- How to graph linear, quadratic, and rational functions
- How to solve linear, quadratic, and rational equations

Instructor Notes for Chapter 6

Introduction to Chapter 6

This chapter introduces basic properties of rational functions and expressions, as well as algebraic operations involving rational expressions and solutions of rational equations.

Systems of equations, both linear and nonlinear, are then considered from the point of view of solution by substitution and graphing. (Solution by elimination is introduced in an exercise.)

It is assumed that students are familiar with operations involving numeric fractions, as well as solving linear and quadratic equations.

Preparation Reading

This brief introductory section presents a context for creating a model using a rational expression. Students should become aware that the concept of a ratio or fraction can serve as the basis for modeling a variable quantity. The Reflect and Discuss question is intended to stimulate student thinking about real contexts that are based on ratios of two quantities.

Section 6.1

This section requires one class period.

Fettuccini seems to work best for this activity. Because it's flatter and more brittle than spaghetti or other pastas, it tends to break more cleanly and not bend as much. This activity works best when one student holds down the ends of the pasta on the desk, or whatever movable horizontal surfaces are available. Otherwise, the pasta tends to bend considerably, resulting in less reproducible data.

Let a rubber band hang from the center of the pasta. Use an opened paper clip or binder clip to attach a small plastic bag or other container to the rubber band so that pennies can be added one at a time until the pasta breaks. The weight of the bag and rubber band/clip will probably be negligible compared to the weight of a penny.

Students are not expected to find an analytical model but should observe a nonlinear decrease in breaking force as length increases. (Theoretically, it should be an inverse variation $F = k/L$.) They should also observe qualitative features of the scatter plot and think about what makes sense at the extremes of the graph. Students also are asked to consider why linear and quadratic functions would not be good models for the data.

As an alternative to the "low-tech" use of a bag of pennies, a CBL with a force probe can be used to break the pasta and gives good results. When using the force probe, it is easiest to break the pasta by pulling upward on the probe.

Extend the Activity

Question 12 involves further experimentation with slight alterations of the context. If time permits, some students may be interested in trying the suggested variations.

Questions 13 and 14 may be used for class discussion or assigned as homework.

Section 6.2

Two days are suggested for this section. Discovery 6.1 and vertical asymptotes can be examined on the first day, with a second day devoted to considering horizontal asymptotes.

Exercises 1–3, 6, and 8–17 can be assigned after vertical asymptotes are discussed. Exercises 4, 5, 7, and 18–23 require a background in horizontal asymptotes.

Only the most fundamental features of rational functions are discussed: domain and excluded values and asymptotic behavior. The primary goal is for students to distinguish the qualitative characteristics of rational functions from those of linear and polynomial functions and to see why they are more appropriate for modeling certain types of real-world behaviors. There is considerable emphasis on critical thinking regarding the suitability of rational models.

A thorough treatment of rational functions and their graphs, as would be appropriate in a precalculus course, is not intended here. Although students should be able to locate vertical asymptotes by examining a function's domain for excluded values, they are not expected to quantitatively evaluate limits to find horizontal asymptotes. Manual construction of graphs can allow estimation of horizontal asymptotes, but such observations are greatly facilitated by the use of technology—either computer graphing software or graphing calculators.

Discovery 6.1

The constant of 40,000 in the beam strength function is an approximation. It varies with wood quality and could be two or three times as large for oak or ash beams. It may also be worth noting to students that what is referred to as a 2" x 10" beam actually has a cross section of 1-1/2" x 9-1/2" due to finishing of the wood surface.

A rational function is defined as a ratio of two polynomial functions. If students have difficulty later on in seeing that something like $y = 1 + \dfrac{1}{x}$ meets this definition, remind them that $1 + \dfrac{1}{x}$ can be rewritten as a single fraction $\dfrac{x+1}{x}$.

Students should be encouraged to explain how excluded values for a rational function are naturally related to the problem domain in this Discovery and in examples that follow in the section.

Vertical Asymptotes

Students should not be led to think that an excluded value always leads to a vertical asymptote. The "hole" in a graph that results when a rational function contains identical factors in its numerator and denominator is discussed in Exercise 6 (although the term *indeterminate form* is not mentioned).

Example 1

The 125 that appears in the function represents the luminous intensity in candela (cd) of a 100-watt light bulb. The luminous intensity of a commercial light bulb is usually labeled (on the box) in lumens. The relationship between lumens and candela is 1 candela = 4π lumens.

Discovery 6.2

The alcohol elimination function may need some explaining for low blood alcohol levels. For example, if only 2 grams of alcohol are in the system, the function shows that 3-1/3 grams are eliminated in an hour. Clearly no more than 2 grams can be eliminated in such a situation. For values of a between 0 and 6 grams, the function is not a good model for the actual amount eliminated. Because it is the asymptotic behavior of the function that is being investigated here, the discrepancy does not affect the conclusion drawn by the example.

Example 4

In addition to using a graph to identify a horizontal asymptote, this example shows that it is possible for a graph to *cross* an asymptote before exhibiting asymptotic behavior.

This example also mentions the difficulty of obtaining a good graph of a rational function on a graphing calculator, a subject that is explored further in Exercise 4. Most graphing calculators do not recognize excluded values unless the window is chosen in such a way that an excluded value is located exactly on a pixel of the screen. For this reason, computer graphing programs are a preferred technology for this section. Students may need to be reminded that spurious vertical lines are often graphed in the approximate locations of asymptotes due to the limitations of calculator technology. Although the use of dot mode rather than connected mode can eliminate this problem, it is not a perfect solution.

Special Notes on Exercises

Exercises 1 and 2 involve critical thinking about the properties of rational functions and their use as models. It is important that they be discussed in class. One possible strategy is to assign each exercise to half of the students and have them make presentations to the class as a whole. In Exercise 1, students may wonder about the very large equilibrium levels involved. For the doses used (near 15 grams), it would take a long time for the equilibrium levels to be reached.

⇒ Exercise 3 is an examination of inverse variation from a graphical point of view.

❏ Exercise 4 examines the peculiarities and limitations of graphing rational functions on a graphing calculator.

⇒ Exercise 5 is an investigation of oblique asymptotes. (Although identifying the equation of the asymptote in the second part of the exercise is probably beyond most students, its equation is $L = 2D + 13\pi$.)

⇒ Exercise 6 looks at rational functions that have removable discontinuities (or "holes"), rather than vertical asymptotes, at excluded values. An in-depth treatment of this topic is not appropriate at this level.

Exercise 7 is analogous to Activity 6.1 but involves buckling a simulated vertical column rather than breaking a beam. It differs in that the function involves an inverse square variation instead of a simple inverse variation. The complete problem was first investigated by Euler, and the Euler Load P required to buckle a column of length L is given by $P = \dfrac{\pi^2 EI}{L^2}$, where E is the modulus of elasticity of the column's constituent material and I is the moment of inertia of the column's cross-sectional shape. In collecting the data, it is important that the spaghetti be pressed vertically, with no horizontal component of the force. Very short (a few centimeters) strands will not buckle but will break instead. (Short structural columns fail by crushing before they buckle.)

This exercise can be ended with (d), providing an alternate context to the opening chapter Activity 6.1 and Discovery 6.1. However, depending on available time and student ability, the last half of the activity is an opportunity to engage in some in-depth model building with rational functions. (Note: Exercise 3 is a prerequisite to this part of the exercise.)

Exercises 8–10 involve evaluation of rational functions, whereas Exercise 11 includes construction of a graph from a table of values.

Exercises 12–14 provide additional practice on graphing.

Exercises 15–21 ask students to identify excluded values.

❏ In Exercises 22–27 ask students to use either graphing calculators or computers to identify horizontal asymptotes. They are also asked to identify vertical asymptotes.

Section 6.3

One day is suggested for this section.

Throughout the remainder of this chapter, the terms *rational expression* and *algebraic fraction* are used interchangeably. The decision to refer to rational expressions informally as algebraic fractions was made because this terminology often makes more sense to the student.

Discovery 6.3

Many different criteria can be used to measure packaging efficiency, but this section focuses on only one criterion: the ratio of the surface area of the package to its volume. Prior to beginning this Discovery, it may be necessary to review how to find the surface area and the volume of a rectangular prism.

The Discovery begins with a concrete example and moves to writing a rational expression in order to model the situation. Students are asked in items 6–10 to explore the equivalence of two rational expressions, one a simplified form of the other.

Exploring Rational Expressions

A rational expression is defined here as the quotient of two polynomials. If students have forgotten how to identify a polynomial, remind them that a polynomial can be defined (informally) as an algebraic expression made up of terms where the variables are raised to whole-number exponents.

In this text, if a student is asked to *simplify* a rational expression, it means that the student is to reduce the expression to lowest terms by making sure that no factors other than 1 are common to the numerator and denominator of the expression.

In this section and in Section 6.4, examples provide students with reminders on how to simplify, multiply, divide, add, and subtract numeric fractions. These examples occur just prior to corresponding work with algebraic fractions. Making this connection is very helpful for most students.

Multiplying and Dividing Rational Expressions

If students ask about why we invert and multiply when dividing fractions, you might want to give them the following explanation:

$\dfrac{a}{b} \div \dfrac{c}{d}$ can be rewritten as the complex fraction $\dfrac{\frac{a}{b}}{\frac{c}{d}}$.

To eliminate the denominator of this fraction, you can multiply the numerator and denominator by $\dfrac{d}{c}$.

$$\frac{\frac{a}{b} \cdot \frac{d}{c}}{\frac{c}{d} \cdot \frac{d}{c}} = \frac{\frac{a}{b} \cdot \frac{d}{c}}{1} = \frac{a}{b} \cdot \frac{d}{c}.$$

So any problem involving the division of two fractions can be rewritten as an equivalent multiplication problem.

Special Notes on Exercises

The purpose of Exercise 1 is to help students recognize that a rational expression can be used to represent an average, in this case an average hourly wage.

Exercise 2 has students examine two expressions to see if they are equivalent. This exercise reiterates the importance of examining a rational expression for values for which it is undefined prior to simplifying the expression.

Exercise 3 revisits the context of packaging efficiency. This time students are asked to examine two shapes (a cube and a sphere) that fit into a cube with an edge of x cm. The result may be both surprising and possibly misleading, as the ratio of surface area to volume is the same for both solids. This problem will be revisited once again in Section 6.4, Exercise 3, when students are asked to find the ratio for a cylinder. And again, it is the same as for the sphere and the cube. At that point, posing a question as to whether it is true for all solids that fit into the cube might be appropriate. If students examine a square pyramid with a base of x cm x x cm and a height of x cm, they will find that its ratio is not the same.

❏ Exercise 4 explores using a graphing calculator or computer to simplify rational expressions.

Exercise 5 revisits the context of blood alcohol elimination that was introduced in Discovery 6.2.

⇒ Exercise 6 asks students to use the quadratic formula to solve a quadratic equation, and then examines how to simplify solutions when common factors appear in the numerator and denominator.

⇒ In Exercise 7, students explore multiplying binomials that are conjugate pairs. They use the information they gather to rationalize the denominators of expressions that have radicals in binomial denominators.

Exercise 8 is an open-ended project that has students explore three methods of apportionment. All three methods are divisor methods in which rational functions can play a part. Listed below are several excellent references on the topic:

Balinski, Michel, and H. Peyton Young. *Fair Representation: Meeting the Ideal of One Man, One Vote*. (New Haven, CT: Yale University Press, 1982.)

COMAP. *For All Practical Purposes: Introduction to Contemporary Mathematics*. 4th ed. (New York: W. H. Freeman, 1997.)

Crisler, N., Patience Fisher, and Gary Froelich. *Discrete Mathematics through Applications*. 2nd ed. (New York: W. H. Freeman, 1999.)

Eisner, Milton P. *Methods of Congressional Apportionment*. (Lexington, MA: COMAP, 1982.)

In Exercises 9–12, students are asked to write an expression for an average or a ratio and to evaluate the expression for specific values of the variable.

Exercise 13 gives a rational expression in context and asks student to evaluate it for specific values of the variables.

Exercises 14–23 provide practice with reducing rational expressions.

Exercises 24–29 ask students to find values for which a given expression is undefined.

Exercise 30 is a rational expression, contextual in nature. Students are asked to rewrite it in simplified form.

Exercise 31 provides students with a product of two rational expressions and asks the students to find the factors.

Exercises 32–43 provide practice with multiplying and dividing rational expressions. In Exercises 40–43, students may need to be reminded that a complex fraction is simply another way of indicating the division of two fractions.

Section 6.4

Two days are suggested for this section. Discovery 6.4 and Solving Rational Equations can be completed on the first day. On the second day, the remainder of the section, which focuses on adding and subtracting rational expressions, can be addressed.

Exercises 1, 2, 6, and 8–22 can be assigned after solving rational equations are discussed.

Exercises 3, 4, 5, 7, and 23–35 require background in adding and subtracting fractions.

Discovery 6.4

This Discovery revisits the context of the structural beam and asks students to find the length of a beam that can support a certain force. When answering the question, students begin to see the need to solve rational equations. They also find that one way to solve the equation is to use multiplication to eliminate the fractions.

Solving Rational Equations

The groundwork that was presented in Discovery 6.4 is followed up here through the use of examples and a summary of the steps used in solving rational equations. Students are cautioned to check their solutions because extraneous solutions may be introduced when multiplying both sides of the equation by a common denominator. Remind students that in Chapter 5 they were first introduced to extraneous solutions when they solved radical equations by raising both sides of an equation to some power.

Adding and Subtracting Fractions with Like Denominators

It is important to continually remind students that expressions and equations are different and that when they are working with them, the procedures differ.

Adding and Subtracting Fractions with Unlike Denominators

Once again, students are reminded of their work with numeric fractions before exploring examples with algebraic fractions.

Have students take note of and possibly discuss the sidebar on Common Denominators.

Figure 6.22 is provided for students who have a hard time remembering the procedures for adding and subtracting fractions in general and, more specifically, rational expressions.

Special Notes on Exercises

In Exercise 1, students are asked to develop a cost function, to use it to create a table of values, and to use the table to graph the function. The asymptotic behavior of the graph is then interpreted.

Exercise 2 presents a context that is familiar to most students. Students are asked to develop an expression for GPA and apply it in a meaningful situation.

Exercise 3 is an extension of Exercise 3 in Section 6.3. The cylinder is examined here rather than in the previous section because students need to be able to add fractions in order to simplify the expression for surface area.

Students are asked to compare the surface area to volume ratio for the cube, sphere, and cylinder. All three models have the same efficiency. If students find this problem interesting and wonder if this is true for all solids that fit into the cube, have them explore the ratio for a square pyramid with a base x cm x x cm and a height of x cm. Students will find that the efficiency is not the same for a square pyramid as it is for the three solids examined here.

The topic of efficiency is continued in Exercise 4, but in this exercise solids of equal volume are examined. Students conclude in this case that the sphere is more efficient than the cube.

Exercise 5 provides a context in which students are asked to solve Ohm's law ($V = IR$) for current I and then for resistance R.

⇒ Exercise 6 asks student to solve a rational equation and note that one of the solutions is extraneous.

In the activity in Exercise 7, students use a magnifying lens to explore the relationship given by the formula $\frac{1}{f} = \frac{1}{d} + \frac{1}{i}$. If possible, use an unfrosted, chandelier-type light bulb for this activity. With it, students find it easier to tell when the image is in focus on the paper or cardboard.

Exercises 8–10 provide contextual situations in which students solve equations to answer specific questions.

In Exercise 11, students explore and solve a work problem.

Exercises 12–18 provide practice with solving equations and checking for extraneous solutions.

Exercises 19–22 ask students to solve formulas for a specified variable.

In Exercises 23 and 24, students are given two fractions and are asked to identify possible common denominators.

In Exercises 25–27, students add fractions in various contextual situations. Exercise 27 requires extensive algebraic manipulation and can be a challenge to most students. Take care before assigning it to all students.

Exercises 28–35 provide practice with adding and subtracting rational expressions.

Section 6.5

One or two days can be devoted to Discovery 6.5 and the general substitution method, with another day to focus on linear systems. However, some instructors may choose to emphasize the elimination method presented in Exercise 5 in preference to solving linear systems by substitution.

Exercises 4 and 5 require discussion of linear systems, but all others can be assigned at any time after Discovery 6.5 has been completed.

Discovery 6.5

Ski jumps for international competition involve a takeoff angle of $11°$ below horizontal, but a horizontal takeoff is used here. Also, in order to slow jumpers down more quickly, runouts at the bottom of the hill are usually curved rather than straight.

This Discovery can be done without technology, but the computations, especially creating the required tables, may be somewhat tedious. And the graphical method is not very useful without some form of graphing technology.

It is important that students understand that finding the solution of a system of equations is equivalent to finding points of intersection of the graphs of the equations.

Example 18

This example provides a culminating extension to the contextual explorations that began with Activity 6.1 and continued throughout the chapter.

Example 19

This is a typical mixture problem and requires careful study if students are to understand the modeling aspect. Otherwise, it is a straightforward substitution problem, albeit with messy numbers. The term *aggregate* when applied to concrete refers to any particulate matter mixed in with the cement. It may be sand (hence a "sand mix") or larger gravel particles (usually called "concrete mix").

Special Notes on Exercises

It is recommended that Exercises 1 and 2 be assigned, as they involve modeling and equation writing. Exercise 1 also provides a foundation for study of linear programming and optimization in subsequent courses (although for linear programming, the solution of a system of linear inequalities, rather than equations, is usually involved).

⇒ Exercise 3 revisits the simplified GPS problem that was introduced in Exercise 3 of Section 5.6.

⇒ Exercise 4 discusses linear systems for which unique solutions do not exist (inconsistent and dependent systems).

Exercise 5 illustrates the elimination method as an alternative to the substitution method for solving linear systems. For students who will be introduced to matrix methods and/or Cramer's rule in subsequent courses, this exercise should be completed.

Exercise 6 requires students to think in general terms about graphs of linear, rational, and quadratic functions and how they relate to each other.

Exercise 7 is an investigation of the golden mean, which involves solving a nonlinear system. Students who are especially interested in this topic (e.g., architecture majors) can be encouraged to explore the occurrence of golden rectangles in design. Spaghetti is a good substitute for straws in this activity.

Exercises 8–44 are practice problems on solving systems, some linear and some nonlinear. Exercises 8–23 are contextual, and most involve either some modeling or rewriting; Exercises 24–42 are non-contextual and straightforward. Instructors may wish to suggest that students solve these both algebraically and graphically.

❏ Students are instructed to solve Exercises 43–44 using graphs. Accurate solutions can be found using technology.

Chapter Review

The breakdown of exercises by section is:

Chapter Summary:	Exercise 1
Sections 6.1 and 6.2:	Exercise 2
Section 6.3:	Exercises 3–14
Section 6.4:	Exercises 15–19
Section 6.5:	Exercises 20–22

Instructor Notes

Chapter 7: Probability

Goals of the Chapter

- To understand the meaning of probability references in the news and everyday occurrences
- To become familiar with some typical workplace uses of probability
- To understand sampling as the basis of inference
- To be aware of potential uses and abuses of statistics

What You Will Learn

Section 7.1

- To calculate the probability of a simple event

Section 7.2

- To calculate the probability of a compound event
- To express probabilities in different ways
- To know the difference between a deterministic model and a probabilistic model
- To use the Basic Multiplication Principle in counting

Section 7.3

- To obtain probability information from relative frequencies in data
- To find the probability of the complement of an event
- To compare the relative likelihood of common and rare events

Section 7.4

- To find joint probabilities for independent events
- To find joint probabilities for mutually exclusive events

Section 7.5

- To construct a probability distribution
- To create and interpret a graph of a probability distribution
- To find the expected value of a probability distribution

Section 7.6

- To use sample measurements to estimate values of unknown quantities
- To understand the meaning of sampling error

Section 7.7

- To construct a table of random number assignments for a simulation
- To conduct a simulation of an experiment using random numbers
- To generate random numbers using calculators and/or computers

What You Need To Know

Section 7.1

- How to construct a line graph
- How to express fractions as equivalent decimals and percentages

Section 7.2

- How to find the probability of a simple event

Section 7.3

- How to calculate the probability of an event
- How to express probabilities in different forms

Section 7.4

- How to compute probabilities of individual events

Section 7.5

- How to calculate probabilities from data
- How to construct and interpret a histogram

Section 7.6

- How to find a sample mean
- How to draw a frequency histogram

Section 7.7

- How to construct a probability distribution
- How to find the expected value of a probability distribution

Instructor Notes for Chapter 7

Introduction to Chapter 7

The chapter is an introduction to the basic concepts and some of the terminology of probability. It is not intended to be a thorough treatment of the subject but should help students build an intuitive sense of probability. The meanings of such things as equally likely events, theoretical and experimental probabilities, joint probabilities, and confidence intervals are introduced through hands-on activities and visual representations. Jargon and formal definitions (common to many probability texts) are kept to a minimum. Graphs and area models are used throughout the chapter to provide concrete images. Students who complete this chapter should be equipped to understand probability references in the news and to succeed in more comprehensive courses in probability, statistics, finite mathematics, and related subjects. As has been customary throughout the text, the joint themes of theory-driven and data-driven models are explored in the context of probability.

The first five sections of the chapter introduce fundamental probability concepts through experiential and workplace-oriented activities. But the emphasis is on simplicity. For example, joint probabilities using multiplication and addition rules are discussed, but only for independent (Multiplication Rule) and mutually exclusive (Addition Rule) events. Most of the exercises labeled *Investigations* or *Projects and Group Activities* are optional and intended for enrichment purposes. It is here that such things as conditional probabilities and non-mutually exclusive events are considered, as well as some interesting extended applications of probability (several of which are intended primarily for students in technical fields). Instructors who prefer a more in-depth treatment of probability can assign the relevant exercises, while those who wish to stick with a more streamlined introduction to the subject can assign only the *Additional Practice* exercises.

Section 7.6 provides an elementary introduction to statistical inference. Estimation and confidence intervals are introduced conceptually directly from descriptive statistics, without reference to standard deviation. An introduction to the more elusive concept of statistical significance is presented as an exercise in the *Projects and Group Activities* section.

Section 7.7 introduces probabilistic simulation along with random numbers and can be considered optional. But so much of modern business and industry uses computer simulation models in decision making and design that we believe all students should be exposed to its fundamentals. Generation of random numbers with calculators is emphasized along with more traditional physical simulators, and rudiments of simulation with computer spreadsheets are introduced as exercises. A few specific commands for the TI-83 calculator and Microsoft Excel are discussed. Consequently, some portions of this section may need to be altered or omitted if other technologies are used.

The nature of probability problems and activities based on experiment dictates that many answers depend on chance. In such cases, the answers given are merely typical examples. Students' results may show considerable variation.

Some activities and exercises in the chapter rely on the use of spinners to generate data. Spinner templates are provided as handouts. A simple way to construct spinners involves gluing the template to thin cardboard, then spinning a paper clip around a pencil point in the center of the spinner, as in **Figure 1.**

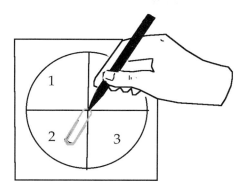

Figure 1

Several activities and exercises refer to "tossing" a coin. It is assumed that students understand that this means "flipping" a coin. If a coin is merely tossed without being flipped, whether it lands heads or tails will not be randomized but will often depend on initial orientation. Exercise 4 of Section 7.3, on the other hand, requires that students "spin" a penny rather than toss it. Surprisingly, the probability of heads in this case is less than 1/2.

There are a number of good Internet resources for statistics and statistical data, as well as links to such sites. A few of these are:

- The Journal of Statistical Education, http://www.amstat.org/publications/jse/

- Math Forum at Swarthmore College, http://www.mathforum.org

- CHANCE Database, http://www.geom.umn.edu/docs/education/chance/

- The Math Archives, http://archives.math.UTK.edu

- Data Surfing, http://it.stlawu.edu/~rlock/datasurf.html#Individual

- Statistical Resources on the Web, http://www.lib.umich.edu/libhome/Documents.center/stats.html

(A word of caution: URLs for Web sites are subject to change, but there usually will be a link to the new site from the old one.)

Preparation Reading

The chapter begins with anecdotal statements that implicitly or explicitly invoke probability to demonstrate how probability enters routine conversations. Reference is made to some of the concepts that will be explored in the chapter, as well as the connection between probability and statistics.

Students are asked to list some common and rare occurrences in order of likelihood. Although some of the probabilities listed are well defined, others are estimates, in some cases averaged from several sources.

Section 7.1

The activity in this section uses graphs to visualize probability as a limiting value of relative frequency. It requires one class period.

Quality control is introduced as a workplace application of probability. This context is revisited later in the chapter. It is the focus of Discovery 7.5 in Section 7.6 and also appears in exercises in other sections.

Activity 7.1

Students should already have an intuitive sense of the 50-50 chances involved in tossing a coin. This activity examines what happens in reality when a coin is tossed (flipped). In a small number of trials, large deviations from 50% heads (or tails) will often be observed. Only by repeating the coin-tossing experiment many times will students see a trend toward 50%. By examining many sets of 10 tosses, students see both the variability in small samples and the overall trend after many repetitions. Instructors may wish to point out the meaning of a "fair" coin, which is implicitly assumed here. Throughout the chapter, materials such as coins and dice will be assumed to be "fair" and not biased or weighted.

Section 7.2

This section requires one class period.

Discovery 7.1

This investigation illustrates why all the equally likely outcomes of an experiment must be identified if correct probabilities are to be found. The situation of joint controls for a bumper car ride, while somewhat contrived, provides an example in which students' intuitions are not likely to reveal those probabilities until all configurations of the two controls are examined. This activity introduces the first use of an area model for visualizing probability concepts. A visual model is used both here and in Example 1 to aid in counting (although tree diagrams are discussed in Exercise 1) and will be adapted later in the chapter for direct computation of probabilities.

Examples 2 and 3

Counting is limited to the use of the Basic Multiplication Principle, as the use of combinations and permutations formulas is deemed beyond the scope of developmental mathematics. However, a first exploration of more involved counting techniques is contained in Exercise 2.

Deterministic and Probabilistic Models

Students are accustomed to thinking that formulas they encounter in their coursework are absolutely correct. However, many (if not most) formulas used by engineers and other professionals have some degree of uncertainty because of inherent variation in their input values. This section concludes with a brief mention of this state of affairs.

Special Notes on Exercises

Exercises 1–6 can be assigned according to instructor preference.

⇒ Exercise 1 introduces tree diagrams as an aid in counting and is a prerequisite to a few exercises later in the chapter.

Exercise 2 examines more advanced counting techniques (without calling them by name) in the context of the probabilities of winning state lotteries.

⇒ Exercise 3 discusses the concept of odds and its relationship to probability.

Exercise 4 is a geometrical look at errors in rounding numbers. It is adapted from the Mathematical Sciences Education Board publication *High School Mathematics at Work: Essays and Examples for the Education of All Students* (National Academy Press, 1998).

Exercises 5 and 6 further explore graphs as tools for estimating probabilities from relative frequencies. Exercise 5 uses the context of genetic combinations. Exercise 6 is an activity involving tossing a penny on a grid and continues the use of geometrical models.

Exercises 7–36 involve practice in computing theoretical probabilities, identifying equally likely outcomes, counting, and distinguishing between deterministic and probabilistic models.

Section 7.3

This section focuses on experimental (empirical) probabilities as determined from data.

Although the section might be completed in one day, it could be split between two days, with the second devoted to complements.

Exercises 1–16 can be assigned after Discovery 7.2 and the definition of experimental probability.

Exercises 17–27 involve probabilities of complements of events.

Discovery 7.2

It is suggested that this discovery be done as a whole-class activity. It is analogous to a component of time-and-motion studies common in manufacturing measurements of workplace efficiency. Students are asked to perform a simple act of dealing cards while being timed with a stopwatch. The purpose of the activity is to formulate probability statements concerning a situation for which theoretical probabilities cannot be determined. Some students may try to achieve very fast dealing times; if this results in sloppy dealing, it may be worth suggesting that the group collectively decide whether a "deal" is valid. There will usually be a few people out of a group of 20 or more who take longer than 30 seconds to deal the cards into 12 piles. If not, the threshold may have to be set at a shorter time. The analysis of the data introduces language like "no more than," "at least," and so on, which students often have difficulty translating into numerical equivalents. This type of language will be used in various places throughout the chapter.

Special Notes on Exercises

Exercise 1 examines a few cases of especially rare events and is designed to give students a better sense of the meaning of very small probabilities.

⇒ Exercise 2 involves an important application of probability in industry, which is the reliability of manufactured products. It also provides a background (though not essential) for Discovery 7.3 in the next section.

⇒ Exercise 3 compares discrete and continuous variables.

Exercise 4 is an experimental investigation of probabilities related to spinning a penny. Interestingly, a spun penny does not land heads or tails with equal frequency. Rather, a tails-up landing is more likely than a heads-up one.

Exercise 5 is an open-ended experiment to explore the probabilities associated with whether a thumbtack lands point up or point down.

Either of Exercises 4 or 5 could also be assigned as group activities.

Exercise 6 requires the entire class to collect data about themselves and so has some built-in interest. The reason for the separation into younger and older student groups is to prepare students for using the data to explore conditional probability in Exercise 5 of Section 7.4. If the class is quite variable in age, the dividing line between younger and older could be at about 25 years or so, but if it's more homogeneous, the age distinctions may not be worthwhile. In such a case, another category such as whether a student lives on campus or works part or full time could be used to separate the class into two groups. If Exercise 5 of Section 7.4 will not be assigned, the breakdown into two groups is not necessary.

Exercises 7–27 involve practice in computing probabilities from data and in finding probabilities of complements of events.

Section 7.4

Joint probabilities for multiple events are explored in this chapter. The Multiplication and Addition Rules for Probabilities are introduced.

One day should be devoted to Discovery 7.3 and the Multiplication Rule for independent events. A second day could be used for the Addition Rule for mutually exclusive events. (Conditional probabilities and non-mutually exclusive events are considered only in exercises.)

Exercises 1, 2, 5–7, and 9–25 can be assigned after the Multiplication Rule has been discussed.

Exercises 3, 4, 8, and 26–29 require the Addition Rule.

Discovery 7.3

Reliability of systems is an important concept in engineering. The area models drawn in this discovery provide a concrete visualization of why the probability of occurrence of two independent events is found by multiplying their individual probabilities. In contrast with Discovery 7.1, in which area diagrams were used merely to enumerate outcomes, the areas are used here in a quantitative sense to represent probability values. Notice that although the system configuration discussed in item 8 involves an "exclusive or" type of event, the use of the area diagram eliminates the need for bringing up the Addition Rule at this point.

Special Notes on Exercises

Exercises 1 examines probabilities associated with the *Challenger* shuttle disaster.

Exercise 2 discusses coincidences and the "Birthday Problem," which has a result that is surprising to most people. Although not mentioned in the exercise, it can be shown through similar reasoning that it requires a group of 253 people in order for the probability that 2 of them share a particular birthday to reach 50%.

⇒ Exercise 3 discusses non-mutually exclusive events and the use of area models in finding their probabilities.

In Exercise 4, a tree diagram is used to illustrate an answer; it may be helpful for students to have completed Exercise 1 in Section 7.2. This is the first use of a tree diagram to examine events that are not equally likely, which may raise some questions.

⇒ Exercise 5 examines conditional probability in the context of risk factors for disease. Note that there is no implication of causation here: Baldness is not shown to be a cause of heart problems, but it is statistically linked to some extent. The complete study controlled for age, smoking patterns, and so on, so the actual risk factors are not identical with the conditional probabilities.

Exercise 6 examines the notion of false positives in testing.

Exercise 7 presents a role-playing activity in which students investigate the possible savings when samples are pooled in drug testing.

Exercises 8–29 involve practice in calculating joint probabilities.

Section 7.5

Probability distributions are discussed, as well as a few probability models for non-numerical data that don't meet a strict definition of probability distribution based on values of a random variable. The expected value (or mean) of a probability distribution is also introduced.

One day each should be spent on probability distributions and expected value.

Exercises 1, 3, and 6–22 can be assigned after Discovery 7.4 and Example 12.

The remaining exercises require expected value.

Discovery 7.4

This activity is identical to the type of traffic study that is done by highway engineers in preparation for design of new roads, traffic lights, expressway interchanges, and so on. It may not be feasible to count cars at every college, though. If this is not a possibility, some good alternatives may include the following:

- Counting the number of calls received at the college's telephone switchboard each minute.

- Counting the number of students that enter the Registrar's office in different five-minute intervals.

- Counting the number of prospective students per day who visit the Admissions Office. (That office may be able to supply such data.)

- Counting the number of people that go through the cafeteria line each minute.

If none of these activities is feasible, students can be given the sample data table in item 2 and asked to complete the discovery using those data. The data analysis involves constructing histograms. If students have not previously studied histograms, provide them with Handout 7.4. Although the histograms can be made by hand, either calculator or computer spreadsheet programs can be used. Probabilities (relative frequencies) can be given in percentages rather than in decimal form.

In item 8, the 8-car limit may have to be changed if traffic counts are particularly high or low.

Special Notes on Exercises

⇒ Exercise 1 examines the uniform probability distribution.

⇒ Exercise 2 discusses expected monetary value as a special case of expected value. Part (d) is based on the Massachusetts state lottery. The jackpot payoff is in reality a variable amount that depends on the number of tickets submitted.

Exercise 3 introduces a binomial distribution, without naming it, in the context of quality control inspection. Exercise 1 of Section 7.2 (tree diagrams) is a prerequisite.

Exercise 4 (replacement analysis) may be quite difficult for some students, but it is an interesting management application involving probability.

Exercises 5 and 6 are good follow-up activities involving construction of a theoretical and an experimental probability distribution. Exercise 6 is essentially based on quality control measurements for salt packets. If enough packets were used, the distribution should approach a normal distribution. The limits discussed in parts (g)–(i) correspond roughly to two standard deviations above and below the mean. This activity requires a balance accurate to at least 0.01 grams; otherwise, the masses of packets will not show much variability.

Exercises 7–22 involve constructing and interpreting probability distributions.

Exercises 23–29 involve expected value calculations.

Section 7.6

This section contains a very basic introduction to statistical inference based on sampling. The fundamental idea of a confidence interval is examined, without a formal or rigorous definition. Exercise 5 then introduces the notion of statistical significance.

Only one day is needed for this section, although a second could be devoted to Exercise 5.

Discovery 7.5

This experiment requires preparation by the instructor. For each group, a few dozen coffee stirrers should be cut up into pieces varying in size between 2 and 8 centimeters. These are all placed in a sack or other container, and students draw samples of five pieces at a time to measure their lengths. Unless very large numbers of pieces are used, each sample should be returned to the bag after being measured to avoid changing the true mean length of all pieces during the course of the activity. (In actual quality control testing of large lots, sampling would be done without replacement.)

Although confidence intervals for estimation would usually be constructed based on single samples, that would require computation of a standard deviation. We avoid this complication by examining many samples and inferring the size of a confidence interval from the distribution of the sample means. The development presented here is meant only to suggest the meaning of a confidence interval and is not to be construed as a practical method for finding one.

Example 14

The given margin of error of 1.3 ppm is based on the sample standard deviation of 2.54, and a student's *t* value of 1.796 for 90% confidence is based on a sample with 11 degrees of freedom.

Special Notes on Exercises

Exercise 1 examines opinion polls in greater depth than as discussed within the section. A method for approximating the margin of error from the sample size is given, which can be used only for 95% confidence interval determination. It comes from using the formula $E = z_{\alpha/2} \sqrt{\dfrac{pq}{n}}$, where $z_{\alpha/2} = 1.96$ for 95% confidence and p and q both equal 0.5, for a conservative 50% default percentage in favor of the poll question.

Exercise 2 examines the consequences of nonrandom sampling.

⇒ ❑ Exercise 3 introduces sample standard deviation for those instructors who wish to discuss it. Part (c) gives a formula for calculating margin of error that is only valid when the sampling distribution of the mean is normal; this occurs for large sample sizes or for situations when the underlying distribution of the random variable is known to be normal.

Exercise 4 invites students to do their own research on polling. It requires the completion of Exercise 1.

⇒ Exercise 5 allows students to explore how often an event with a known probability of occurrence of 1/2 can happen with a relative frequency that differs from 1/2 by fairly large amounts. The concept of statistical significance is examined.

Exercises 6–11 are drill problems on writing confidence intervals, and 12–16 require more thoughtful understanding of the concepts discussed in the section.

Section 7.7

This section can be considered optional, especially if students don't have calculators with random number generators. It introduces simulations based on probability, which are commonly used in many workplaces.

The section can be completed in one day (Discovery 7.6), with students exploring various exercises on their own.

Discovery 7.6

Students explore a variety of simulators to create a virtual free throw contest between two professional basketball players. As alternatives, students could select other players or base the simulations on their own free throw percentages. However, it is advisable to keep the simulation simple by using round numbers (e.g., 90%, 80%, 70%) for the percentages.

In item 6, the random number generation can be repeated by merely pressing the Enter key.

Note that computers and calculators generate what are called "pseudorandom numbers," based on algorithms that nevertheless randomize the displayed digits. A new calculator will be programmed to use the same "seed" number as all other identical calculators. Therefore, if two students with identical calculators use their **randInt(** commands for the first time, they will get the same random numbers. This may be confusing for some students. The seed number can be manually changed (consult the calculator manual).

Special Notes on Exercises

Exercises 1 explores attributes of various kinds of simulators.

Exercise 2 is a critical-thinking question dealing with the assignment of random numbers.

❏ Exercise 3 is an in-depth investigation of using a calculator to generate random numbers.

❏ Exercise 4 discusses the use of a computer spreadsheet to conduct simulations. The discussion of generating random numbers is valid for older versions of Excel. More recent versions include a "random integer" command, which may have to be specially loaded.

❏ Exercise 5 involves programming a calculator to perform a simulation.

Exercise 6 is a simple simulation that examines the consequences of China's family planning policy of encouraging married couples to have children until a boy is born.

Exercise 7 invites students to research a baseball player's statistics and to design a simulator for that player's batting abilities. Some good resources, besides local newspapers, are:

- *Baseball: The Biographical Encyclopedia*, by David Pietrusza, et al.
- *The Major League Handbook,* by Bill James
- *The Sporting News Baseball Guide*
- *Total Baseball: The Official Encyclopedia of Major League Baseball,* by John Thorn et al (eds.)
- Several Internet sites, including www.baseball-reference.com and The Baseball Archive at www.baseball1.com. These both contain links to other sites.

Alternatively, students may wish to choose a different sport and design a simulation more appropriate to it.

Exercises 9–16 involve practice in designing and executing simulations.

Chapter Review

The breakdown of exercises by section is:

Chapter Summary:	Exercise 1
Section 7.1:	Exercises 2–3
Section 7.2:	Exercises 4–6
Section 7.3:	Exercises 7–11
Section 7.4:	Exercises 12–15
Section 7.5:	Exercises 16–19
Section 7.6:	Exercises 20–21
Section 7.7:	Exercises 22–23

Instructor Notes

Chapter 8: Trigonometry

Goals of the Chapter

- To investigate similar triangles and use their properties to solve real-world problems

- To relate angles and their trigonometric ratios

- To apply trigonometric ratios to determine unknown distances and angles in a variety of contexts

What You Will Learn

Section 8.1

- To identify similar triangles

- To use the properties of similar triangles to solve problems

Section 8.2

- To classify triangles by their sides and by their angles

- To use the converse of the Pythagorean theorem to determine whether a triangle is a right triangle

- To identify similar polygons

- To use the properties of similar polygons to solve problems

Section 8.3

- To find the sine, cosine, and tangent of an acute angle

- To use the sine, cosine, and tangent of an angle to find a missing side of a right triangle when given one side and an acute angle

Section 8.4

- To solve a right triangle

- To use right triangle trigonometry to solve real-world problems

Section 8.5

- To determine the sign of the sine, cosine, and tangent of a non-acute angle

- To solve real-world problems that can be modeled with oblique triangles

What You Need to Know

Section 8.1

- How to find the ratio of two numbers
- How to solve proportions
- How to use a protractor to measure an angle
- How to find lengths of line segments

Section 8.2

- How to use the Pythagorean theorem
- How to identify similar triangles

Section 8.3

- How to write proportions for the corresponding sides of similar triangles
- How to solve proportions

Section 8.4

- How to find the sine, cosine, and tangent of a given angle
- How to use the sine, cosine, and tangent of an angle to find a missing side of a right triangle when given one side

Section 8.5

- How to find the sine, cosine, and tangent of a given acute angle

Instructor Notes for Chapter 8

Introduction to Chapter 8

This chapter is designed to meet the needs of many different curricula and instructors. It is possible that some or all of the sections are appropriate. Similarity is introduced and explored in Section 8.1. Section 8.2 provides a basic review of several geometric terms dealing with triangles and the Pythagorean theorem. Sections 8.3 and 8.4 focus on right triangle trigonometry, and the final section (8.5) explores non-acute angles and oblique triangles.

Preparation Reading

This reading helps make the connection between the familiar (the Pythagorean theorem) and something most likely unfamiliar (trigonometry). It poses a question about finding distances that can't be measured directly and talks briefly about people who might use trigonometry in their lives.

Section 8.1

This section is designed to be completed in one day.

Prior to the activity, a brief history of the Leaning Tower of Pisa is presented. This context was chosen for two reasons: the situation is rich in geometry and the solution to the leaning problem has made news headlines recently. The three questions posed here are solved within the chapter.

Activity 8.1

Students will need copies of Handout 8.1 and protractors for this activity. If protractors are not available, use Blackline Master 8.1 to make transparencies of protractors, which can be given to the students.

This activity provides students with the opportunity to discover what specific attributes of triangles cause them to have the same shape. If students experience difficulty in finding the pairs of triangles with the same shape in Handout 8.1, it might be helpful to have them cut out the eight triangles and match them up. Another option is to give students a transparency of the handout. Students can then place the transparency on top of a paper copy to help them match up the similar pairs.

A word of caution: As students measure the angles of the triangles in the handout, error may be introduced. (Sizes of figures may also have changed slightly during the production of the handout.) If the measures of the angles are off 1 or 2 degrees, you may need to ask students to consider those angles the same size for the purposes of this activity. Students may experience similar errors as they examine the ratios of the lengths of the corresponding sides.

In item 4, students are asked to compute the ratios of the lengths of corresponding sides. Ask students what they think makes sides corresponding. Make a point that corresponding sides are those opposite equal angles.

In the Extend the Activity section, item 11 returns to the original problem posed in the Preparation Reading. To help students understand the solution to this problem, it may be necessary to remind them that the sum of the angles of a triangle is 180°. Also caution students that this "shadow" method works outside because the sun is so far away that the rays of light are considered parallel. Inside, with artificial light, it's a different story because the triangles formed by the objects and shadows will not be similar unless the objects are the same size and the same distance from the bulb.

Section 8.2

This section requires one class period.

Because the material in this section is mainly a geometry review, it might be possible to do one or two Investigations during class time.

The first part of this lesson provides a review of some specific geometric concepts and terms. While discussing isosceles and equilateral triangles, point out that isosceles triangles are not limited to those with *exactly* two equal sides. Hence, any triangle that is equilateral is also isosceles.

Special Notes on Exercises

Exercise 1 presents a problem in which the Pythagorean theorem must be applied before setting up a proportion.

In Exercise 2, students apply the converse of the Pythagorean theorem.

⇒ Exercise 3 uses the converse of the Pythagorean theorem to determine if a triangle is acute, right, or obtuse.

Exercise 4 applies the concept of similarity to pinhole cameras.

In Exercise 5, students use similar figures to determine the diameter of the sun. This investigation can only be done on a sunny day when students can go outside.

Exercise 6 provides another method (in addition to the "shadow" method) of using similar triangles to indirectly measure the height of an object. The project in Exercise 14 asks students to actually try one of these methods to measure an object of unknown height. In Exercise 6, you may need to remind students that when light reflects off a flat surface such as a mirror, the angle of incidence is equal to the angle of reflection (see **Figure 1**).

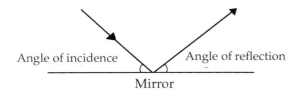

Angle of incidence   Angle of reflection

Mirror

Figure 1

Exercise 7 asks students to verify a relationship given to them in Chapter 6. This exercise makes the connection between the distances observed in the magnifying glass investigation (Section 6.4, Exercise 7) and our study of similar triangles. This problem is one that could be done in class, as the algebra in (c) may be a bit challenging for students.

⇒ In Exercise 8, remind students that many calculators have a feature that changes DMS to decimal degrees, and visa versa. Have them check their manuals.

❏ ⇒ Exercises 9 and 10 provide an opportunity to use a geometric drawing utility to help in exploration. If one is available, have students construct 45°-45°-90° and 30°-60°-90° triangles and observe the lengths of the sides. Handout 8.1 (used in Activity 8.1) can also be used because there are two 45°-45°-90° and two 30°-60°-90° triangles that can be used rather than having students create their own.

The project in Exercise 12 asks students to explore the many different proofs of the Pythagorean theorem. A good source for this information is the Web site www.cut-the-knot.com, which has 29 proofs.

❏ If a geometric drawing utility such as Geometer's Sketchpad is available, have students explore properties of similar triangles by completing Exercise 13.

Exercises 15–42 provide skill practice. Be sure to assign Exercise 36 because this is the first time students are confronted with three sides that don't represent a triangle. You may want to take a few minutes with this exercise to talk about why it doesn't make a triangle.

Exercise 43 asks students to think about different ways of viewing similar triangles and writing proportions for their sides. This exercise leads into the next section, where students will see all three of these forms.

Section 8.3

This section may take one or two days depending on how Discoveries 8.1 and 8.2 are used. Discovery 8.1 should be done as an in-class activity. Discovery 8.2 can be completed outside class.

Discovery 8.1

This exploration provides an opportunity for students to put into practice what they learned about similar triangles in the previous sections. Examining similar triangles and setting up proportions from figures in a book differs greatly from looking at physical situations, determining if the triangles are similar, and then writing and solving the proportions.

In this activity, it very important that the meter stick not move, so you may want to have students secure the top and bottom of the stick with pieces of masking tape.

Discovery 8.2

This discovery activity asks students to measure the sides of several right triangles on Handout 8.2 and to calculate the sine, cosine, and tangent of each of the acute angles.

This activity can be done as described in this Discovery, or it can be replaced with Exercise 9 if you prefer to use a geometric drawing utility such as Geometer's Sketchpad or Cabri. The activities are identical otherwise. Note that both introduce complementary angles.

Item 7 asks students to check the values in their table. If students have calculators with a table feature, suggest that they use this feature to check as shown in **Figure 2**.

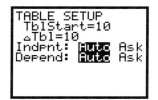 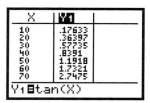 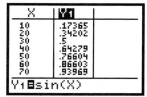 

Figure 2

Sine, Cosine, and Tangent Summary

This is provided as a summary and quick reference for students.

Special Notes on Exercises

Exercise 1 revisits the problem posed in the Preparation Reading and solved using similar triangles in Section 8.1. This time students are asked to reconsider the problem given a side and an angle.

In Exercise 2, students return to one of the initial problems posed regarding the Leaning Tower.

⇒ Exercise 3 introduces the cosecant, secant, and cotangent ratios.

⇒ Exercise 4 returns to the topic of special triangles and builds on two previous exercises (9 and 10 from Section 8.2). For students who completed those, this exercise is appropriate.

⇒ Exercise 5 should be assigned because it introduces the term *angle of elevation*. Problems of this type are scattered throughout the remainder of this chapter.

Exercise 6 is an important exercise for students enrolled in some technical courses, as it introduces the use of sine bars and sine plates to establish angles.

⇒ Exercise 8 explores the topic of radian measure. In this exercise, a strip of paper and a paper plate are needed. For this investigation, adding machine tape works very well. For additional ideas on this investigation, see "Experiencing Radians," by Patrick J. Eggleton, in *Mathematics Teacher* (NCTM), Vol. 92, No. 6 (Sept. 1999), p. 468.

❑ As mentioned earlier, Exercise 9 can be used as a replacement for Discovery 8.2. Or if students need additional help, it can be used to reinforce that activity.

Exercise 10 gives students a right triangle and asks them to identify the hypotenuse, the side opposite a specific angle, and the side adjacent to a specific angle.

Exercises 11–18 provide practice using trigonometric ratios to find missing sides of right triangles.

Exercises 19–23 ask students to explore the connection between the sine and cosine.

Exercise 24 uses the Greek letter θ (theta). If students did not do Exercise 8, you may want to mention that Greek letters are often used to denote angles; in this case, it's the letter θ.

Section 8.4

This section is designed to be completed in one day, with most of the class time spent on making and using a clinometer.

Discovery 8.3

As with Discovery 8.1, having students actually take measurements and do calculations to determine the height of something differs from having them look at a picture of a flag pole or a tower in a textbook. It is very important that students "experience" this activity. If you find it cumbersome to make a clinometer, a simpler device can be made from which the desired angle of elevation can be found directly. This device is made by attaching a straw to a protractor in such a way that the straw can rotate (see **Figure 3**). The major drawback of this device is that the zero edge of the protractor must be held horizontal as the top of the object is sighted through the straw.

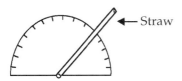

Figure 3

Finding Acute Angles

Up to this point, students have been asked only to find missing sides of triangles. In this section, they are asked to use the calculator to find missing angles. No formal mention is made of inverse functions or of reading $\sin^{-1}$ as "inverse sine."

To help check for understanding, you may wish to go over at least one of the examples (Examples 6–9) in class.

Special Notes on Exercises

The final question regarding the Leaning Tower of Pisa is answered in Exercise 1. If students express interest in what is currently happening with the tower, encourage them to use the Internet, as hundreds of sites are available.

Exercises 2–5 provide a variety of real-world contexts such as building codes, forestry, and agricultural surveying.

The group activity in Exercise 6 has purposely been left open except for the visual directions on how to build a hypsometer. Students are expected to figure out how to use it to get the desired indirect height.

In Exercise 7, to find out when the moon is half full, have students consult the Weather Channel, the Internet, or an almanac. In (a) of this project, the estimation of θ is difficult because the angle is so close to 90° (actually 89.83°). If several members of the class are working on this project, they may wish to average the individual estimates of θ.

Exercises 8–14 ask students to solve right triangles in contextual settings.

⇒ It is important to assign Exercise 9 because it introduces the term *angle of depression.* Problems dealing with angles of depression will appear throughout the remainder of this chapter.

Exercise 12 introduces the concept of the *angle of repose.* Many students will find this problem of interest.

Exercises 15–19 ask students to solve right triangles from information given to them.

⇒ To solve the problems in Exercises 20–21, students will need to solve a system of two equations in two unknowns. Problem 20 will be revisited in Section 8.5, Exercise 20, when the Law of Sines is introduced.

Section 8.5

How (and if) this section is used depends on the instructor's curriculum. This section can be easily broken into two parts so that if only the section on finding trigonometric ratios of non-right angles is needed, that can be accomplished easily.

Part 1: Finding the trigonometric ratios of non-right angles

This includes the exposition in the section and Exercises 1, 2, 3, 4, and 9–13.

Part 2: The Law of Cosines and the Law of Sines

The Law of Cosines is introduced in Exercise 5, and the Law of Sines is introduced in Exercises 6 and 7. Practice is provided in Exercises 14–23.

Special Notes on Exercises

If the introduction to the cosecant, secant, and cotangent ratios was skipped (Exercise 3 in Section 8.3), you may want to skip Exercise 1(b) and Exercise 3.

⇒ Exercise 4 introduces the term *quadrantal angles* and asks students to find the sine, cosine, and tangent of these angles.

⇒ The purpose of Exercises 5, 6, and 7 is to make students aware that trigonometry can be used to solve oblique triangles. It is not intended to achieve mastery.

In Exercise 5, if students find it difficult to remember all three of these forms, have them look for the following pattern: Select any side of the triangle and write the Pythagorean formula with this side as the hypotenuse. Then subtract twice the product of the other two sides and the cosine of the angle opposite the side with which you started.

Exercise 8 is a fairly complicated problem, but it may be of high interest to people who use GPS receivers.

Exercises 14–18 require the Law of Cosines to solve.

In Exercise 14, it might be necessary to remind students that a bearing is an angle measured clockwise from north.

Exercises 19–23 require the Law of Sines to solve.

Exercise 20 revisits a problem from Section 8.4, Exercise 20, where it was solved using a system of two equations. Here the Law of Sines is used.

Draw students' attention to the problem in Exercise 22. Ask them how they know there is only one solution. Redrawing the triangle in a different position, as in **Figure 4,** may help them if they have trouble seeing this.

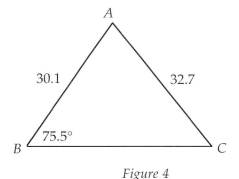

Figure 4

Chapter Summary and Review

For instructors who use only parts of this chapter, the Chapter 8 Review exercises break down in the following manner:

Chapter Summary: Exercise 1

Section 8.1 and 8.2:   Exercises 2–5

Sections 8.3 and 8.4:   Exercises 6–13

Section 8.5: (Part 1):   Exercises 14 and 15

              (Part 2):   Exercises 16 and 17

Chapter 5—Quadratic Functions and Radicals

Goals of the Chapter

- To solve quadratic equations using analytical and graphical methods
- To model real phenomena with quadratic functions
- To simplify radicals and solve radical equations
- To use fractional exponents to represent radicals

Preparation Reading

Recall that a linear equation in two variables has a graph that is a straight line. In such an equation, one of the variables (the dependent variable) is said to be a function of the other (the independent variable). A large variety of situations can be modeled with linear functions, such as weight/volume relationships, the design of housing developments, the relationship between powerboat registrations and manatee deaths in Florida, frictional force, people's heights and forearm lengths, and the elasticity of a spring. For example, **Table 5.1** shows data relating body length L to head circumference H for newborn female babies.

Length (in)	Head Circumference (in)
17.75	12.75
18.25	13.00
19.00	13.25
19.75	13.50
20.00	13.75
20.50	14.00
20.75	14.25

Table 5.1

Figure 5.1 shows a scatter plot of the data from Table 5.1. It also shows the graph of a linear regression line, or least-squares line, based on the data. The equation of the regression line is $H = 0.47L + 4.37$. This linear model provides a good description of the data.

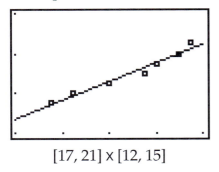

[17, 21] x [12, 15]

Figure 5.1

However, linear models are sometimes insufficient for describing data. **Table 5.2** contains data on the stopping distance D required for a vehicle to stop for various initial speeds S.

Speed (mph)	Stopping Distance (ft)
20	42
30	73.5
40	116
50	173
60	248
70	343
80	464

Table 5.2 Source: U.S. Bureau of Public Roads.

Stopping distance varies with speed in a way that a linear model cannot adequately explain (see **Figure 5.2**).

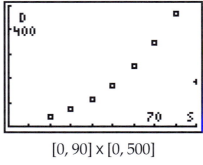

[0, 90] x [0, 500]

Figure 5.2

Polynomial functions of degree two or higher sometimes provide good models for nonlinear types of behavior. The simplest nonlinear polynomial function is a quadratic, or second-degree function, having the general form $y = ax^2 + bx + c$. In this chapter we will explore a number of situations that can be modeled with quadratic functions, as well as the algebraic and graphical properties of these functions.

The analysis of quadratic functions frequently involves square roots. These and other kinds of radicals (for example, cube roots) are found in a variety of applications. Some of the properties of radicals, as well as methods for solving equations containing radicals, will also be examined in this chapter.

Reflect and Discuss *(SE page 3)*

Why might an equation be helpful in modeling a problem involving collected data?

An equation can "smooth out" irregularities in data due to random variation and can be used to predict values for which no data exists, either between available data values or (with caution) beyond the range of the data. Epidemiologists can use models of infectious disease cases to predict the course of the disease. Businesses can make sales projections from models of past sales volume. Engineers can construct calibration curves for instruments from equations based on data.

Section 5.1 Parabolas

What You Need to Know

- How to identify coordinates of points on a rectangular coordinate system
- How to factor polynomials

What You Will Learn

- To describe parabolic shapes with quadratic functions

Materials

- Graph paper
- Bulletin board, cardboard, or thin plywood
- Pushpins or tacks
- Lightweight chain
- Hooks or paper clips
- Thread
- Large metal washers

The Pochuck Quagmire Bridge was completed in 1996 on the Appalachian Trail in New Jersey. It is a pedestrian bridge built out of timber, with "wire rope" suspension cables made of steel. Many such bridges have been built in state and national parks by governmental forest services.

A suspension bridge is the lightest bridge system for a given span and supported weight. This makes suspension bridges good for spanning streams in forests where banks can be unstable. Such bridges can also be constructed without using heavy equipment, which is important because forest trails are often in remote locations.

One of the most common types of bridges, especially for spanning large distances over rivers and harbors, is a suspension bridge. The Golden Gate Bridge (**Figure 5.3**) and the Brooklyn Bridge are two of the most famous ones. There are also many others throughout the United States and the rest of the world.

The most important structural feature of such bridges is the main curved suspension cable that spans the bridge towers. In a large bridge, the main cable is made up of many strands of steel braided together to form a cable with a diameter that can sometimes be greater than one foot.

Throughout this chapter you will examine, explore, and create mathematical models of these curved suspension bridge cables, as well as other real-world situations that exhibit similar characteristics.

Figure 5.3

Activity 5.1 A Suspension Bridge *(SE pages 5–7)*

In this activity you will construct a small model of a suspension bridge cable and then explore its mathematical properties. These properties are essentially the same for all suspension bridges.

Directions for Constructing the Model

- Attach a sheet of graph paper to a vertical piece of cardboard or a bulletin board.

- Pick a point near the upper left-hand corner of the sheet to use as an origin, and draw x- and y-axes from that point. Only the positive part of the x-axis and the negative part of the y-axis will appear on your graph paper (see **Figure 5.4**).

- Label the axes with any convenient scale, using the same scale on both axes.

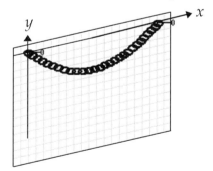

Figure 5.4

- Use a pin or tack to attach a chain exactly at the origin. Then attach another link of the chain somewhere on the x-axis so that the chain hangs in the shape of a suspension bridge cable.

---

1. Record the coordinates of both attachment points and the lowest point in the center of the chain.

1. **A typical result, using 1/4-inch graph paper: Chain is suspended from the points (0, 0) and (42, 0), with the lowest point (vertex) at (21, –13).**

2. Now use small hooks and thread to hang identical weights, such as washers or bunches of paper clips, from the chain at *equally spaced horizontal intervals* (not equally spaced along the curved chain). If you are using a standard $8\frac{1}{2}$ in x 11 in sheet of graph paper, then about seven to nine weights should be sufficient. Did the shape of the chain appear to change much?

> A true parabola will result only from a uniform horizontal loading, that is, where the load is continuously applied to the cable at a constant horizontal rate (measured in pounds per foot or newtons per meter). For a real suspension bridge, as well as for the model constructed here, the load is not continuous but is applied at fixed horizontal intervals. Therefore, the shape is only approximately parabolic.
>
> A cable or chain supporting no load at all has a slightly different shape, called a **catenary,** which has a mathematical form different from that of a parabola. The differences between the two shapes are very small, however.

2. **The shape will probably not change in a measurable way, even though the theoretical shapes of an unloaded chain (a catenary) and of a chain bearing a uniform horizontal load (a parabola) have distinctly different mathematical forms.**

3. When a chain or cable supports a load that is evenly spaced in the horizontal direction, it assumes a shape called a **parabola.** A parabola is always symmetrical across some line, called the **axis of symmetry.** For your chain, the axis of symmetry will be the vertical line that passes through the lowest point of the chain.

   The point where the axis of symmetry and the parabola intersect is called the **vertex** of the parabola. In this case, the vertex is the lowest point on a parabola.

   You can find an equation for your chain's parabolic shape. It will be of the form $y = ax(x - d)$. In this equation, d is the *horizontal distance between the points of attachment of the chain.* Because the equation should be true for any pair of values (x, y) that represent coordinates of a point on the curve, you can find the value of a by substituting the coordinates of the vertex in $y = ax(x - d)$. Determine an equation for your parabola.

3. **Sample answer: For the points given in the answer to item 1, $d = 42$, $x = 21$, and $y = -13$. Then $a \approx 0.0295$, and $y = 0.0295x(x - 42)$.**

4. The equation you have written is an example of a **quadratic function,** which is a **polynomial function** of degree two. It is easier to identify the function as quadratic if you use the distributive property to rewrite your function from item 3. Rewrite your function. Note that it is now in the form $y = ax^2 + bx$. (In this case, the value of b is negative.)

> Recall that a **function** provides a way of finding the value of one variable quantity, called the **dependent variable,** from the values of one or more other quantities, called the **independent variable(s)**. A function is often stated in the form of an equation.
>
> A function is evaluated by substituting a particular value of the independent variable in the equation and finding the corresponding value of the dependent variable. In the Suspension Bridge Activity, x is the independent variable and y is the dependent variable.

4. **Sample answer using the equation found in item 3:** $y = 0.0295x^2 - 1.24x$.

5. You can now check to see whether your function actually describes the curved shape of your chain. Evaluate the function when $x = 10, 15,$ and 30. Check to see if these (x, y) pairs are points on your chain.

5. **The (x, y) pairs should be close to points on the chain.**

6. Does the point $(3, -5)$ lie on your chain? Use your equation from item 4 to justify your answer.

6. **Sample answer: No, the point $(3, -5)$ does not lie on the chain, because the ordered pair $(3, -5)$ is not a solution to the equation $y = 0.0295x^2 - 1.24x$. That is, $-5 \neq 0.0295(3)^2 - 1.24(3)$.**

7. Notice that even when the chain's equation is written in the form $y = ax^2 + bx$, you can still determine the end points and the lowest point of the chain.

   a) The end points each have a y-coordinate of zero so that the x-coordinates must satisfy the equation $ax^2 + bx = 0$. In this equation, the left-hand side can be factored $x(ax + b) = 0$. Then each factor can be set equal to zero ($x = 0$ or $ax + b = 0$), which allows you to solve for x. This follows a general method for solving factorable equations, which will be discussed in Section 5.2.

   Use your equation from item 4 to solve for the endpoints of your chain.

7. a) **Sample answer: $0.0295x^2 - 1.24x = 0$; $x(0.0295x - 1.24) = 0$. Then $x = 0$ or $\left(\dfrac{1.24}{0.0295}\right) \approx 42$.**

   b) Because of the symmetry of the parabola, the lowest point of the chain (the vertex of the parabola) must have an x-coordinate midway between the endpoints. Use your equation from item 4 to solve for the lowest point of the chain.

   b) **Sample answer: The coordinates of the endpoints of the chain are $(0, 0)$ and $(42, 0)$. The x-value midway between 0 and 42 is 21, and $0.0295(21)^2 - 1.24(21) \approx -13$, so the lowest point (vertex) is $(21, -13)$.**

Extend the Activity *(SE pages 8–9)*

8. Use the information in **Figure 5.5** to answer the questions.

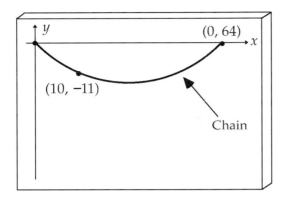

Figure 5.5

   a) What is the equation of the parabola in the form $y = ax(x - d)$?

8. **a) To find a, we know one (x, y) pair (10, −11) and that $d = 64$. So
   −11 = a(10)(10 − 64); a = 11,540 = 0.0204. Hence, $y = 0.0204x(x − 64)$.**

   b) Use the distributive property to rewrite your equation from part (a) in the form $y = ax^2 + bx$.

   b) From part (a), we have the equation $y = 0.0204x(x − 64)$. Using the distributive property: $y = 0.0204x^2 − 1.28$.

   c) Use the equation from either part (a) or part (b) to find the coordinates of the vertex.

   c) The x-coordinate of the vertex is 32. When $x = 32$, $y = 0.0204(32)(32 − 64)$, or $y = −20.9$. The coordinates of the vertex are (32, −20.9).

   d) Describe where the axis of symmetry in the parabola in Figure 5.5 is located.

   d) It is a vertical line passing through the vertex (32, −20.5). The equation of the axis is $x = 32$.

   e) Is the point (40, −19.6) on the parabola? Explain.

   e) Yes, (40, −19.6) lies on the parabola because the ordered pair satisfies the equation: −19.6 = 0.0204(40)(40 − 64).

9. In the original activity, the origin of the coordinate system was placed at one of the points of attachment of the chain. What if we had chosen to place the origin at the vertex of the parabola?

   Repeat the steps in Directions for Constructing the Model, hanging the chain in exactly the same position on the paper as you did in the original activity. Only this time, draw the axes so that the origin is located at the vertex of your parabola (see **Figure 5.6**).

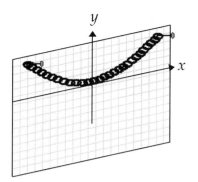

Figure 5.6

a) The equation of the parabola with respect to the new origin is $y = ax^2$, where a is the same number you found in item 3 of the activity. (Note that the value of a depends only on the shape of the parabola, not its position on the axes.) Determine the new equation for your parabola.

9. a) $y = 0.0295x^2$

   b) Verify the accuracy of your equation from (a) by comparing ordered pairs that satisfy the function for this new coordinate system to ordered pairs of coordinates for points on the chain.

   b) The vertex is now at (0, 0), etc. The ordered pairs that satisfy the function should be close to the points on the chain.

10. It is usually the case that the simplest equation for a curve will result if the origin of the coordinate system is placed at a central point of symmetry. But there may be a reason to use another point for the origin. For example, if the roadbed of a bridge were located some distance below the lowest point of the main cable, it might be more convenient to measure everything from an origin located at the height of the road.

   Determine an equation for your chain's parabolic shape with respect to an origin located exactly 5 units below the lowest point of the chain. Why do you think this type of equation might be helpful for the builders of a suspension bridge?

10. **Because such a location for the point (0, 0) would increase the y-coordinate of every point on the parabola by 5, the new equation would be $y = 0.0295x^2 + 5$. For a suspension bridge, this type of equation would give the height of the main cable above the road, which would also give the lengths of all the vertical cables connecting the roadbed to the main cable.**

Section 5.2 Quadratic Functions and Their Graphs

What You Need to Know

- How to factor polynomials
- How to find the square root of a number

What You Will Learn

- To identify the properties of a quadratic function
- To sketch the graph of a quadratic function
- To find the vertex of the graph of a quadratic function
- To find the equation of a parabola when the axis of symmetry is the y-axis
- To find the real solutions to quadratic equations using factoring, graphing, and the quadratic formula
- To interpret the meaning of vertices and horizontal intercepts of real-world quadratic models

Materials

- Small wooden ball
- Large graph paper and cardboard or plywood
- Watercolor paint or washable ink
- Scissors
- Handout 5.1

The functions we examined in Section 5.1 are examples of quadratic functions. Recall that a quadratic function is a polynomial function of degree two. This means that the function contains a term in which the highest power of the independent variable is two, and it can be written in the **standard form** $y = ax^2 + bx + c$, where $a \neq 0$. The domain of a quadratic function includes all real numbers, and the graph is a smooth curve called a parabola.

In Discovery 5.1 we will investigate the properties of quadratic functions as we explore their graphs.

Discovery 5.1 Graphs of Quadratic Functions *(SE pages 11–13)*

In Activity 5.1 we explored several quadratic functions and found that the domain of a quadratic function includes all real numbers and that the graph is a smooth, U-shaped curve called a parabola. Recall that the lowest (or highest) point on the parabola is called the vertex and that the vertical line passing through the vertex is called the axis of symmetry.

When sketching the graph of a quadratic function, it is helpful to first locate the vertex. For a function written in standard form $y = ax^2 + bx + c$, the x-coordinate of the vertex is $-\dfrac{b}{2a}$. After finding the coordinates of the vertex, create a table of values with several x-values smaller and several x-values greater than the x-value of the vertex.

1. a) Consider the graph of $y = x^2 + 2x - 5$. Find the coordinates of the vertex.

1. **a) The x-value of the vertex is $-\dfrac{b}{2a} = -\dfrac{2}{2(1)} = -1$. The y-value of the vertex is**

$(-1)^2 + 2(-1) - 5 = -6$. **So the coordinates of the vertex are the ordered pair $(-1, -6)$.**

b) Complete a table of values similar to **Table 5.3**.

x	y
−4	
−3	
−2	
−1	
0	
1	
2	

Table 5.3

b)

x	y
−4	3
−3	−2
−2	−5
−1	−6
0	−5
1	−2
2	3

c) Graph the parabola by plotting the points from the table and connecting them with a smooth curve.

c)

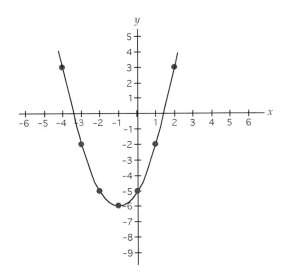

In items 2–4 we will investigate the effects of the constants a, b, and c on the graphs of quadratic functions by examining three special cases. We will consider the graphs of $y = ax^2 + bx + c$ when $b = 0$ and $c = 0$ ($y = ax^2$), when $b = 0$ ($y = ax^2 + c$), when $c = 0$ ($y = ax^2 + bx$), and when none of the constants are equal to zero ($y = ax^2 + bx + c$). In order to explore the characteristics of these graphs, it is helpful to use either a graphing calculator or a computer.

2. Consider the graphs of quadratic functions whose general form is $y = ax^2$.

a) Because $b = 0$ for this function, what are the coordinates of the vertex of each graph for any value of a? Explain.

2. a) (0, 0) or the origin. When b = 0, the x-coordinate of the vertex is $-(0/2a) = 0$. And when $x = 0$, $y = 0$.

b) Using a window [–10, 10] × [–10, 10], graph $y = ax^2$ for $a = 5, 3, 1,$ and 0.5 on the same set of axes. What do you observe as the values of a become smaller? Name three things these four parabolas have in common.

b) As the value of a gets smaller, the parabola gets wider. The vertex and the axis of symmetry are the same for each parabola. Each graph opens upward.

c) Using a window [–10, 10] × [–10, 10], graph the functions when $a = 5, -5, 2,$ and -2. What do you observe about the graphs when $a > 0$? When $a < 0$?

c) **When $a > 0$, the parabola opens upward. When $a < 0$, the parabola opens downward.**

d) What is the domain of the function $y = ax^2$? What is the range of the function when $a > 0$? What is the range of the function when $a < 0$?

d) **The domain of the function is all real numbers. When $a > 0$, the range is numbers greater than or equal to 0. When $a < 0$, the range is numbers less than or equal to 0.**

3. Consider the graphs of quadratic functions whose general form is $y = ax^2 + c$.

a) To explore the effect that the value of a has on the graph, on the same set of axes graph $y = ax^2 + 3$ for various values of a. What happens when $a > 0$? What happens when $a < 0$?

3. a) **The vertex is located at (0, 3). When $a > 0$, the graph opens upward. When $a < 0$, the graph opens downward.**

b) To explore the effect that the value of c has on the graph, on the same set of axes graph $y = 2x^2 + c$ for various values of c. (Hint: Don't forget to let $c = 0$ and to use values less than 0.) What do you observe about the general shape of each of the graphs? What do you observe about the axes of symmetry? What do you observe about the vertices?

b) **The vertex is located at (0, c). The general shapes remain unchanged. The axis of symmetry is the y-axis for each graph. The vertex of each graph lies on the y-axis.**

c) Because $b = 0$ for the function $y = ax^2 + c$, what is the x-coordinate of the vertex of the graph for any values of a or c? What is the y-coordinate?

c) **The x-coordinate of the vertex is 0. The y-coordinate is c.**

d) How many x-intercepts does the graph have when a is positive? How many when a is negative?

d) **When a is positive and $c < 0$, there are two x-intercepts. When $c = 0$, there is one. And when $c > 0$, there are no x-intercepts.**

When a is negative and $c > 0$, there are two x-intercepts. When $c = 0$, there is one. And when $c < 0$, there are no x-intercepts.

e) How many y-intercepts does the graph have? What are their coordinates?

e) **There is only one y-intercept, the point (0, c).**

4. Consider the graphs of quadratic functions whose general form is $y = ax^2 + bx$. Graph $y = ax^2 + bx$ for various non-zero values of a and b.

a) When does the graph open upward? When does it open downward?

4. a) The graph opens upward when $a > 0$ and downward when $a < 0$.

b) For each graph, what do you notice about its axis of symmetry? What do you notice about the x-intercepts?

b) The axis of symmetry varies, but it is not the y-axis in any of the graphs. In each graph, one of the x-intercepts is the origin.

5. Consider the graph of $y = x^2 - 4x + 1$. Before graphing this function, consider the observations you've made about graphs of quadratic functions in items 2–4, then make conjectures about the following:

a) Does the graph of the function open upward or downward? Explain.

5. a) Because $a > 0$, the graph opens upward.

b) What are the coordinates of the vertex of the graph?

b) The x-coordinate of the vertex is $-\dfrac{b}{2a} = -\dfrac{(-4)}{2(1)} = 2$. The y-coordinate is -3. Thus, the coordinates of the vertex are $(2, -3)$.

c) What is the equation of the axis of symmetry of the graph?

c) The vertical line $x = 2$.

d) What is the domain of the function? What is the range?

d) The domain is all real numbers. The range is all numbers greater than or equal to -3.

e) Graph the function to see if your conjectures for (a)–(d) were correct.

e) Answer depends on student conjectures.

6. Now consider the functions $y = 2x^2 + 3x + 6$ and $y = -3x^2 + 12x + 6$. Before graphing each, predict how you think they will differ. Then graph each to see if you were correct. (Hint: Consider how the graphs open, the locations of the vertices, and the axes of symmetry.)

6. Predictions will vary. Some things that should be predicted are that neither function has an axis of symmetry that is the y-axis, the x-coordinate for the vertex of $y = 2x^2 + 3x + 6$ is $-\dfrac{3}{4}$, the x-coordinate for the vertex of $y = -3x^2 + 12x + 6$ is $-\dfrac{12}{6} = 2$, and the graph of $y = 2x^2 + 3x + 6$ opens upward, whereas the graph of $y = -3x^2 + 12x + 6$ opens downward.

In Discovery 5.1 we found that the graphs of quadratic functions in the form $y = ax^2 + bx + c$ have the following properties:

- There is a single maximum or minimum point called the vertex that lies on the axis of symmetry. The x-coordinate of the vertex is $-\dfrac{b}{2a}$, and the axis of symmetry is the line $x = -\dfrac{b}{2a}$.

- The graph of a quadratic function is a U-shaped curve called a parabola. When a is positive, the parabola opens upward. When a is negative, the parabola opens downward.

- The domain of a quadratic function includes all real numbers. For parabolas that open upward, the range includes all numbers greater than or equal to the y-value of the vertex. For parabolas that open downward, the range includes all numbers less than or equal to the y-value of the vertex.

- There is exactly one y-intercept, but there may be zero, one, or two x-intercepts.

Quadratic Equations

As we have seen, quadratic functions can be used to model the shapes of bridge cables. Structural arches and the paths of rockets can also be modeled by quadratic functions. In such cases, we might ask questions like "What is the highest point on the arch?" or "When does the rocket hit the ground?" To help answer these questions, it becomes necessary to solve quadratic equations.

> A **quadratic equation** in x can be written in the form $ax^2 + bx + c = 0$, called its **standard form.** A quadratic equation can have no more than two real solutions but may have only one or none.

Several different methods can be used to find the solutions to a quadratic equation. In this section we will examine three of these methods: factoring, graphing, and using the quadratic formula.

Solving Quadratic Equations by Factoring

In item 7(a) of Activity 5.1 we made use of factoring to solve an equation. Justification for this method is provided by the **Zero Product Property.**

> **Zero Product Property**
> If a and b are numbers and $a \cdot b = 0$, then a or b (or both) is equal to 0.

Consider the equation $3x^2 + 12x = 0$. Factoring the left side of the equation, we get $3x(x + 4) = 0$. From the Zero Product Property, we know that if $3x(x + 4) = 0$, then $3x = 0$ or $x + 4 = 0$. If $3x = 0$, then $x = 0$. And if $x + 4 = 0$, then $x = -4$. We then conclude that $x = 0$ or $x = -4$. We say that 0 and -4 are the solutions to the equation $3x^2 + 12x = 0$. These solutions are also called **roots** of the equation or **zeros** of the polynomial function $y = 3x^2 + 12x$.

The Zero Product Property can be applied in a similar manner when solving any quadratic equation that can be factored.

Example 1

Solve by factoring:

a) $2x^2 + 10x - 48 = 0$

b) $H^2 - 7H = 18$

Solutions:

a)
$2x^2 + 10x - 48 = 0$	Original equation
$2(x^2 + 5x - 24) = 0$	Factor common equation.
$2(x + 8)(x - 3) = 0$	Factor trinomial.
$x + 8 = 0$ or $x - 3 = 0$	Set factors equal to 0.
$x = -8$ or $x = 3$.	Solve for x.

Note that in the solution we did not set the factor 2 equal to 0 because $2 \neq 0$.

b) First the equation $H^2 - 7H = 18$ must be rewritten in standard form where one side of the equation is 0.

$H^2 - 7H - 18 = 0$	Standard form
$(H - 9)(H + 2) = 0$	Factor.
$H - 9 = 0$, or $H + 2 = 0$	Set factors equal to 0.
$H = 9$ or $H = -2$.	Solve for H.

Example 2

When a highway passes over a hill, engineers often design the road's vertical curve in the shape of a parabola. This is done to make it easier for the driver's eye to follow the changing slope of the road. **Figure 5.7** shows a drawing of the cross section of a hill with a road passing over it.

A typical survey for such a road would measure from an origin placed at the point where the straight rise of the road on one side of the hill meets the start of the curve.

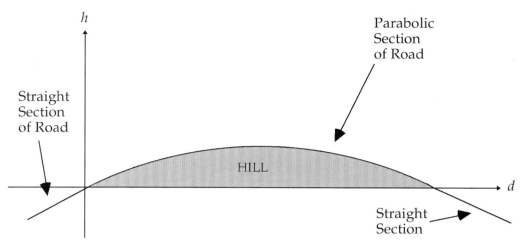

Figure 5.7

The equation of the curved portion of this road is $h = 0.082d - 0.00025d^2$, where h is the height of the road above the origin of every point on the curve and d is the horizontal distance from the origin (all distances measured in feet).

1. Assume that the curve starts and stops at the same elevation. Find the coordinates of the points where the curve meets the straight sections of the road.

2. Find the horizontal distance between the ends of the curved portion of the road.

3. Find the vertex of the parabolic portion of the road. What do the coordinates of the vertex indicate in this context?

Solution:

1. Because the ends of the curved portion are at the same elevation and the left end is at the origin, both ends are at points with h-coordinates of 0. The d-coordinates of these points must then satisfy the equation:

$0.082d - 0.00025d^2 = 0$	Original equation
$d(0.082 - 0.00025d) = 0$	Factor.
$d = 0$ or $d = 328.$	Solve for d.

   The coordinates of the points are $(0, 0)$ and $(328, 0)$.

2. The distance between the two points is 328. So the horizontal distance from one end of the curve to the other is 328 feet.

3. The vertex of the parabola has a d-coordinate of $-\dfrac{b}{2a} = -\dfrac{0.082}{2(-0.00025)} = 164.$
   When $d = 164$, $h \approx 6.7$, so the coordinates of the vertex are $(164, 6.7)$.

   These coordinates indicate that the highest point, or summit, of the road is located a distance d (164 feet) from the origin. Again, note that this is halfway between the ends of the curved sections of the road. The height of the road h at that point is 6.7 feet above the ends of the curve.

Using the Quadratic Formula to Solve Quadratic Equations

Not all quadratic expressions are factorable. In fact, only a small percentage of quadratic functions that are used to model real-world situations can be factored. Even when the factoring method cannot be used to solve a quadratic equation, it is possible to find exact solutions.

Once an equation is written in standard form $y = ax^2 + bx + c$ and the constants a, b, and c are identified, the **quadratic formula** can be used to find the solutions to the equation. (This formula is derived in Exercise 11.)

The Quadratic Formula

The solutions to a quadratic equation that is written in standard form $y = ax^2 + bx + c$ are given by the formula $x = \dfrac{-b \pm \sqrt{b^2 - 4ac}}{2a}$.

This one formula represents two solutions:

$$x = \frac{-b + \sqrt{b^2 - 4ac}}{2a} \quad \text{and} \quad x = \frac{-b - \sqrt{b^2 - 4ac}}{2a}$$

Example 3

Use the quadratic formula to solve the equation $5x^2 + 11x + 4 = 0$.

Solution:

The equation $5x^2 + 11x + 4 = 0$ is in standard form with $a = 5$, $b = 11$, and $c = 4$.

Hence, $x = \dfrac{-b \pm \sqrt{b^2 - 4ac}}{2a} = \dfrac{-11 \pm \sqrt{11^2 - 4(5)(4)}}{2(5)} = \dfrac{-11 \pm \sqrt{41}}{10}$.

This expression represents the two solutions $x = \dfrac{-11 + \sqrt{41}}{10} \approx -0.460$

and $x = \dfrac{-11 - \sqrt{41}}{10} \approx -1.740$. Remember that whereas $x = \dfrac{-11 + \sqrt{41}}{10}$ and

$x = \dfrac{-11 - \sqrt{41}}{10}$ are exact solutions of the equation, -0.460 and -1.740 are only approximate solutions.

Example 4

Use the quadratic formula to solve the equation $9t^2 - 4 = 0$.

Solution:

The equation $9t^2 - 4 = 0$ is in standard form with $a = 9$, $b = 0$, and $c = -4$. Hence,

$$t = \frac{-b \pm \sqrt{b^2 - 4ac}}{2a} = \frac{-(0) \pm \sqrt{0^2 - 4(9)(-4)}}{2(9)} = \frac{\pm\sqrt{144}}{18}.$$ The two solutions are

$$t = \frac{\sqrt{144}}{18} = \frac{12}{18} = \frac{2}{3} \text{ and } t = \frac{-\sqrt{144}}{18} = \frac{-12}{18} = -\frac{2}{3}.$$

Example 5

Use the quadratic formula to solve the equation $3r^2 = 5r + 6$.

Solution:

The equation must first be rewritten in standard quadratic form as $3r^2 - 5r - 6 = 0$. Now we can see that $a = 3$, $b = -5$, and $c = -6$.

$$r = \frac{-b \pm \sqrt{b^2 - 4ac}}{2a} = \frac{-(-5) \pm \sqrt{(-5)^2 - 4(3)(-6)}}{2(3)} = \frac{5 \pm \sqrt{97}}{6}.$$ The two solutions are

$$r = \frac{5 + \sqrt{97}}{6} \approx 2.475 \text{ and } r = \frac{5 - \sqrt{97}}{6} \approx -0.808.$$

The quadratic formula can be used to solve any quadratic equation in one variable, including those that can be solved by other methods. For example, the equation $2x^2 + 10x - 48 = 0$, from Example 1, can be solved using the formula to give $x = \dfrac{-b \pm \sqrt{b^2 - 4ac}}{2a} = \dfrac{-10 \pm \sqrt{10^2 - 4(2)(-48)}}{2(2)} = \dfrac{-10 \pm \sqrt{484}}{4}$, which reduces to the solutions 3 and −8. But because the equation is factorable, it may be quicker to solve this equation by factoring.

Identifying an equation as quadratic may require careful inspection or even rewriting the equation in a different form. The equation $(2 - 3x)^2 = 15$ can be rewritten as $9x^2 - 12x - 11 = 0$. It is a quadratic equation in either form, but the second form may be easier to recognize.

On the other hand, $x^2 = (5 - x)^2$ is not a quadratic equation, because it simplifies to $10x - 25 = 0$, which is a linear equation. If all terms are collected on one side of an equation and written in polynomial form, the correct degree of an equation can be identified.

Using Graphs to Solve Quadratic Equations

Approximate solutions to quadratic equations can be found by constructing and analyzing graphs. Although the following method applies even for hand-drawn graphs, a graphing calculator or other graphing technology is required for accuracy.

Example 6

Use a graphing utility to solve the equation $5x^2 + 11x + 4 = 0$. (This is the same equation that was solved using the quadratic formula in Example 3.)

Solution:

Enter the function $y = 5x^2 + 11x + 4$ on the function screen of a graphing calculator. By choosing an appropriate window, a complete graph of the function, including the intercepts and the vertex, can be shown.

As an aid in choosing the window, it is often helpful to find the vertex of the parabola. For this graph, the vertex is $\left(-\dfrac{11}{10}, -\dfrac{41}{20}\right)$, which is close to the point $(-1, -2)$. **Figure 5.8** uses the window $[-3, 1] \times [-3, 5]$.

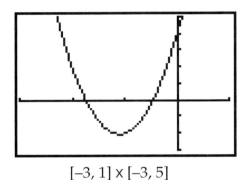

$[-3, 1] \times [-3, 5]$

Figure 5.8

The graph shows that one of the x-intercepts lies between $(-2, 0)$ and $(-1, 0)$, while the other lies between $(-1, 0)$ and $(0, 0)$. The x-coordinates of these intercepts are the solutions of the equation $5x^2 + 11x + 4 = 0$. In order to examine the first of these more closely, we can zoom in on the intercept to get a more accurate reading of its location.

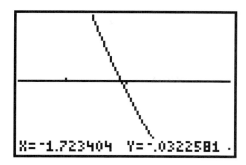

Figure 5.9

Figure 5.9 is the result of one such zooming process. Using the TRACE command, we can see that the graph crosses the *x*-axis at approximately $x = -1.74$, so this approximates one of the solutions of the equation. If a more accurate estimate is required, the zooming process can be repeated.

Example 7

Figure 5.10 is a sketch of a computer circuit board that is a 10.6 cm by 6.2 cm rectangle.

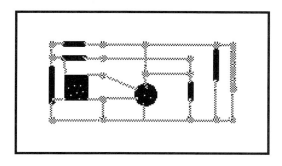

Figure 5.10

The circuit components are to be printed onto a rectangular area in the interior of the board, covering 90% of the board's area and leaving a border of uniform width. Determine the required width of the border.

Solution:

Begin by drawing a picture (see **Figure 5.11**).

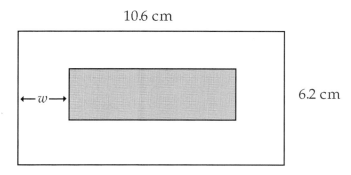

10.6 cm

$\leftarrow w \rightarrow$

6.2 cm

Figure 5.11

Then write a "word equation." That is, write a statement using the word *equals*, which indicates the relationship between the area of the printed part and the area of the board. One possibility is:

The area of the printed portion equals 90% of the total area of the board.

Rewriting the problem in this way allows us to focus on the mathematical formulation of the problem:

(length of printed portion)(width of printed portion) = 0.90(65.72 cm^2) = 59.148 cm^2

Replace the length and width of the printed portion with expressions in terms of w:

$(10.6 - 2w)(6.2 - 2w) = 59.148$.

This equation is quadratic. In standard quadratic form, it is

$4w^2 - 33.6w + 6.572 = 0$.

Use the quadratic formula to find the values of w that make the equation true:

$$w = \frac{-(-33.6) \pm \sqrt{(-33.6)^2 - 4(4)(6.58)}}{2(4)} \approx 0.20 \text{ cm or } 8.20 \text{ cm}.$$

A border width of 8.20 cm makes no sense because it is larger than the width of the board, so 0.20 cm is the only feasible solution.

As was illustrated in Example 7, there are times when a quadratic equation that is used to model a real-world situation has a solution(s) that makes no physical sense. Making note of the problem domain prior to solving the equation helps in making decisions about which solutions are feasible.

Exercises 5.2

I. Investigations *(SE pages 21–25)*

1. Consider the function $y = -3x^2$.

   a) From looking at the equation, what do you know about the graph of the function? In other words, what is the shape of the graph, where is the vertex, where is the axis of symmetry, and how does the graph open?

1. **a) Because the equation is of the form $y = ax^2$, the graph is a parabola with its vertex at the origin. The axis of symmetry is the y-axis, and the graph opens downward.**

   b) When $x = 2$, what is the value of the function?

   b) When $x = 2$, $y = -12$.

   c) From the information in (a) and (b), sketch a graph of $y = -3x^2$.

   c) Because the point (2, –12) is on the graph, from symmetry across the y-axis, we know that the point (–2, –12) is also on the graph.

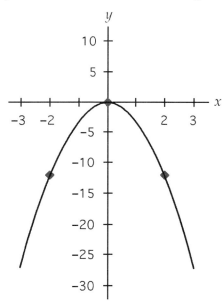

2. The Golden Gate Bridge in San Francisco is a suspension bridge in which the two towers are approximately 1280 m apart (see **Figure 5.12**). The two main suspension cables are attached to the tops of the towers 213 m above the mean high water level in San Francisco Bay, and the lowest point of each cable is 67 m above the water.

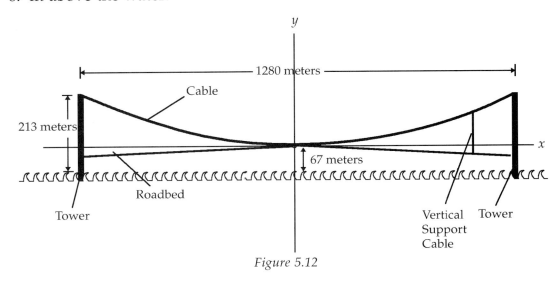

Figure 5.12

a) As we saw in the activity in Section 5.1, the main cables of a suspension bridge hang in the shape of a parabola. Find an equation for the shape of one of the cables of the Golden Gate Bridge. (Assume that the origin is placed at the center of the cable, as shown in Figure 5.12.)

2. **a) If the origin is placed at the center of the cable (the vertex of the parabola), the equation is of the form $y = ax^2$. One end of the cable is at the point (640, 146). Then $a(640)^2 = 146$, and $a = 0.0003564$. The equation of the cable is $y = 0.0003564x^2$.**

b) The roadbed of the bridge is not flat but about 6 meters higher in the center than at the towers. (This is partly to allow for downward deflection when the bridge is supporting a load.) If the center of the roadbed is 67 m above the mean high water level, find an equation for the height of the roadbed at any point. Assume that the roadbed also has a parabolic shape, and use the same coordinate system as in part (a).

b) **The end of the road (at one of the towers) has coordinates (640, –6). Then from $y = ax^2$, we have $a(640)^2 = -6$, and $a = -1.46 \times 10^{-5}$. The equation for the roadbed is $y = -0.0000146x^2$.**

c) Vertical support cables are used to connect the roadbed to the suspension cables. (One is shown in Figure 5.12.) Use your equations from (a) and (b) to find the length of a vertical support cable that is located 300 m from one of the towers.

c) One such cable is located at an x-coordinate of (640 – 300) = 340. When $x = 340$, the distance from the cable to the horizontal axis is given by the equation $y = 0.0003564x^2$ or $y = 0.0003564(340)^2 \approx 41.2$ m. The distance from the horizontal axis to the roadbed is given by the equation $y = -0.0000146x^2$ or $y = -0.0000146(340)^2 \approx -1.7$ m. So the total length of the cable is approximately 41.2 + 1.7 = 42.9 m.

3. The Gateway Arch in St. Louis, Missouri, is 630 ft high and 630 ft wide at its outermost extremes (see **Figure 5.13**). Although its shape is actually a curve called a catenary, it can be approximated by a parabola.

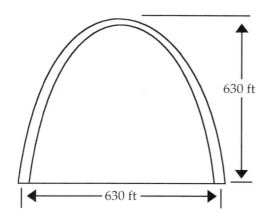

Figure 5.13

a) Assume that the origin of the coordinate system is located at the highest point of the arch. What are the coordinates of the point at the lower right-hand edge of the arch?

3. a) The point at the lower right-hand edge of the arch has coordinates (315, –630).

b) What is the equation of the outer edge of the arch?

b) The general equation for a parabola with its vertex at the origin is $y = ax^2$. Substituting these coordinates for x and y in the equation gives $-630 = a(315)^2$, from which $a = -630/(315)^2 \approx -0.00635$. The equation is $y = -0.00635x^2$.

c) What is the equation of the arch if the origin is located on the ground directly beneath the center of the arch?

c) Every point on the parabola will have a y-coordinate exactly 630 greater than in (b). The equation is $y = -0.00635x^2 + 630$.

d) What is the equation of the arch if the origin is located on the ground at the leftmost edge of the arch? (See item 3 of Activity 5.1.)

d) The equation will be of the form $y = ax(x - d)$. The distance d is 630 ft, and a is the same number as in (b) and (c). The equation is $y = -0.00635x(x - 630)$ or $y = -0.00635x^2 + 4x$.

e) Why might someone involved in construction of the arch have a preference for one of the equations in (b), (c), and (d)?

e) $y = -0.00635x^2$ is the simplest mathematical form. Both of the forms $y = -0.00635x^2 + 630$ and $y = -0.00635x^2 + 4x$ have the advantage that the y-values represent the heights of various points on the arch and might be more useful in the construction of such an arch.

4. Not every U-shaped graph is a parabola. There are other polynomial functions that can have U-shaped graphs that differ from parabolas.

   a) On the same set of axes, graph the functions $y = x$, $y = x^2$, $y = x^3$, $y = x^4$, and $y = x^5$.

4. a) Labeling the functions a–e respectively,

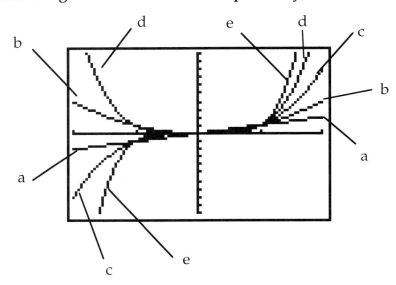

$[-2, 2] \times [-10, 10]$

b) What are your observations about the graphs of these functions? Write a paragraph to summarize your observations.

b) All but $y = x$ have curved graphs. Every graph passes through (0, 0) and (1, 1). (0, 0) is the only intercept for every graph. As each of the graphs moves away from 0 horizontally, it either increases without any limit or decreases without any limit; the higher the power, the faster the increase (or decrease). Graphs of even-powered functions are U-shaped, with the y-axis as an axis of symmetry, and pass through (–1, 1). Graphs of odd-powered functions are symmetric with respect to the origin. That is, for every point (x, y) on the graph, $(-x, -y)$ is also on the graph. Hence, the odd-powered functions pass through (–1, –1).

c) From your observations in (b), what do you expect that the graph of
$y = x^6$ would look like? What about $y = x^7$?

c) **The graph of $y = x^6$ should be a relatively narrow U-shaped curve
passing through (0, 0), (1, 1), and (–1, 1) and symmetric with respect to
the y-axis. The graph of $y = x^7$ should be a steep curve showing
symmetry with respect to the origin and passing through (0, 0), (1, 1),
and (–1, –1).**

Because each of the polynomials investigated in this exercise contains only one
term, they form a special class of functions called **power functions.** More general
polynomial functions (that is, those containing more than one term) can have
more complicated shapes with many changes of direction.

5. The formula $d = 16t^2$ gives the distance, d, in feet that an object travels in t
seconds after it is dropped. This formula ignores the effect of air resistance.

a) Use this formula to complete **Table 5.4,** with the third column
representing the distance the object fell in the previous half second, which
measures the speed of the object.

Time After Object is Dropped (s)	Distance the Object Has Fallen (ft)	Distance Fallen in Previous Half-Second (ft)
0	0	******
0.5	4	4
1	16	12
1.5		
2		
2.5		
3		
3.5		
4		
4.5		
5		

Table 5.4

5. a)

Time After Object is Dropped (s)	Distance the Object Has Fallen (ft)	Distance Fallen in Previous Half-Second (ft)
0	0	* * * * * * *
0.5	4	4
1	16	12
1.5	36	20
2	64	28
2.5	100	36
3	144	44
3.5	196	52
4	256	60
4.5	324	68
5	400	76

b) What do you notice about successive values in the third column of the table? What does this tell you about the speed of the falling object?

b) Each number is 8 units greater than the previous one. The speed is a linear function of time.

c) Add a fourth column to your table and label it "Change in Distance Fallen per Half-Second." Subtract the entries in the third column to obtain these values.

c)

Time After Object is Dropped (s)	Distance the Object Has Fallen (ft)	Distance Fallen in Previous Half-Second (ft)	Change in Distance Fallen per Half-Second (ft)
0	0	*******	*******
0.5	4	4	*******
1	16	12	8
1.5	36	20	8
2	64	28	8
2.5	100	36	8
3	144	44	8
3.5	196	52	8
4	256	60	8
4.5	324	68	8
5	400	76	8

d) The acceleration of any object that continues to move in one direction is defined as the change in speed per unit time. By looking at the fourth column of your table, what can you say about the acceleration of the object in this problem?

d) **Its acceleration is constant.**

e) The values in the third column of the table can be called the **first differences** of the function values. Those in the fourth column are then called **second differences.** The patterns you have noticed here are general characteristics of quadratic functions. How would you summarize these characteristics?

e) **The first differences of a quadratic function increase (or decrease) at a constant rate. The second differences are constant.**

6. A model rocket rises vertically so that its height above the ground (in feet) is given by $h = 300t - 16t^2$, with time measured in seconds.

a) At what time will the rocket return to the ground? (Hint: What is the value of h when the rocket is on the ground?) Write an appropriate equation for t, solve it, and interpret your solution(s).

6. a) $h = 0$ **at ground level. The equation becomes** $300t - 16t^2 = 0$**, which can be solved by factoring:** $4t(75 - 4t) = 0$**, so** $t = 0$ **or** $18\dfrac{3}{4} = 18.75$ **seconds.** $t = 0$ **when the rocket is launched, so it returns to the ground after 18.75 s.**

b) Considering your answer to part (a), what is the problem domain for this situation? (That is, what are the possible values of time during the flight of the rocket?)

b) $0 \le t \le 18.75$ **s; times less than 0 or greater than 18.75 s make no sense.**

c) At what time will the rocket be at a height of 1000 ft? Again, write and solve an equation and interpret the solution(s).

c) $300t - 16t^2 = 1000$**, or** $16t^2 - 300t + 1000 = 0$**. Solutions are** $t = 4.34$ **s and** $t = 14.41$ **s. The smaller time is when the rocket is rising through the 1000 ft height, and the larger is when it is falling through 1000 ft after reaching its peak height.**

d) Write the quadratic equation for the time when the rocket would be at a height of 2000 ft.

d) $300t - 16t^2 = 2000$**, or** $16t^2 - 300t + 2000 = 0$**.**

e) If you try to solve this equation using the quadratic formula, a difficulty arises. What is it?

e) $(b^2 - 4ac) = -38{,}000$**. The square root of a negative number is not a real number, so there is no real number solution for** t**.**

f) How can you interpret this result?

f) The rocket will not reach a height of 2000 ft.

g) It is possible for a quadratic equation to have no real solutions. You can identify such equations by examining the quantity $(b^2 - 4ac)$, which is called the **discriminant** of the quadratic equation because it allows you to discriminate among the types of possible solutions. If the discriminant is positive, there will be two real and distinct solutions, as you saw in (b). If the discriminant is negative, as in (d), there will be no real solutions. What if the discriminant equals zero? See if you can determine for what rocket height $(b^2 - 4ac)$ would equal zero for this problem, then find the corresponding time and interpret the result.

g) $(b^2 - 4ac) = 0$ **if** $(-300)^2 - 4(16)(h) = 0$**, which will occur for** $h = 1406.25$ **ft. The corresponding quadratic equation is** $300t - 16t^2 = 1406.25$**, which has only one solution,** $t = 9.375$ **s. This is the time at which the rocket will be at its maximum height.**

7. To solve the quadratic equation $4x^2 - 16 = 0$, we can use factoring, the quadratic formula, or graphical methods. However, another method could be used.

   a) Solve the equation for x^2.

7. **a) $4x^2 = 16$; $x^2 = 4$.**

   b) You should now have an equation that gives the value of x^2. Solve the equation by taking the square root on both sides. There are two numbers that have a square of 4. What are they?

   b) 2 and –2

   Remember that positive real numbers have two square roots—one is positive and the other is negative. Although using this square root property is an acceptable method for solving quadratic equations of this type, it is necessary to be careful in order to avoid overlooking negative solutions.

   c) Solve $3x^2 - 62 = 13$ using square roots.

   c) $3x^2 = 75$; $x^2 = 25$; $x = 5$ or –5.

   d) Solve $7x^2 - 91 = 0$ using square roots.

   d) $7x^2 = 91$; $x^2 = 13$; $x = \sqrt{13}$ or $-\sqrt{13}$, which are approximately 3.606 and –3.606.

   e) Solve $23x^2 - 40 = 0$ using square roots.

   e) $23x^2 = 40$; $x^2 = \dfrac{40}{23}$; $x = \sqrt{\dfrac{40}{23}}$ or $-\sqrt{\dfrac{40}{23}}$.

   f) Solve $x^2 + 7 = 0$ using square roots.

   f) $x^2 = -7$; $x = \pm\sqrt{-7}$, which is not a real number. There are no real solutions to $x^2 + 7 = 0$.

8. A golf ball hit by Tiger Woods from an elevated tee follows a trajectory given by the equation $h = -0.0030x^2 + 0.87x + 2.5$ until it hits the fairway. For this function, x represents the horizontal distance of the ball from the tee (in yards) and h is the height of the ball measured from the level of the flat fairway (also in yards). Air resistance is ignored.

   a) Construct a graph of the function representing the ball's trajectory. If you use a graphing calculator, choose an appropriate window so that the entire trajectory is displayed.

8. a)

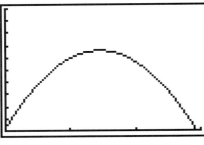

[0, 300] **×** [0, 100]

b) What is the elevation of the tee?

b) At the tee $x = 0$, so $h = 2.5$ and the height of the ball on the tee is 2.5 yards.

c) What is the height of the ball above the fairway after it has traveled 100 yards horizontally from the tee?

c) When $x = 100$, $h = -0.0030(100)^2 + 0.87(100) + 2.5 = 59.5$ yd.

d) How far does the ball travel horizontally before it hits the fairway?

d) When the ball hits the fairway, $h = 0$, so the equation $-0.0030x^2 + 0.87x + 2.5 = 0$ must be solved. Using the quadratic formula, the solutions are $x \approx 292.8$ and $x \approx -2.8$. The positive result is the only one that makes sense for the length of the drive, so the ball travels about 293 yards horizontally.

e) What is the maximum height reached by the ball?

e) The maximum height occurs at the vertex of the parabola that is the graph of the ball's trajectory. The vertex has an x-coordinate of $\dfrac{-b}{2a} = \dfrac{-0.87}{2(-0.0030)} = 145$. The h-coordinate of the vertex is $-0.0030(145)^2 + 0.87(145) + 2.5 = 65.575$, so the maximum height of the drive is about 66 yards above the fairway.

9. A small startup manufacturing company intends initially to make a single product. It would be helpful to know what production level is necessary in order for the company to cover its costs and break even. The company has done a market analysis and found that if it charges $15 for each item, it can expect to sell 800 items per week, but the sales will drop to 600 items per week if the price rises to $20, with a further drop-off of 200 fewer items sold for each $5 price rise.

a) Write a function that expresses the number of items sold per week N as a function of price p, assuming that the relationship is linear.

9. a) The function is of the linear form $N = mp + b$. Here $\dfrac{800 - 600}{15 - 20} = \dfrac{200}{-5} = -40$ units per dollar. The point-slope form of the equation is $(N - 800) = (-40)(p - 15)$, which simplifies to $N = -40p + 1400$.

b) The weekly revenue, or income, equals the number of items sold per week multiplied by the price of one item. This relationship can be expressed as $R = Np$. Write R in two ways: as a function of p only and as a function of N only.

b) $R = (-40p + 1400)(p) = -40p^2 + 1400p$. Alternatively, solve $N = -40p + 1400$ for p to get $p = -0.025N + 35$, from which $R = (N)(-0.025N + 35) = -0.025N^2 + 35N$.

c) The company has fixed weekly overhead costs of \$5600, and each item costs \$6 to produce. Write a function expressing total weekly cost as a function of N.

c) $C = 6N + 5600$.

d) The company will break even when weekly revenue is large enough to cover total weekly cost, that is, for $R = C$. How many items must be sold per week in order for the company to break even? Interpret your answer. You may find a graph of revenue and cost helpful.

d) $R = C$ means that $-0.025N^2 + 35N = 6N + 5600$. In standard quadratic form, this is $-0.025N^2 + 29N - 5600 = 0$. Solutions are 245 and 915. Graphs of revenue and cost as functions of number of items sold show two break-even points, for sales of 245 or 915 items per week. At the lower of these values, revenue and cost both equal about \$7070. (There will be a profit for all sales levels between these two values.)

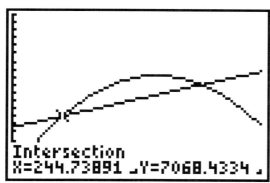

e) What price should the company charge in order to break even?

e) Assuming the lower ($N = 245$) production level, $p = -0.025N + 35 = 28.875$, so the price is \$28.88.

II. Projects and Group Activities *(SE pages 26–28)*

10. Materials: small wooden ball, watercolor paint or washable ink, large graph paper, and cardboard or plywood

    If an object is thrown into the air in any direction other than vertical, the object's trajectory (that is, its path through space) will have the shape of a parabola, provided that air resistance can be ignored. It is difficult to record the path of an actual thrown object, however, unless some kind of multiple flash photography is used (see Exercise 7 of Section 5.3). However, it is possible to observe the path directly by altering the circumstances somewhat. Instead of throwing a ball into the air (a situation called free fall), a ball can be rolled along a tilted surface.

    a) Attach a large sheet of graph paper to cardboard or a piece of plywood. Tilt the board, and roll a small wooden ball near the top of the board so that it reaches its highest point *after* being released (see **Figure 5.14**). Experiment with different angles of tilt of the board and initial speeds of the ball in order to produce a trajectory that approximately "covers" the board. What board angle did you use?

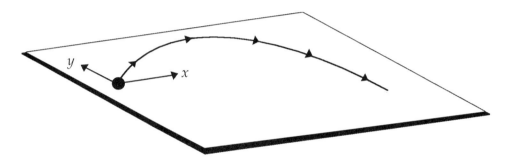

Figure 5.14

10. a) **A board tilted at 10°–20° with the horizontal works well.**

    b) Now coat the ball with paint or ink, and roll it along the board so that the trajectory of the ball is recorded on the graph paper. Mark an origin at the point of release of the ball, and draw in a (horizontal) *x*-axis and a *y*-axis perpendicular to it. Label an appropriate scale on each axis.

    b) **The graph should resemble Figure 5.14.**

    c) The curve of the ball's path should be approximately parabolic. Find the vertex and the *x*-intercepts of the parabola.

    c) **Answers will vary. For one sample trial, the vertex is (26, 13) and the *x*-intercepts are (0, 0) and (52, 0). This is based on using 1/4-inch graph paper, with each 1/4 inch as one unit.**

d) Because the curve passes through the origin, its equation will have the form $y = ax^2 + bx$. See if you can determine the equation of the parabola, using the same procedure as in Activity 5.1. Check your equation by evaluating your function for a few values of x to see if it predicts the correct y-coordinates on the curve of the ball's path.

d) The equation is of the form $y = ax(x - d)$, where $d = 52$. Substituting coordinates of the vertex, we have $a(26)(26 - 52) = 13$ and $a = -0.0192$. The equation is $y = -0.0192x(x - 52)$ or $y = -0.0192x^2 + 0.998x$. This should predict correct positions of points on the trajectory to within the nearest unit.

e) Roll the ball again and try to release it so that its initial direction is horizontal (that is, in the direction of the x-axis). Where is the vertex of the parabola for this path?

e) The vertex is at the point of release.

f) See if you can determine an equation for this path.

f) Because the vertex is at the origin and the y-axis is the axis of symmetry, the parabola's equation is of the form $y = ax^2$. If the coordinates of one point on the ball's path are measured, the value of a can be determined by substituting them in the equation. If, for example, the trajectory passes through $(45, -18)$, we have $a(45)^2 = -18$, from which $a = -0.00889$ and $y = -0.00889x^2$.

11. Materials: Handout 5.1, scissors

The quadratic formula is a generalized method for solving a quadratic equation in the standard form $ax^2 + bx + c = 0$. In this activity you will show how any quadratic equation can be rewritten in a special factorable form. You can then use that form to develop the quadratic formula.

a) Consider the equation $x^2 + 6x + 9 = 0$. Solve the equation by factoring.

11. a) $(x + 3)^2 = 0$; $x + 3 = 0$, so $x = -3$.

The perfect-square expression $(x + 3)^2$ can be visualized with the area model in **Figure 5.15.**

x
1
1
1
x
x^2
x
x
x
1
x
1
1
1
1
x
1
1
1
1
x
1
1
1

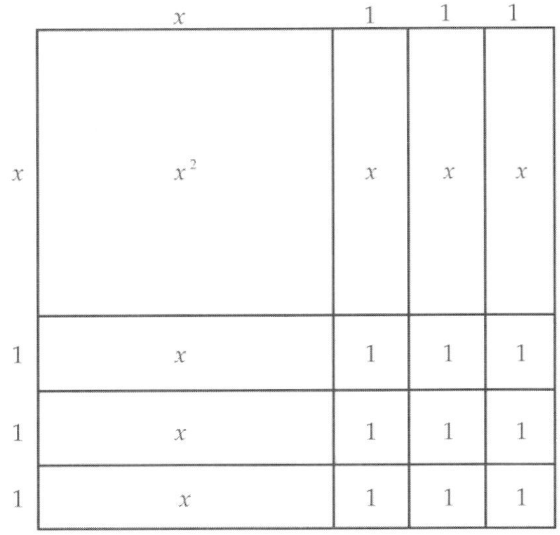

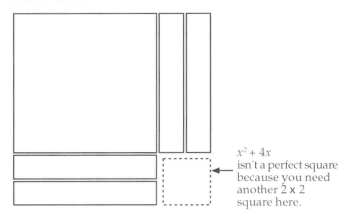

Figure 5.15

Each side of the large square has length $(x + 3)$, and the square is subdivided into blocks that, when combined, make up the terms in the expansion of $(x + 3)^2$. Notice that there are nine unit squares, which taken together represent the constant term in the expression $x^2 + 6x + 9$.

b) Now consider the expression $x^2 + 4x$. This can be thought of as part of a perfect-square trinomial. To find the constant term that would complete the trinomial, cut out an x-by-x square and four 1-by-x rectangles from Handout 5.1 and arrange them in the manner of **Figure 5.16,** forming an incomplete larger square. By determining the area of the missing region, you can complete the square by adding this missing area value to $x^2 + 4x$. Then factor the trinomial.

$x^2 + 4x$
isn't a perfect square
because you need
another 2 x 2
square here.

Figure 5.16

b) Missing area is 4; $x^2 + 4x + 4 = (x + 2)^2$.

c) Now repeat (b), using an x-by-x square and ten 1-by-x rectangles to complete the square of $x^2 + 10x$.

c) 25 completes the square; $x^2 + 10x + 25 = (x + 5)^2$.

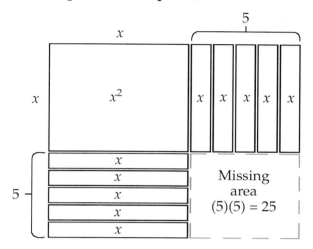

d) Experiment with other expressions of the form $x^2 + kx$, and see if you can make a general statement about finding the number that completes the square of $x^2 + kx$.

d) $\left(\dfrac{k}{2}\right)^2 = \dfrac{k^2}{4}$ completes the square.

Now consider the quadratic equation $x^2 + 6x + 7 = 0$. The left-hand side is not factorable. If 7 is subtracted from both sides, the result is $x^2 + 6x = -7$. Now we add 9 to both sides of the equation, **completing the square** on the left-hand side:

$x^2 + 6x + 9 = 2$

Factoring the trinomial, we get:

$(x + 3)^2 = 2$

Finally, an arrow diagram (**Figure 5.17**) can be used in reverse to solve the equation:

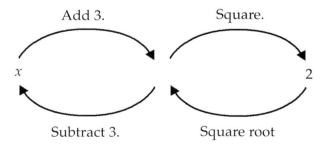

Figure 5.17

The square roots of 2 are $\sqrt{2}$ and $-\sqrt{2}$.

Subtract 3 from each root to find the values of x, which are $\sqrt{2}-3$ and $-\sqrt{2}-3$, or approximately –1.596 and –4.414, respectively.

e) Use this method to solve $x^2 + 10x - 8 = 0$.

e) Add 8 to both sides of the equation: $x^2 + 10x = 8$. Exercise 11(c) showed that 25 completes the square on $x^2 + 10x$, so add 25 to both sides: $x^2 + 10x + 25 = 33$. The factored form is $(x + 5)^2 = 33$, which can be solved with an arrow diagram:

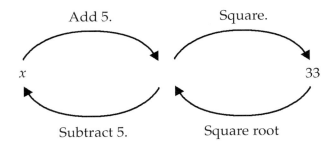

Add 5. Square.

x 33

Subtract 5. Square root

The solutions are $\sqrt{33}-5$ and $-\sqrt{33}-5$.

f) You are now ready to derive the quadratic formula. Start with the standard quadratic form $ax^2 + bx + c = 0$. Rewrite the equation using rules of algebra so that the left-hand side has the form $x^2 + kx$.

f) $ax^2 + bx = -c$

$$x^2 + \frac{b}{a}x = -\frac{c}{a}.$$

g) Complete the square on the left-hand side the same way you would if the coefficient of x were a number.

g) $x^2 + \frac{b}{a}x + \frac{b^2}{(2a)^2} = -\frac{c}{a} + \frac{b^2}{(2a)^2}.$

h) Finally, factor the left-hand side and use an arrow diagram to solve the equation. (Hint: The right-hand side can be rewritten as $\dfrac{b^2 - 4ac}{(2a)^2}$, as we will see in Chapter 6.)

h) $\left(x + \dfrac{b}{2a}\right)^2$

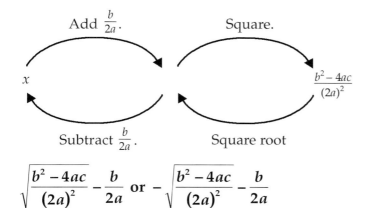

$\sqrt{\dfrac{b^2 - 4ac}{(2a)^2}} - \dfrac{b}{2a}$ or $-\sqrt{\dfrac{b^2 - 4ac}{(2a)^2}} - \dfrac{b}{2a}$

This is equivalent to the familiar form $x = \dfrac{-b \pm \sqrt{b^2 - 4ac}}{2a}$.

III. Additional Practice *(SE pages 29–32)*

For 12–15, write an equation for the parabola.

12. The parabola in **Figure 5.18** passes through (5, –1). Its vertex is at the origin.

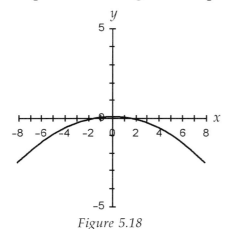

Figure 5.18

12. $a(5)^2 = -1$, so $a = -\dfrac{1}{25} = -0.04$, **and the equation is** $y = -\dfrac{1}{25}x^2$ **or** $y = -0.04x^2$.

13. The parabola in **Figure 5.19** passes through (4, 80) and (0, 0).

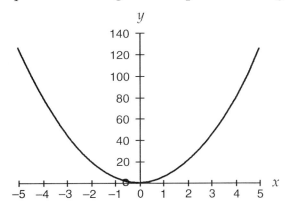

Figure 5.19

13. $a(4)^2 = 80$, so $a = 5$, and the equation is $y = 5x^2$.

14. Write an equation for the parabola that passes through the point (2, 7) and has a vertex at the origin.

14. **Because the vertex is at the origin and the axis of symmetry is the y-axis, the parabola's equation is of the form $y = ax^2$. The coordinate pair (2, 7) satisfies this equation, so $a(2^2) = 7$. Therefore, $4a = 7$, and $a = \dfrac{7}{4}$. The equation of the parabola is $y = \dfrac{7}{4}x^2$ or $y = 1.75x^2$.**

15. Write an equation for the parabola that crosses the x-axis at the points (0, 0) and (12, 0) and passes through the point (–2, 14). (Hint: See item 3 in Activity 5.1.)

15. **We know that because one of the x-intercepts is the origin, the equation can take the form $y = ax(x - d)$. To find a, $14 = a(-2)(-2 - 12)$. $a = 0.5$. An equation for the parabola is $y = 0.5x(x - 12)$ or $y = 0.5x^2 - 6x$.**

16. A mirrored reflector for a searchlight has a parabolic cross section (see **Figure 5.20**). The diameter across the rim of the reflector is 1.50 m, and the depth of the reflector is 0.25 m. Find an equation for the parabolic cross section, with the origin placed at the bottom of the reflector.

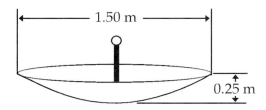

Figure 5.20

16. **The equation is of the form $y = ax^2$. Substitute the coordinates of the point (0.75, 0.25) to find $a = \dfrac{4}{9} \approx 0.444$. The equation is $y = \dfrac{4}{9}x^2$.**

17. Construct a graph of $y = 3x^2 - 4$.

17.

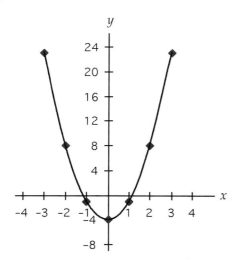

18. Construct a graph of $y = -0.3x^2$.

18.

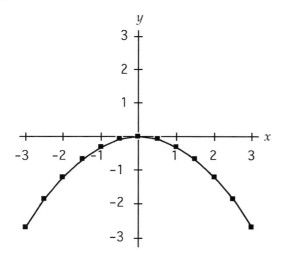

19. Construct a graph of $y = -x^2 + 6x - 13$.

19.

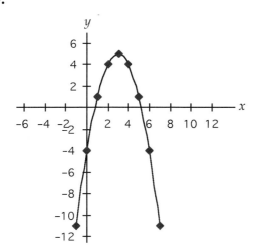

20. Construct a graph of $y = 2x^2 - 4x + 1$.

20.

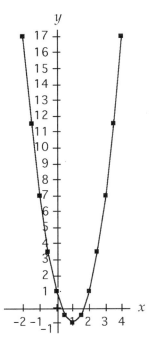

For 21 and 22, determine whether the given point is on the graph of the given quadratic function.

21. $y = 5x^2 + 2x - 7$; $(-1, -14)$.

21. $5(-1)^2 + 2(-1) - 7 = -4$, so the point is not on the graph.

22. $y = -2x^2 - x + 9$; $(3, -12)$.

22. $-2(3)^2 - (3) + 9 = -12$, so the point is on the graph.

23. The Zero Product Property holds true for any number of factors. If the product of any number of factors is 0, then at least one of the factors is equal to 0. Use this property to solve the following polynomial equations.

a) $x^3 + 5x^2 + 4x = 0$. (Hint: First factor the left side of the equation.)

23. a) First, factor: $x^3 + 5x^2 + 4x = x(x^2 + 5x + 4) = x(x + 1)(x + 4) = 0$. Then from the Zero Product Property, we know that either $x = 0$, $x + 1 = 0$, or $x + 4 = 0$. So $x = 0, -1,$ or -4.

b) $x^3 - 49x = 0$.

b) First, factor: $x(x^2 - 49) = x(x + 7)(x - 7) = 0$. Then from the Zero Product Property, we know that either $x = 0$, $x + 7 = 0$, or $x - 7 = 0$. So $x = 0, -7,$ or 7.

c) $x^4 - 4x^2 = 0$.

c) First, factor: $x^2(x^2 - 4) = x^2(x - 2)(x + 2) = 0$. Then from the Zero Product Property, we know that either $x = 0$, $x - 2 = 0$, or $x + 2 = 0$. So $x = 0, 2,$ or -2.

24. If a ball is thrown vertically upward at an initial speed of 50 mi/hr (73 ft/s), its height h above the point of release (measured in feet) at any time t (measured in seconds) is given by $h = 73t - 16t^2$. At what time will the ball return to its point of release?

24. **Solve the equation for $h = 0$: $73t - 16t^2 = t(73 - 16t) = 0$. $t = 0$, 4.56. Because $t = 0$ is the time at which the ball is thrown, $t \approx 4.6$ s is the time at which it returns.**

25. The power output P (measured in watts) of an electrical generator is given by $P = EI - RI^2$, where E represents generator voltage (in volts), R represents internal resistance (in ohms), and I represents current (in amperes). Find the current that will produce 80 watts of power when $E = 24$ volts and $R = 1$ ohm.

25. **The equation becomes $80 = 24I - I^2$, or $I^2 - 24I + 80 = (I - 20)(I - 4) = 0$. $I = 4$ or 20, so a current of either 4 amperes or 20 amperes will produce 80 watts of power.**

26. The parabolic archway in **Figure 5.21** has an equation $y = 6x - 0.5x^2$, where both x and y are measured in feet from the lower left corner of the base of the arch. How wide is the arch (a) at its base and (b) at a point 5.0 ft above the ground?

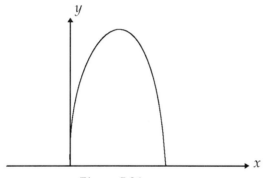

Figure 5.21

26. a) $6x - 0.5x^2 = x(6 - 0.5x) = 0$

   $x = 0$ or 12, so the arch is 12 ft wide.

   b) $6x - 0.5x^2 = 5.0$

   $-0.5x^2 + 6x - 5.0 = 0$

   $x = 0.9$ and 11.1, so the width of the arch at this level is $(11.1 - 0.9) = 10.2$ ft.

27. The main cable of a suspension bridge has the equation $h = 0.0025x^2 - 1.6x + 80$, where h represents the height of the cable above the roadway and x measures the horizontal distance across the bridge starting from the left tower (see **Figure 5.22**). At what positions will the vertical support cables be 50 feet long?

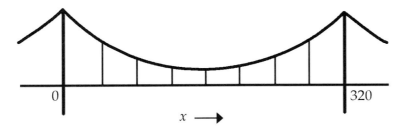

Figure 5.22

27. $0.0025x^2 - 1.6x + 80 = 50$

$0.0025x^2 - 1.6x + 30 = 0$

$x = 19$ or 621. The problem domain is $0 < x < 320$; otherwise, the cables would not be within the central span of the bridge. So 19 feet is the only feasible answer.

28. A long sheet of aluminum, 10 inches wide, is to be folded twice to create a rectangular gutter (see **Figure 5.23**). Find the value of x that will make a gutter with the largest possible carrying capacity, as measured by the cross-sectional area.

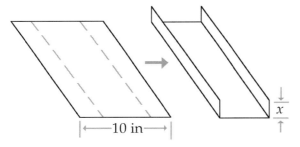

Figure 5.23

28. Area = (length of cross section)(width of cross section)

$A = (10 - 2x)(x)$

Zeros of the area function occur for $x = 0$ and 5, so the maximum area occurs for $x = 2.5$ inches.

29. Based on market surveys and cost analysis, the manager of a coffee shop has determined that its weekly profit P on sales of large cups of regular coffee is a function of the selling price p and is given by $P = -1000x^2 + 2170x - 712$. ($P$ and p are both measured in dollars.)

a) What price(s) would result in zero profit for the shop?

29. a) $-1000x^2 + 2170x - 712 = 0$; $x = 0.40$ or 1.77.

Either a price of \$0.40 or \$1.77 would result in zero profit.

b) What price should be charged to produce maximum profit?

b) Maximum profit will result when $x = \dfrac{-b}{2a} = \dfrac{-2170}{2(-1000)} \approx \$1.09$. (This is the x-coordinate of the vertex of the graph of the quadratic function.)

For 30–33, determine whether the given function is quadratic.

30. $(r - 2)^2 = 2r^2 - (r + 6)^2$.

30. The equation simplifies to $-4r + 4 = -12r - 36$, which is not quadratic, because it lacks a term in r^2.

31. $v^2 + (4 - v)^2 = 0$.

31. $v^2 + 16 - 8v + v^2 = 0$; $2v^2 - 8v + 16 = 0$.

The equation is quadratic.

32. $8x^3 + x = (2x + 1)^3$.

32. $8x^3 + x = 8x^3 + 12x^2 + 6x + 1$; $x = 12x^2 + 6x + 1$.

The equation is quadratic.

33. $5y^3 - y(5y - 1)(y + 1) = y(2 - 4y) + 3$.

33. $5y^3 - 5y^3 - 4y^2 + y = 2y - 4y^2 + 3$; $y = 2y + 3$.

The equation is not quadratic because its degree is 1.

34. Find the x- and y-intercepts and vertex of the parabola whose equation is $y = x^2 - 7$. Then use those points to draw a sketch of its graph.

34. $(0)^2 - 7 = -7$, so the y-intercept is $(0, -7)$.

The x-intercepts occur where $x^2 - 7 = 0$, or $x = 2.646$ or $x = -2.646$. The x-intercepts are $(2.646, 0)$ and $(-2.646, 0)$. The vertex occurs where $x = -(0)/2 = 0$ and $y = (0)^2 - 7 = -7$, so the vertex is $(0, -7)$.

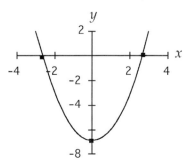

35. Find the x- and y-intercepts and vertex of the parabola whose equation is $y = x^2 - 4x + 1$. Then use those points to draw a sketch of its graph.

35. $(0)^2 - 4(0) + 1 = 1$, so the y-intercept is $(0, 1)$.

The x-intercepts occur where $x^2 - 4x + 1 = 0$, or $x = 0.268$ or $x = 3.732$. The x-intercepts are $(0.268, 0)$ and $(3.732, 0)$. The vertex occurs where $x = \dfrac{-(-4)}{2} = 2$ and $y = (2)^2 - 4(2) + 1 = -3$, so the vertex is $(2, -3)$.

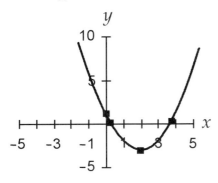

36. Find the x- and y-intercepts and vertex of the parabola whose equation is $y = x^2 - 8x - 9$. Then use those points to draw a sketch of its graph.

36. $(0)^2 - 8(0) - 9 = -9$, so the y-intercept is $(0, -9)$.

The x-intercepts occur where $x^2 - 8x - 9 = (x - 9)(x + 1) = 0$, or $x = -1$ or $x = 9$. The x-intercepts are $(-1, 0)$ and $(9, 0)$.

The x-coordinate of the vertex is $\dfrac{-(-8)}{2} = 4$, and the y-coordinate is $(4)^2 - 8(4) - 9 = -25$, so the vertex is $(4, -25)$.

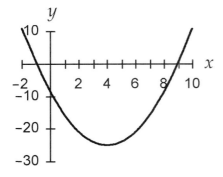

37. Find the x- and y-intercepts and vertex of the parabola whose equation is $y = x^2 - 7x + 10$. Then use those points to draw a sketch of its graph.

37. $(0)^2 - 7(0) + 10 = 10$, so the y-intercept is (0, 10).

The x-intercepts occur where $x^2 - 7x + 10 = (x - 5)(x - 2) = 0$, or $x = 5$ or $x = 2$. The x-intercepts are (5, 0) and (2, 0).

The x-coordinate of the vertex is $\dfrac{-(-7)}{2} = 3.5$, and the y-coordinate is $(3.5)^2 - 7(3.5) + 10 = -2.25$, so the vertex is (3.5, -2.25).

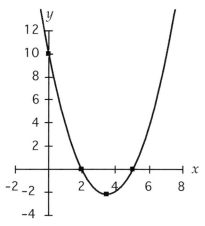

For 38–45, solve the given quadratic equations by factoring if possible; otherwise, use the quadratic formula. If the equation has no real solutions, write "No real solutions." Round any irrational answers to three decimal places.

38. $x^2 + 10x + 4 = 0$.

38. Not factorable; $x = -9.583, -0.417$.

39. $x^2 + 4x = 21$.

39. $x^2 + 4x - 21 = 0$

$(x + 7)(x - 3) = 0$

$x = -7, 3$.

40. $4d^2 - 9d = 25$.

40. Not factorable; $d = -1.616, 3.866$.

41. $9k^2 = 144$.

41. $9k^2 - 144 = 0$

$(3k + 12)(3k - 12) = 0$

$k = -4, 4$.

42. $-7s^2 + 12s + 1 = 0$.

42. Not factorable; s = -0.080, 1.794.

43. $x^2 + 5x + 7 = 0$.

43. No real solutions

44. $(3 - L)^2 = 6$.

44. $9 - 6L + L^2 = 6$

$L^2 - 6L + 3 = 0$

Not factorable; $L = 0.551, 5.449$.

45. $2w^2 - 3w - 21 = 0$.

45. Not factorable; $w = -2.576, 4.076$.

For 46 and 47, use a graphical method to solve each quadratic equation, accurate to three decimal places.

46. $7x^2 + 12x + 4 = 0$.

46.

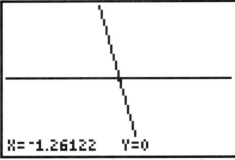

Graphing the function and zooming in on the leftmost intercept gives a value $x = -1.261$, which approximates one solution to the equation. (It may be necessary to zoom in several times in order to achieve three decimal place accuracy.) A similar process gives the other solution $x = -0.453$.

47. $18x^2 - 23x - 37 = 0$.

47.

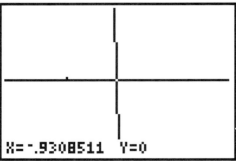

Graphing and zooming shows that the leftmost x-intercept is at $x = -0.931$, so this is one approximate solution to the equation. In a similar manner, the other solution is found to be $x = 2.209$.

Section 5.3 Modeling Data with Quadratic Functions

What You Need to Know

- How to use linear regression to find a line of best fit

What You Will Learn

- To use quadratic regression to construct models of data

- To choose between linear and quadratic models for data

Materials

- CBL with ultrasonic motion sensor

- Basketball

- Beach ball

- Ruler

- Compass

- Handout 5.2 (optional)

Quadratic Models

Recall that the term **mathematical model** refers to any use of mathematics to describe the behavior of some system. A good mathematical model can be used to predict how the system will work under conditions for which measured data are not available, either to forecast future behavior or merely to fill in gaps in data. Mathematical models can consist of equations, graphs, tables, diagrams, or combinations of all of these.

A wide variety of situations can be modeled by linear functions. But when data relating two quantities cannot be described by a linear function, it becomes necessary to look for a nonlinear model. One possible nonlinear model is a quadratic function.

Discovery 5.2 Modeling the Amount of Lumber in a Log *(SE pp. 33–35)*

Materials: ruler, compass

When trees are cut for lumber, the amount of lumber that can be produced from any log depends on the diameter and length of the log. A common unit of measure for lumber is a *board foot*. Despite its name, a board foot is a measure of volume. It is defined as the amount of lumber contained in a 1-inch-thick board that has an area of 1 square foot.

- A 6-ft-long 1" x 12" board contains 6 board feet, because its area is $(6 \text{ ft})(1 \text{ ft}) = 6 \text{ ft}^2$ and it is 1 inch thick.

- An 8-ft-long 1" x 6" board contains 4 board feet, because it has an area of $(8 \text{ ft})(\frac{1}{2} \text{ ft}) = 4 \text{ ft}^2$ and it is 1 inch thick.

- A 12-ft-long 2" x 8" piece of lumber contains 16 board feet, because it has an area of $(12 \text{ ft})(\frac{2}{3} \text{ ft}) = 8 \text{ ft}^2$ but its thickness is 2 inches.

1. How many board feet are contained in a 10-ft-long 1" x 4" board?

1. Board area = $(10 \text{ ft})(\frac{1}{3} \text{ ft}) = \frac{10}{3} \text{ ft}^2$. Because the thickness is 1", there are $(1)\left(\frac{10}{3}\right) = \frac{10}{3}$, or $3\frac{1}{3}$ board feet in the board.

2. Draw a circle on a sheet of paper. The circle is a smaller scale representation of the cross section of a log. Choose a suitable scale to represent 1 inch, and measure the diameter of the log using that scale.

2. Convenient scales to use, depending on the type of ruler available, might be $\frac{1}{4}$ inch = 1 inch or 0.5 cm = 1 inch.

3. Now draw in parallel lines representing the edges of 1-inch boards that could be cut from the log, as in **Figure 5.24**.

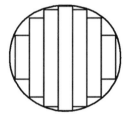

Figure 5.24

3. Students' circles will vary in size, but lines should be equally spaced according to the scale chosen.

4. Assume that you are looking at the ends of the boards that could be cut from the log. If the log is 1 foot long, how many board feet of lumber are contained in this log?

4. If Figure 5.24 corresponds to an 8-inch-diameter log, typical measurements might yield the following board widths: one 8-inch board, two 7-inch boards, two $6\frac{1}{2}$-inch boards, and two 4-inch boards for a total of 43 inches, or 3.6 feet, of width. Because the boards are all 1 foot long and 1 inch thick, this equals 3.6 board feet of lumber.

5. Draw several more circles of varying sizes, and repeat items 3 and 4 for each one. Record data on log diameter and amount of lumber (in board feet) in a table. For an extra data pair, include a diameter of 0 inches.

5. **Sample data:**

Log Diameter (in)	Amount of Lumber (board ft)
0	0
4.5	1.0
8	3.6
13	9.8
21	27.5

6. Draw a scatter plot of the data. Note that the amount of lumber depends on log diameter. Therefore, log diameter is the independent variable, and the amount of lumber is the dependent variable.

6.

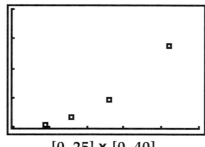

[0, 25] x [0, 40]

7. Does the scatter plot have a linear form? Would a linear function be a good model for the relationship between log diameter and amount of lumber? Explain.

7. **The points on the scatter plot do not lie along a straight line; therefore, a linear function is not an appropriate model for the data.**

8. To determine whether a quadratic function is a good model for the data, use the quadratic regression feature of your calculator to find the best-fitting quadratic function. Graph it along with the scatter plot. Does it appear to be a good fit?

8.

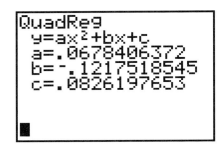 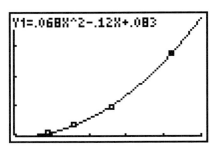

[0, 25] x [0, 40]

The quadratic function provides a very good fit to the data.

9. Write an equation expressing the amount of lumber L in a 1-foot-long log, measured in board feet, as a function of the log's diameter d.

9. $L = 0.068d^2 - 0.122d + 0.083$

Although your quadratic function for the lumber data in Discovery 5.2 is an example of a data-driven model, it is possible to use geometrical reasoning to show that this situation can be expected to produce a quadratic model.

There are many situations for which there is no apparent theoretical reason to expect that one particular model will describe the data better than another. In such cases, it is desirable to explore two or more possible models in search of one that best accounts for any variation in the data. Unless the pattern in the data is clearly nonlinear, as in Discovery 5.2, it is customary to begin by examining a linear model because it is the simplest possibility.

Discovery 5.3 Choosing the Best Model (SE pages 35–36)

A common rule of thumb for determining a person's ideal body weight says that a 5-foot-tall person should weigh 100 pounds and that you should add 5 pounds for each extra inch of height. This rule assumes that there is a simple linear relationship between height H and ideal weight W. In this activity you will compare two models to determine which one provides the better description of the connection between height and weight.

Table 5.5 lists ideal weights for "medium-framed" men of different heights, as determined from lowest mortality for each height by an insurance company. (Each weight is actually the median of a range of values, and all data are for fully clothed men.)

Height (in)	Weight (lb)
62	136
63	138
64	140
65	142.5
66	145
67	148
68	151
69	154
70	157
71	160
72	163.5
73	167
74	171
75	174.5
76	179

Table 5.5 Source: MetLife.

1. Construct a scatter plot of the data. Note that height is the independent variable.

1.

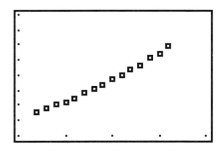

[60, 80] x [120, 200]

2. Find a linear model for the data by determining a line of best fit, using linear regression. Graph the regression line on the scatter plot, and describe what you see.

2. $W = 3.07H - 56.5$; the line seems to follow the data fairly closely.

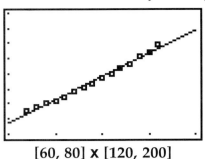

LinReg
 y=ax+b
 a=3.066071429
 b=-56.45892857

[60, 80] x [120, 200]

3. A **residual plot** can be helpful in assessing the adequacy of a model for describing data relating two variables. To create a residual plot, add two more columns to Table 5.5.

Label the new column "Predicted Weight (lb)." In this column, list the weight values predicted from your linear model.

Label the second new column "Residual (lb)." Then calculate the residuals (or errors) based on your model. Recall that residuals are the differences between the actual and predicted weights. For this example, residual = $W_{actual} - W_{predicted}$. List these in the second new column of the table.

3. **Sample answer:**

Height (in)	Weight (lb)	Predicted Weight (lb)	Residual (lb)
62	136	134	2
63	138	137	2
64	140	140	0
65	142.5	143	−0.5
66	145	146	−1
67	148	149	−1
68	151	152	−1
69	154	155	−1
70	157	158	−1
71	160	161	−1
72	163.5	164	−0.5
73	167	167	0
74	171	170	1
75	174.5	173.5	1
76	179	177	2

(Values in this table are rounded.)

4. Now construct a residual plot. Is there a pattern to the residuals, or do they seem to be random?

4.

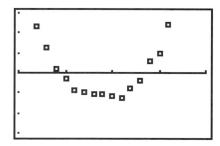

[60, 80] x [–3, 3]

There is a definite U-shaped pattern to the residual plot. This suggests that although the size of the residuals is small, the linear model does not provide a good description of the data.

5. Now find a quadratic model for the data, and plot its graph. Does it seem to fit the data better than the graph of the line from item 2?

5. $W = 0.0728H^2 - 6.99H + 289$

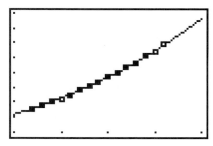

[60, 80] x [120, 200]

Yes, the quadratic regression curve follows the data more closely than the linear model did.

6. Add two more columns to Table 5.5 as you did with the linear model. Again, list predicted weights based on the quadratic model, as well as the respective residuals. Plot the residuals and describe the pattern.

6.

Height (in)	Weight (lb)	Predicted Weight (lb)	Residual (lb)
62	136	136	0
63	138	138	0
64	140	140	0
65	142.5	143	−0.5
66	145	145	0
67	148	148	0
68	151	151	0
69	154	154	0
70	157	157	0
71	160	160	0
72	163.5	164	−0.5
73	167	167	0
74	171	171	0
75	174.5	175	−0.5
76	179	179	0

(Values in table are rounded.)

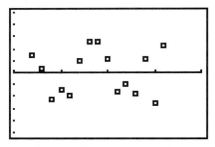

[60, 80] **x** [−0.5, 0.5]

The residuals are very small and randomly scattered. The quadratic model describes the data well. (But see Exercise 5 for a more extended analysis.)

As we have seen in Discovery 5.3, examining the patterns in residual plots is a useful way of choosing between two possible models (in this case, a linear model and a quadratic model). But occasionally, different models may appear to describe data equally well. In such cases, it's wise to try to collect more data in an attempt to make a clear choice between the models. After further research, if a linear and a nonlinear model both have small-valued and random residual plots, researchers often choose the linear model unless theory suggests a nonlinear one. And there may be commonsense reasons for preferring one model over another, as discussed in the following example.

Example 8

Table 5.6 shows data on the growth in the use of the instant messenger e-mail system of an Internet Service Provider (ISP) over a three-year period. The instant messenger system, which allows real-time online "chat" among groups of friends, is believed to be used primarily by teenage Internet users.

Month	Number of Instant Messages Sent Each Day (millions)
December 1996	50
April 1997	94
April 1998	225
April 1999	430
February 2000	651

Table 5.6 Source: America Online.

Find an appropriate mathematical model for these data.

Solution:

Figures 5.25 through **5.28** show several possible regression models for the data, along with their respective graphs and residual plots. The independent variable here is time, measured in months (with December 1996 being month 1).

Linear model: $N = 15.5t + 8.05$.

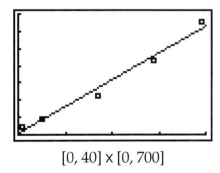

[0, 40] x [0, 700]

Figure 5.25a

[0, 40] x [−50, 50]

Figure 5.25b

Quadratic model: $N = 0.223t^2 + 6.733t + 47.860$.

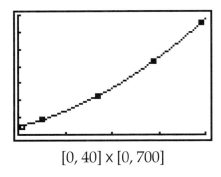

[0, 40] × [0, 700]

Figure 5.26a

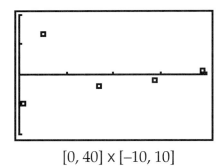

[0, 40] × [−10, 10]

Figure 5.26b

Third-degree model: $N = 0.00172t^3 + 0.118t^2 + 8.337t + 44.317$.

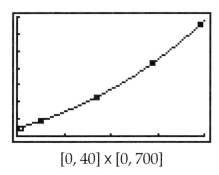

[0, 40] × [0, 700]

Figure 5.27a

[0, 40] × [−10, 10]

Figure 5.27b

Fourth-degree model: $N = -0.000269t^4 + 0.02333t^3 - 0.429t^2 + 12.890t + 37.516$.

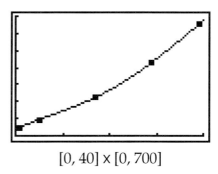

[0, 40] x [0, 700]

Figure 5.28a

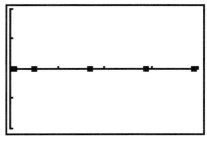

[0, 40] x [−10, 10]

Figure 5.28b

The linear model does not fit the scatter plot well. In addition, its residual plot is U-shaped, indicating that a linear model does not work well here. However, all three of the nonlinear models are reasonably good at accounting for the given data.

The fourth-degree model (a "quartic" function) seems to fit the data exactly. In fact, it can be shown that *any* five points can be exactly modeled by some fourth-degree model. (Similarly, a cubic function can always be found that exactly fits four points, and any three points can be exactly modeled by some quadratic function.) So the apparent perfect fit of the function
$N = -0.000269t^4 + 0.02333t^3 - 0.429t^2 + 12.890t + 37.516$ is deceptive. It is likely that if more data were available, the exactness of the model would disappear. Furthermore, if the fourth-degree model and the quadratic model are compared to see how each would predict future usage, we see the results shown in **Figure 5.29:**

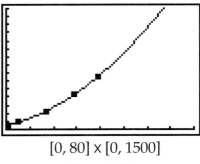

[0, 80] x [0, 1500]

Figure 5.29a

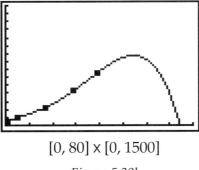

[0, 80] x [0, 1500]

Figure 5.29b

The fourth-degree model is inadequate for forecasting because it predicts an eventual drop off to zero messages. Although one must always be cautious when using any model to extrapolate beyond known data, the quadratic model seems most appropriate for this example.

Exercises 5.3

I. Investigations *(SE pages 41–44)*

1. We saw in Section 5.1 that the primary cables of suspension bridges can be modeled by quadratic functions. **Figure 5.30** shows the primary cable of a suspension bridge. **Table 5.7** lists the heights above the road surface of points on the cable at several horizontal distances from the left-hand supporting tower.

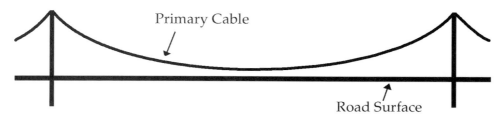

Primary Cable

Road Surface

Figure 5.30

Distance from Left-Hand Tower (ft)	Height of Cable (ft)
0	92
100	21
200	8
300	52

Table 5.7

    a) Use quadratic regression to find an equation that models the shape of the suspension cable.

1. **a) If the origin is located at the point where the left-hand tower meets the road surface, then the equation is $y = 0.002875x^2 - 0.9955x + 91.95$.**

    b) Use your equation to find the height of the lowest point on the cable.

    b) The lowest point on the cable is at the vertex of the parabola that is the graph of the equation. The x-coordinate of the vertex is
$$-\frac{b}{2a} = -\frac{-0.9955}{2(0.002875)},$$ **or $x \approx 173$ ft. The height of the cable at that point is** $0.002875(173)^2 - 0.9955(173) + 91.95 \approx 5.8$ **ft.**

    c) What is the span of the bridge between the towers?

    c) The lowest point of the cable is midway between the towers, so the towers are 2(173 ft) = 346 ft apart.

2. The data in **Table 5.8** show the distances required for a car to stop when braking from different initial speeds. (This table is the same as Table 5.2 in the Preparation Reading section at the beginning of this chapter.) **Figure 5.31** shows a scatter plot of the data.

Speed (mph)	Stopping Distance (ft)
20	42
30	73.5
40	116
50	173
60	248
70	343
80	464

Table 5.8 Source: U.S. Bureau of Public Roads.

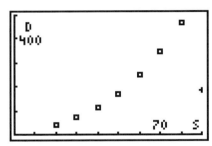

$[0, 90] \times [0, 500]$

Figure 5.31

a) Find a linear regression equation (least-squares line) for the data expressing stopping distance D as a function of speed S. Graph the line on the scatter plot.

```
LinReg
 y=ax+b
 a=6.917857143
 b=-137.3928571
```

2. a) $D = 6.92S - 137.$

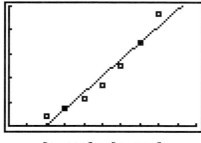

[0, 100] x [0, 500]

b) Add two columns to Table 5.8. In the first, list predicted stopping distances based on your linear model. In the second, list the residuals based on comparing the linear model to the data: residual = $D_{\text{actual}} - D_{\text{predicted}}$. Then make a residual plot and assess the appropriateness of the linear model.

b)

Speed (mph)	Stopping Distance (ft)	Predicted Distance (ft)	Residual (ft)
20	42	1	41
30	73.5	70	3.5
40	116	139	−23
50	173	209	−36
60	248	278	−30
70	343	347	−4
80	464	416	48

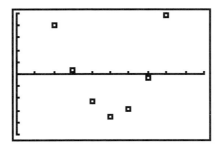

[0, 100] x [−50, 50]

The residuals are large and form a U-shaped pattern. The linear model is not a good model for the data.

c) To see if a quadratic model better describes the data, first perform a quadratic regression on the data, and graph the resulting quadratic function on the scatter plot. Write the regression equation expressing stopping distance D as a function of speed S.

c) $D = 0.0888S^2 - 1.96S + 49.1$.

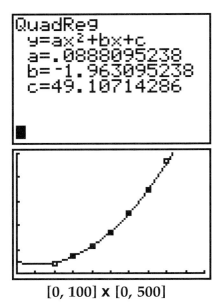

[0, 100] x [0, 500]

d) To test the model, examine a residual plot of the data with quadratic form and decide whether the quadratic regression equation appears to adequately describe the data.

d)

Speed (mph)	Stopping Distance (ft)	Predicted Distance (ft)	Residual (ft)
20	42	45	−3
30	73.5	70	3.5
40	116	113	3
50	173	173	0
60	248	251	−3
70	343	347	−4
80	464	460	4

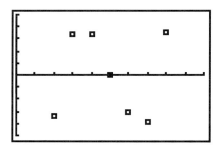

[0, 100] x [−5, 5]

The quadratic regression equation appears to adequately describe the data because the residuals are small and randomly scattered.

e) The original data table did not include an entry for speeds below 20 mph. Does the quadratic model appear to make sense for speeds near 0 mph?

e) No, the model predicts a stopping distance of 49 ft for a speed of 0, which is greater than the observed stopping distance for a speed of 20 mph.

f) How could you refine the model to reflect your answer to (e)?

f) Include a stopping distance of 0 ft for a speed of 0 mph before performing the regression.

3. When young children begin to talk, they usually start with small words. The lengths of words increase with the age of the child. Researchers measure the change in word length using "mean length of utterance" (mlu). **Table 5.9** shows results of observation on twelve children of various ages.

Child	Age (months)	mlu
1	20	2.0
2	22	2.2
3	23	2.1
4	28	2.0
5	31	2.3
6	34	2.3
7	36	2.9
8	43	2.7
9	46	3.5
10	49	3.2
11	53	5.2
12	57	5.8

Table 5.9

a) Make a scatter plot of the data. Explain your choice of independent variable (on the horizontal axis) and dependent variable (on the vertical axis).

3. a) Word length should depend on the age of the child, so it is the dependent variable.

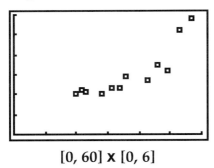

[0, 60] **x** [0, 6]

b) Use a calculator or computer to find both linear and quadratic regression equations for the data, and graph each on the scatter plot.

b) Linear: $M = 0.08744A - 0.20389$.

Quadratic: $M = 0.00397A^2 - 0.21358A + 4.91092$.

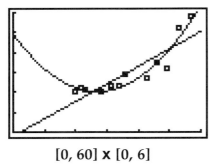

[0, 60] **x** [0, 6]

c) Construct residual plots for both models, and explain why the quadratic model is better than the linear model at describing the data in the table.

c) Linear model:

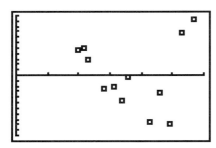

[0, 60] **x** [−1.1, 1.1]

Quadratic model:

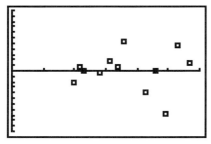

[0, 60] **x** [−1.1, 1.1]

The residuals for the linear model increase with the age of the child and form a U–shaped pattern. The residuals for the quadratic model are generally smaller and more randomly scattered.

d) Use both models to predict the mlu of a child on his or her fifth birthday. Do either or both of the predictions make sense? Explain.

d) The fifth birthday corresponds to an age of sixty months. At this age, the linear model predicts an mlu of 5.0, whereas the quadratic model predicts 6.4. Although both make sense, it might be expected that the mlu at age sixty months would exceed that of any of the children in the sample. Only the quadratic model predicts this.

e) Use both models to predict the mlu of a child on his or her first birthday. Do either or both of the predictions make sense? Explain.

e) At twelve months of age, the linear model predicts an mlu of 0.85, whereas the quadratic model predicts 2.9. It would be unlikely to expect very young children to speak as well as predicted by the quadratic model (as well as three-year-olds). The linear model is probably a better predictor for very young children.

4. During the 1991 Persian Gulf War, one of the high-tech weapon systems used by American forces was a portable radar that quickly detected the origin of enemy artillery fire, allowing rapid response. The radar determines a path for a projectile by sampling coordinates of points on the path. It then uses the mathematical model to determine the source of the fire.

Suppose that the radar, at the origin of the coordinate system, detects an artillery shell headed toward it. The coordinates of three points in the projectile's path (shown in **Figure 5.32**) are (126, 31), (164, 38), and (207, 25). All distances are in meters.

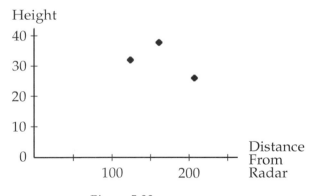

Figure 5.32

a) Trajectories of projectiles such as artillery shells can be approximated by quadratic functions. Use quadratic regression to find an equation for the path of the shell.

4. a) $y = -0.00601x^2 + 1.926x - 116$

b) How well does your model fit the data? Explain.

b) The function is graphed along with the scatter plot:

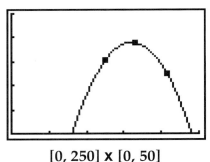

[0, 250] x [0, 50]

The residual plot shows that the quadratic model is a perfect fit:

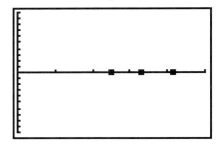

c) Use your model to determine the location from which the artillery shell was fired. (Hint: Assume the shell was fired from the ground, $y = 0$.)

c) Assuming that the shell was fired from the ground, its location can be found by setting y equal to zero in the equation for the trajectory and solving for x:

$-0.00601x^2 + 1.926x - 116 = 0$.

Solutions are 80 and 240. Because the shell is heading toward the origin, the rightmost x-intercept on the graph is the firing point, at (240, 0).

5. As discussed in Example 8, many calculators have several options for fitting regression equations to data. Consider again the weight/height data from Discovery 5.3. Although a quadratic model appeared to be a better fit than a linear model, there are other polynomial functions that can be tried. If your calculator has the capability, find third-degree (cubic) and fourth-degree (quartic) polynomial regression models for the weight/height data. By examining graphs of the regression equations and residual plots, determine which model best describes the data. Also discuss which one might be the best to use to predict weights for heights that are less than or greater than those in the data set.

5. Cubic regression: $W = 0.00006285H^3 + 0.0598H^2 - 6.09H + 268$.

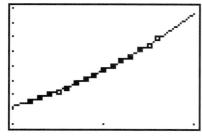

[60, 80] x [120, 200]

Residual plot:

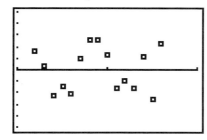

[60, 80] x [−0.5, 0.5]

Quartic regression: $W = 0.0005814H^4 - 0.1604H^3 + 16.64H^2 - 766H + 13{,}316$.

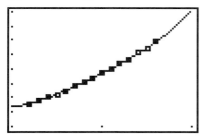

[60, 80] x [120, 200]

Residual plot:

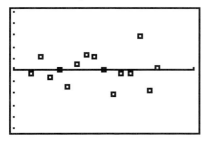

[60, 80] x [−0.5, 0.5]

The cubic model has a residual plot that is almost identical to that for the quadratic model in Discovery 5.3. The residuals for the quartic model are generally smaller than those for any of the other models, so it appears to describe the data best.

The quadratic model might be a better predictor, though, because the quartic model would eventually show decreasing weights as heights increase beyond the given data. The quadratic model would continue to increase.

II. Projects and Group Activities *(SE pages 45–46)*

6. Materials: CBL with ultrasonic motion sensor, or CBR, attached to a graphing calculator; basketball; large, light ball like a beach ball

   In this activity you will examine the behaviors of different objects falling through air. In particular, you will examine how the distance through which a ball falls under the influence of gravity varies with time.

   a) Set up the CBL or CBR to record data on the height of the basketball as it falls toward the sensor. (Have someone catch the ball just before it hits the sensor to avoid damaging the sensor.) Check your CBL program to see how often it records data (suggested setting: every 0.02 seconds). Collect the data and record the values for height and time in calculator lists, or in a separate table. Note: You may have to discard several data pairs both at the beginning and the end of your data set so that you can examine only the data representing the free fall of the ball toward the sensor.

6. a) **This graph is a scatter plot of raw data for the basketball.**

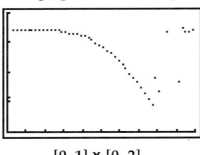

   [0, 1] x [0, 2]

   After deleting the data pairs for times before 0.25177 s (before the ball was dropped) and after t = 0.77181 s (after the ball bounced for the first time), the remaining data and scatter plot are:

Time (s)	Height (m)	Time (s)	Height (m)	Time (s)	Height (m)	Time (s)	Height (m)
0.25177	1.709	0.39179	1.6102	0.53179	1.3621	0.67181	0.90442
0.27178	1.6991	0.41179	1.6025	0.55178	1.3072	0.69181	0.82759
0.29178	1.6903	0.43179	1.5762	0.5718	1.248	0.7118	0.73649
0.31179	1.6837	0.45179	1.5355	0.5918	1.1821	0.73181	0.64429
0.33179	1.6596	0.47179	1.496	0.6118	1.1174	0.75181	0.55319
0.35179	1.6343	0.49179	1.4521	0.6318	1.0504	0.77181	0.4577
0.37179	1.6409	0.51179	1.406	0.65181	0.97138		

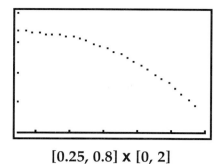

[0.25, 0.8] x [0, 2]

b) Find a quadratic model for the data, and discuss the accuracy of the model.

b) $h = -4.88t^2 + 2.63t + 1.34$; **the function fits the data almost perfectly.**

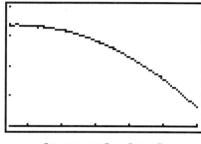

[0.25, 0.8] x [0, 2]

The residuals are small and randomly scattered. The quadratic model adequately describes the data.

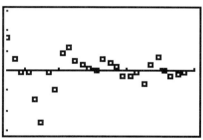

[0.25, 0.8] x [−0.03, 0.03]

c) Now repeat the experiment with a beach ball (or another very large, light ball), and again record the height and time data in a table.

c) Scatter plot of raw data (for a beach ball):

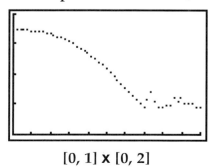

[0, 1] x [0, 2]

After deleting the data pairs for times before 0.05803 s (before the ball was dropped) and after $t = 0.69805$ s (after the ball bounced for the first time), the remaining data and scatter plot are:

Time (s)	Height (m)	Time (s)	Height (m)	Time (s)	Height (m)	Time (s)	Height (m)
0.05803	1.7287	0.23804	1.6255	0.41804	1.3039	0.59804	0.78478
0.07803	1.7276	0.25804	1.6003	0.43804	1.2546	0.61804	0.71673
0.09803	1.7243	0.27804	1.5729	0.45804	1.2052	0.63804	0.64758
0.11803	1.7177	0.29805	1.5421	0.47804	1.1514	0.65804	0.57624
0.13804	1.709	0.31805	1.5048	0.49804	1.0954	0.67805	0.5038
0.15804	1.698	0.33805	1.4664	0.51804	1.0372	0.69805	0.43684
0.17804	1.6804	0.35805	1.4313	0.53804	0.97577		
0.19804	1.6662	0.37804	1.394	0.55804	0.9143		
0.21804	1.6464	0.39804	1.35	0.57804	0.85064		

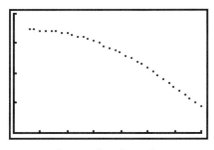

[0, 0.7] x [0, 2]

d) Find a quadratic model for this data set, and discuss its accuracy.

d) $h = -3.06t^2 + 0.254t + 1.73$.

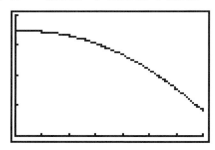

[0, 0.7] x [0, 2]

Although the residuals are small, they form a pronounced **S**-shaped pattern. Another type of model may be necessary.

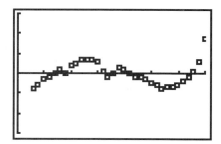

[0, 0.7] x [−0.03, 0.03]

   e) What explanation can you give for the differences you noticed in the accuracy of the quadratic models for these two trials?

   e) Air resistance is more of a factor for the lighter ball and causes the data to depart from the quadratic model.

7. **Figure 5.33** is a multiple-flash photograph of a golf ball that has been thrown into the air. (The photo was created in a darkened room by illuminating the ball with a strobe light that flashed once every one-eighth of a second after the ball was thrown.) The scale on the left side of the photo measures the height of the center of the ball.

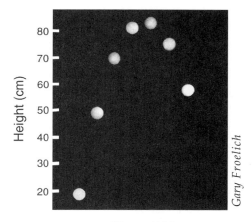

Figure 5.33

   a) Using the scale on the photograph, estimate the height of the ball at each flash and record the data in a table similar to **Table 5.10.**

Time (s)	Height (cm)
0	
0.125	
0.25	
0.375	
0.5	
0.625	
0.75	

Table 5.10

7. a)

Time (s)	Height (cm)
0	19.0
0.125	48.7
0.25	69.5
0.375	80.7
0.5	82.4
0.625	74.8
0.75	57.5

b) Create a scatter plot of the data.

b)

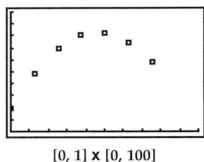

[0, 1] x [0, 100]

c) Find an appropriate model for the data, and make a residual plot to test it. Write an equation expressing the height h of the ball above the point of release as a function of time t.

c)

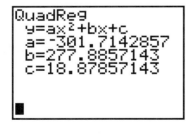 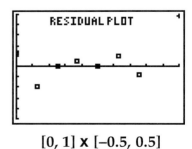

[0, 1] x [0, 100]   **[0, 1] x [−0.5, 0.5]**

The quadratic regression curve seems to fit the data well, and the residuals are small and randomly scattered. The equation is
$h = -302t^2 + 278t + 18.9.$

d) Use your quadratic model to find the height of the ball at times of 0.20 seconds and 0.70 seconds.

d) When $t = 0.20$ seconds, the height is
$-302(0.20)^2 + 278(0.20) + 18.9 = 62.4$ cm. **When $t = 0.70$ seconds, the height is** $-302(0.70)^2 + 278(0.70) + 18.9 = 65.6$ cm.

e) Again using your model, predict the time when the ball will be at height 0 cm.

e) $-302t^2 + 278t + 18.9 = 0$, $t = -0.064$ and 0.984. The ball is at height 0 at 0.064 seconds before the first flash of the strobe light and at 0.984 seconds after the first flash.

f) Determine the maximum height of the ball.

f) The maximum height occurs at the vertex of the parabola, which occurs at $t = -\dfrac{b}{2a} = -\dfrac{278}{2(-302)}$, or about 0.46 seconds. The maximum height is $-302(0.46)^2 + 278(0.46) + 18.9 = 82.9$ cm.

III. Additional Practice *(SE pages 46–52)*

For 8–10,

a) Make a scatter plot of the data.

b) Find the linear regression equation and make a residual plot for that equation.

c) Find the quadratic regression equation and make a residual plot for that equation.

d) Use the residual plots to decide which equation best describes the data.

8. The data in **Table 5.11** represent two characteristics of the ponderosa pine: the diameter measured at chest height and the usable volume.

Diameter (in)	Usable Volume (1000 in^3)
36	276
28	163
28	127
41	423
19	40
32	177
22	73
38	363
25	81
17	23
31	203
20	46
25	124
19	30
39	333
33	269
17	32
37	295
23	83
39	382

Table 5.11

8. a)

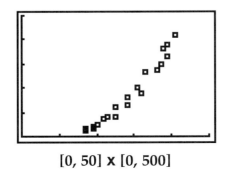

[0, 50] x [0, 500]

b)

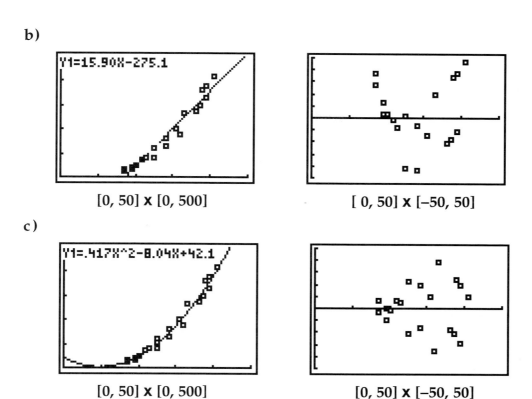

[0, 50] x [0, 500]   [0, 50] x [−50, 50]

c)

[0, 50] x [0, 500]   [0, 50] x [−50, 50]

d) The quadratic regression function seems to describe the data slightly better because the quadratic residuals are somewhat smaller in general, and more randomly scattered, than the linear residuals. The quadratic function $V = 0.417d^2 - 8.04d + 42.1$ appears to be the better model over the range of the data. However, because its graph curves upward for small values of d, it should not be used to predict usable volume for small-diameter trees (below 15 in).

9. The data in **Table 5.12** represent stress produced in a sample of polypropylene plastic due to stretching. The strain measures stretching as a fraction of original length of the sample (in inches per inch). Stress is force per unit of cross-sectional area (in pounds per square inch).

Strain (in/in)	Stress (psi)
0.01	495
0.02	865
0.03	1240
0.04	1600
0.05	1975
0.06	2340
0.07	2720
0.08	3090
0.09	3460

Table 5.12

9. **a)**

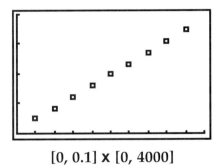

[0, 0.1] x [0, 4000]

b)

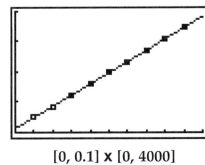

[0, 0.1] x [0, 4000]

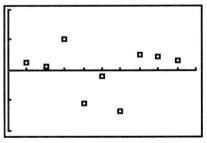

[0, 0, 1] x [−100, 100]

c)

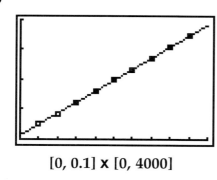
[0, 0.1] x [0, 4000]

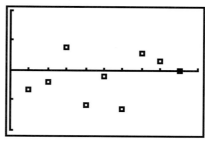
[0, 0, 1] x [−100, 100]

d) **Both the linear and quadratic models seem to describe the data well. The residual plots are similar. A clear choice of models cannot be made solely on the basis of residuals and graphs within the range of the data. This is a situation for which theoretical considerations would be important in choice of model.**

10. The data in **Table 5.13** represent holiday (November–December) retail sales totals between 1988 and 1997.

Year	Holiday Sales (1000s of dollars)
1988	107
1989	113
1990	113
1991	116
1992	127
1993	135
1994	146
1995	151
1996	156
1997	164

Table 5.13 Source: National Retail Federation.

10. a)

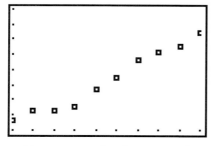
[1988, 1997] x [100, 180]

b)

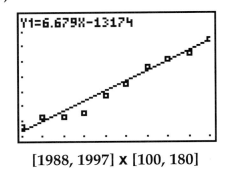 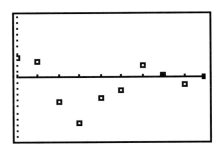

[1988, 1997] x [100, 180]   [1988, 1997] x [−10, 10]

c)

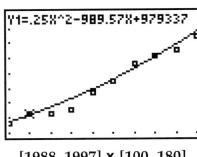 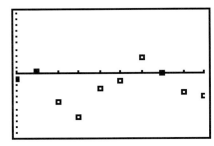

[1988, 1997] x [100, 180]   [1988, 1997] x [−10, 10]

d) Both models seem to describe the data reasonably well, with residual plots being similar. Because the linear model is more accurate for the last two years (1996 and 1997), it might be a better one for forecasting. If the linear model is used, then sales S depends on year y according to $S = 6.679y - 13{,}174$.

11. The cost of mailing a postcard in the United States was one penny for most of the first half of the twentieth century. **Table 5.14** shows the years in which the cost of mailing a postcard changed after 1950.

Year	Cost (cents)
1952	2
1958	3
1963	4
1968	5
1971	6
1974	8
1975	7
1976	9
1978	10
1981	12
1982	13
1985	14
1988	15
1991	19
1998	20

Table 5.14

a) Make a scatter plot of the data.

11. a)

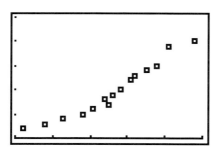

[1950, 2000] x [0, 25]

b) Find a quadratic regression equation for the data.

b) $C = 0.00578769Y^2 - 22.421668Y + 21{,}716.098.$

c) Use the equation to predict the cost of sending a postcard in the year 2015 if the pattern continues.

c) $C = 0.00578769(2015)^2 - 22.421668(2015) + 21{,}716.098.$ **The model predicts the cost of mailing a postcard to be 35 cents in the year 2015.**

d) In what year would you expect the cost of mailing a postcard to rise to 40 cents if the trend continues?

d) $C = 0.00578769Y^2 - 22.421668Y + 21{,}716.098 = 40$, or $0.00578769Y^2 - 22.421668Y + 21{,}676.098 = 0$. **Solutions are** $Y = 1854.46$ **and** **2019.56. 1854 is in the past, so the model predicts that the cost will rise to** **40 cents in the year 2019.**

For 12–16,

a) Make a scatter plot of the data.

b) Find a quadratic regression equation for the data.

c) Use a residual plot to decide if the quadratic regression equation adequately describes the data.

12. **Table 5.15** contains data from a study comparing arsenic levels in people's toenails to arsenic levels in private, unregulated well water that they used for drinking and cooking.

Arsenic in Water (parts per billion)	Arsenic in Toenails (parts per million)
0.87	0.119
0.21	0.118
1.15	0.118
0.13	0.08
0.69	0.158
19.4	0.517
137	2.252
21.4	0.851
17.5	0.269
76.4	0.433
16.5	0.275
0.12	0.135
4.1	0.175

Table 5.15 Source: Dartmouth Hitchcock Medical Center.

12. a)

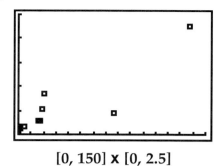

[0, 150] **x** [0, 2.5]

b) If T represents arsenic level in toenails and w represents arsenic level in water, then $T = 0.00009102w^2 = 0.001655w + 0.2032$.

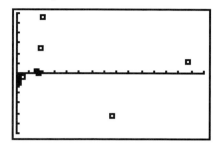

[0, 150] **x** [−0.6, 0.6]

c) The residuals are mostly small and randomly scattered, so the quadratic model may be acceptable. More data is probably needed before any firm conclusions are drawn.

13. **Table 5.16** contains data relating ultimate load strength of structural steel with loss of weight of the steel due to corrosion.

Loss of Weight (%)	Ultimate Load (kilonewtons)
0	183
4	180
5.5	177
11	164
11.5	162
12	160
17.5	146

Table 5.16 Source: American Concrete Institute Structural Journal.

13. a)

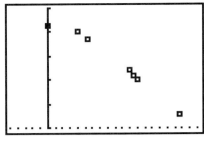

[−5, 20] **x** [140, 190]

b) If *S* represents ultimate strength and *L* represents loss of weight, then $S = -0.06126L^2 - 1.155L + 184$.

c) The residuals are small and randomly scattered, so the quadratic model adequately describes the data.

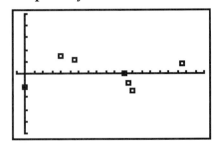

[−1, 20] **x** [−5, 5]

14. **Table 5.17** contains data on Medicare payments to physicians for selected years from 1960 to 1995.

Year	Medicare Payments (billions of dollars)
1960	5.3
1965	8.2
1970	13.6
1975	23.9
1980	45.2
1985	83.6
1990	146.3
1995	201.9

Table 5.17 Source: U.S. Healthcare Financing Administration.

14. a)

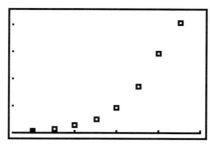

[1955, 2000] x [0, 220]

b) If M represents total Medicare payments and y represents the year, then
$M = 0.23043y^2 - 905.87y + 890308$.

c) The residuals are relatively small and scattered, so the quadratic model adequately describes the data.

[1955, 2000] x [−10, 10]

15. **Table 5.18** contains data on the U.S. balance of trade with China during the 1990s (measured as value of goods imported minus value of goods exported).

Year	Trade Balance (billions of dollars)
1990	−10.4
1991	−12.7
1992	−18.3
1993	−22.8
1994	−29.5
1995	−33.8
1996	−39.5
1997	−49.7
1998	−56.9
1999	−68.7

Table 5.18 Source: Boston Globe/*Associated Press.*

15. a)

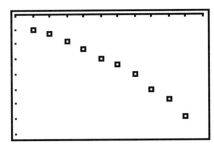

[1989, 2000] x [−80, 0]

b) If T represents trade balance and y represents the year, then
$T = -0.400379y^2 + 1590.77y - 1{,}580{,}112.$

c) The residuals are small and randomly scattered, so the quadratic model adequately describes the data.

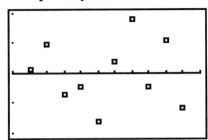

[1989, 2000] x [−2, 2]

16. **Table 5.19** shows the increase in membership of the American Homebrewers Association over an 11-year period.

Year	Number of Members
1983	2600
1984	3300
1985	3600
1986	3900
1987	5500
1988	6400
1989	7400
1990	10,000
1991	12,500
1992	16,400
1993	20,700

Table 5.19 Source: Boston Globe/*Association of Brewers.*

16. a)

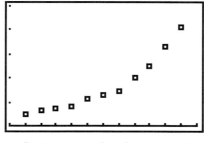

[1982, 1994] x [0, 25000]

b) If N represents number of members and y represents the year, then $N = 219.23y^2 - 869,992y + 863,117,807$.

c) The residuals are relatively small and randomly scattered, so the quadratic model adequately describes the data.

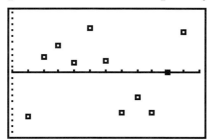

[1982, 1994] x [−1000, 1000]

For 17–18,

a) Make a scatter plot of the data.

b) Find a quadratic regression equation for the data and graph it on the scatter plot.

17. Use **Table 5.20.**

x	−3	−2	−1	0	1	2	3
y	25	10	3	0	1	6	18

Table 5.20

17. a)

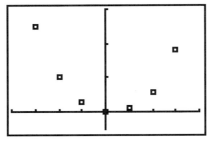

[–4, 4] **x** [–5, 30]

b) $y = 2.41667x^2 - 1.10714x - 0.66667.$

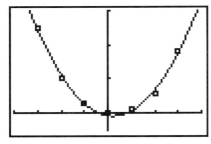

[–4, 4] **x** [–5, 30]

18. Use **Table 5.21.**

x	0.5	1.3	2.8	4.3	6.0
y	3.7	5.2	5.1	4.5	1.1

Table 5.21

18. a)

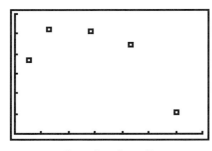

[0, 7] **x** [0, 6]

b) $y = -0.3667x^2 + 1.8857x + 3.0144.$

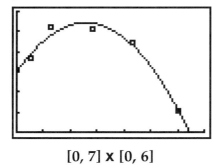

[0, 7] **x** [0, 6]

Section 5.4 Roots and Radicals

What You Need to Know

- How to use the quadratic formula to solve quadratic equations

- How to find the square root of a number

What You Will Learn

- To evaluate square roots and nth roots of numbers

- To simplify radical expressions

- To apply the product and quotient properties of radicals

- To add and subtract radical expressions

Materials

- String

- Washers or other weights

- Meter stick

- Stopwatch

In Section 5.2 we found that quadratic equations can be solved by several methods. The choice of method depends on both the efficiency and accuracy desired. If the factoring method can be used, it is quick and produces exact solutions. Graphing generally produces approximate solutions, and the accuracy of the solution is dependent on the accuracy of the graph.

Even though the quadratic formula can be used to solve any quadratic equation, it is most useful when the equation cannot be factored easily. It allows us to approximate solutions to quadratic equations to any desired accuracy. When an exact answer rather than an approximation is needed, solutions should be left in radical form. For example, when we solve the equation $0.5x^2 - 2x - 23 = 0$, we get the exact solutions $2 \pm \sqrt{50}$.

In this section we will examine numbers written in radical form, explore the properties of radicals, and perform operations on radical expressions.

Evaluating Roots of Numbers

As we use various formulas such as the quadratic formula and the Pythagorean theorem, we find it necessary to take square roots of numbers. Recall that $12^2 = 144$, so the square root of 144 is 12 and is written as $\sqrt{144} = 12$. It is also possible to find roots of numbers other than square roots; for example, the cube root, the fourth root, or the twelfth root.

A number written as $\sqrt[n]{a}$ is said to be in **radical form.** The symbol "$\sqrt{}$" is called the **radical sign,** and the number a is called the **radicand.** The integer n, which must be greater than 1, is called the **index** of the radical. For example, $2^5 = 32$, so the fifth root of 32 is 2. This is written as $\sqrt[5]{32} = 2$. In this case, 32 is the radicand and 5 is the index of the radical.

When the index is 2, the radical is usually written as $\sqrt{a}$ rather than $\sqrt[2]{a}$. Positive numbers have two square roots. For example, $7^2 = 49$ and $(-7)^2 = 49$. This means that both 7 and –7 are square roots of 49.

When a number has two roots, the positive root is called the **principal root.** For the number 49, 7 is the principal square root and –7 is referred to as the **negative root.** In general, the notation $\sqrt{a}$ is used to indicate the principal square root, and $-\sqrt{a}$ indicates the negative square root. Hence, $\sqrt{144} = 12$ and $-\sqrt{144} = -12$.

Using similar reasoning, the **principal nth root** of a number is defined as:

> If a, b, and n are positive numbers and $b^n = a$, then b is called the principal nth root of a and can be written $b = \sqrt[n]{a}$.

Example 9

Evaluate the expression:

a) $\sqrt[3]{27}$

b) $\sqrt{0.36}$

c) $\sqrt[4]{625}$

d) $-\sqrt{121}$

e) $\sqrt[4]{\dfrac{1}{81}}$

f) Find the value of $\sqrt{b^2 - 4ac}$ when $a = 2$, $b = -4$, and $c = -6$.

Solution:

a) $\sqrt[3]{27} = 3$ because $27 = 3^3$.

b) $\sqrt{0.36} = 0.6$ because $0.36 = (0.6)^2$.

c) $\sqrt[4]{625} = 5$ because $625 = 5^4$.

d) $-\sqrt{121} = -11$ because $121 = (-11)^2$.

e) $\sqrt[4]{\dfrac{1}{81}} = \dfrac{1}{3}$ because $\dfrac{1}{81} = \left(\dfrac{1}{3}\right)^4$.

f) $\sqrt{(-4)^2 - 4(2)(-6)} = \sqrt{64} = 8$

Up to this point, we have been examining roots of positive numbers only. What happens when we take a root of a negative number?

Note that $(-2)^3 = -8$, which means that -2 is the cube root of -8, or $\sqrt[3]{-8} = -2$.

Similar reasoning can be used to find odd roots of any number. For example,

$$\sqrt[9]{-1} = -1 \text{ because } -1 = (-1)^9 \text{ and } \sqrt[5]{-\dfrac{1}{32}} = -\dfrac{1}{2} \text{ because } -\dfrac{1}{32} = \left(-\dfrac{1}{2}\right)^5.$$

But what happens when we want to take even roots, such as square roots, of negative numbers? For example, what if we want to find $\sqrt{-9}$? That is, can we find a number whose square is -9. Two possible candidates are -3 and 3. But $(-3)^2$ does not equal -9, and neither does 3^2. Thus, there is no real number x such that $x^2 = -9$.

In fact, we know that if we raise a negative number to an even power, we get a positive number, and the same is true for positive numbers. Hence, we say that $\sqrt[n]{a}$ is undefined if n is even and a is negative.

Example 10
  a) Find all the square roots of 16.

  b) Find all the cube roots of –125.

  c) Find all the fourth roots of –16.

Solution:

  a) Because $4^2 = 16$ and $(-4)^2 = 16$, both 4 and –4 are square roots of 16. The principal root is 4.

  b) Because $(-5)^3 = -125$, the cube root of –125 is –5. (Note that $5^3 = 125$.)

  c) –16 has no fourth roots because the fourth root of a negative number is undefined.

Example 11

The formula $A = 6\sqrt[3]{V^2}$ can be used to find the surface area A of a cube when the volume V of the cube is known. Find the surface area of a cube whose volume is 343 cubic feet.

Solution:

$$A = 6\sqrt[3]{V^2}$$
$$= 6\sqrt[3]{343^2}$$
$$= 6\sqrt[3]{117{,}646}$$
$$= 6(49)$$
$$= 294 \text{ ft}^2.$$

Most graphing calculators provide features that allow the user to find square roots, cube roots, and even nth roots of numbers. For some calculators, these mathematical operations are found under the Math menu (see **Figure 5.34**).

To locate specific instructions for finding roots of numbers, refer to the calculator's manual.

Figure 5.34

Simplifying and Combining Radicals

When using radicals to solve problems, there are times that we are asked to "simplify" expressions containing square roots. This means the following:

- There are no perfect square factors other than 1 in the radicand (under the radical sign).

- There are no fractions in the radicand.

- There are no radicals in the denominator of a fraction.

Two properties of radicals, the **Product Property** and the **Quotient Property**, help simplify expressions that contain square roots.

Properties of Radicals

If a and b are real numbers and n is nonnegative, then

$\sqrt[n]{ab} = \sqrt[n]{a} \cdot \sqrt[n]{b}$ Product Property

$\sqrt[n]{\dfrac{a}{b}} = \dfrac{\sqrt[n]{a}}{\sqrt[n]{b}}$ Quotient Property

---

At the beginning of this section, we gave the solutions to the quadratic equation $0.5x^2 - 2x - 23 = 0$ as $2 \pm \sqrt{50}$. These solutions can be simplified using the product property of radicals to rewrite $\sqrt{50}$:

$$\sqrt{50} = \sqrt{25 \cdot 2} \qquad \text{Factor radicand.}$$
$$= \sqrt{25} \cdot \sqrt{2} \quad \text{Product property}$$
$$= 5\sqrt{2}.$$

Hence, the solutions can be simplified and rewritten as $2 \pm 5\sqrt{2}$.

Example 12

Rewrite each expression in simplified radical form.

a) $\sqrt{300}$

b) $5\sqrt{18}$

c) $\sqrt{20x^3y^4}$

d) $\sqrt{\dfrac{3}{16}}$

e) $\dfrac{1}{\sqrt{3}}$

f) $5\sqrt{2}\left(3\sqrt{10} - 2\sqrt{6}\right)$

> **Take Note**
>
> When simplified, both $\sqrt{3^2}$ and $\sqrt{(-3)^2}$ are equal to 3 because both are equal to $\sqrt{9}$, and the square root symbol $\sqrt{}$ indicates the principal square root. This is not a problem when the radicand contains numbers. But what happens if the radicand contains a variable, for example, $\sqrt{x^2}$? When x represents a number greater than or equal to zero, $\sqrt{x^2} = x$. But if x represents a negative number, then $\sqrt{x^2} = -x$. Hence, we can write $\sqrt{x^2} = |x|$. This assures us that $\sqrt{x^2}$ is a positive number.
>
> In this chapter, when simplifying radical expressions that contain variables, we will assume that the variables are nonnegative unless otherwise noted.

Solution:

a) $\sqrt{300}$ has a perfect square factor in the radicand. To simplify:

$$\sqrt{300} = \sqrt{100 \cdot 3} \qquad \text{Factor radicand.}$$
$$= \sqrt{100} \cdot \sqrt{3} \quad \text{Product property}$$
$$= 10\sqrt{3}.$$

b) $5\sqrt{18}$ has a perfect square factor in the radicand. To simplify:

$$5\sqrt{18} = 5\sqrt{9 \cdot 2} \qquad \text{Factor radicand.}$$
$$= 5\sqrt{9} \cdot \sqrt{2} \quad \text{Product property}$$
$$= 15\sqrt{2}.$$

c) $\sqrt{20x^3y^4}$ has perfect square factors in the radicand. To simplify:

$$\sqrt{20x^3y^4} = \sqrt{4 \cdot 5 \cdot x^2 \cdot x \cdot y^4} \qquad \text{Factor radicand.}$$
$$= \sqrt{4} \cdot \sqrt{5} \cdot \sqrt{x^2} \cdot \sqrt{x} \cdot \sqrt{y^4} \quad \text{Product property}$$
$$= 2xy^2\sqrt{5x}.$$

d) $\sqrt{\dfrac{3}{16}}$ has a fraction in the radicand. To simplify:

$$\sqrt{\frac{3}{16}} = \frac{\sqrt{3}}{\sqrt{16}} \qquad \text{Quotient property}$$
$$= \frac{\sqrt{3}}{4}.$$

e) There is a radical in the denominator of the fraction $\dfrac{1}{\sqrt{3}}$. To simplify, we can multiply the expression by a fraction equivalent to 1. In this case, we choose $\dfrac{\sqrt{3}}{\sqrt{3}}$ because multiplying the denominator by $\sqrt{3}$ results in $\sqrt{3} \cdot \sqrt{3} = \sqrt{9} = 3$.

$$\frac{1}{\sqrt{3}} = \frac{1}{\sqrt{3}} \cdot \frac{\sqrt{3}}{\sqrt{3}} \qquad \text{Multiply by 1 in the form } \frac{\sqrt{3}}{\sqrt{3}}.$$
$$= \frac{\sqrt{3}}{\sqrt{9}} \qquad \text{Product property}$$
$$= \frac{\sqrt{3}}{3}.$$

f) To find the product, use the distributive and product properties:

$$5\sqrt{2}\left(3\sqrt{10} - 2\sqrt{6}\right) = 15\sqrt{20} - 10\sqrt{12} \qquad \text{Distributive property}$$
$$= 15\sqrt{4 \cdot 5} - 10\sqrt{4 \cdot 3} \qquad \text{Factor radicands.}$$
$$= 15\sqrt{4}\sqrt{5} - 10\sqrt{4}\sqrt{3} \qquad \text{Product property}$$
$$= 30\sqrt{5} - 20\sqrt{3}.$$

Example 13

The Americans with Disabilities Act (ADA) of 1990 states that for a ramp to rise 2 feet from ground level to the entrance of a building, the beginning of the ramp must be located at least 24 feet from the building entrance. If a ramp begins exactly 24 feet from the building and rises 2 feet, how long will the ramp be? Give the result both in exact form and in decimal form to the nearest hundredth.

Solution:

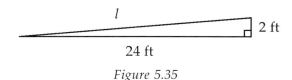

Figure 5.35

Figure 5.35 shows a right triangle. The Pythagorean theorem can be used to calculate the length of the hypotenuse.

$$l = \sqrt{2^2 + 24^2}$$

$$l = \sqrt{580}$$

This gives the exact length, but it is not simplified. Simplifying:

$$\sqrt{580} = \sqrt{4 \cdot 145} = \sqrt{4} \cdot \sqrt{145} = 2\sqrt{145} \text{ feet.}$$

Using a calculator and either $\sqrt{580}$ or $2\sqrt{145}$, the approximate length is 24.08 feet.

Adding and Subtracting Radicals

Recall that when adding or subtracting polynomials, we can only add or subtract like terms. In some ways, adding and subtracting radicals is similar to adding and subtracting polynomials because we can only combine like radicals. Two radical expressions that have the same index and the same radicand are called **like radicals.**

For example, $3\sqrt{6}$ and $\sqrt{6}$ are like radicals, whereas $3\sqrt{5}$ and $\sqrt{6}$ are **unlike radicals.** $3\sqrt{6}$ and $\sqrt[3]{6}$ are also considered unlike radicals because they do not have the same index.

To add or subtract like radicals, we add or subtract their coefficients. For example, $3\sqrt{6} + \sqrt{6} = (3 + 1)\sqrt{6} = 4\sqrt{6}$. As a word of caution, before deciding whether radicals are like or unlike, make sure all radicals are simplified (see Example 14(b)).

Example 14

Perform the indicated operations. Write all answers in simplified radical form.

a) $5\sqrt{5} - 8\sqrt{5} + 6\sqrt{7}$

b) $4\sqrt{2} + 3\sqrt{8} - \sqrt{2}$

c) $2b\sqrt{a^3 b} - 4ab\sqrt{ab} + 3\sqrt{4a^3 b^3} - a\sqrt{ab^3}$

Solution:

a) $5\sqrt{5} - 8\sqrt{5} + 6\sqrt{7} = -3\sqrt{5} + 6\sqrt{7}$ Add like radicals.

b) $4\sqrt{2} + 3\sqrt{8} - \sqrt{2} = 4\sqrt{2} + 3\sqrt{4 \cdot 2} - \sqrt{2}$ Factor radicands.

$$= 4\sqrt{2} + 6\sqrt{2} - \sqrt{2} \quad \text{Simplify.}$$

$$= 9\sqrt{2}. \qquad\qquad\quad \text{Add and subtract like radicals.}$$

c) $2b\sqrt{a^3 b} - 4ab\sqrt{ab} + 3\sqrt{4a^3 b^3} - a\sqrt{ab^3} = 2b\left(a\sqrt{ab}\right) - 4ab\sqrt{ab} + 3\left(2ab\sqrt{ab}\right) - a\left(b\sqrt{ab}\right)$

$$= 2ab\sqrt{ab} - 4ab\sqrt{ab} + 6ab\sqrt{ab} - ab\sqrt{ab}$$

$$= 3ab\sqrt{ab}.$$

Example 15

The area between two concentric circles is called an *annulus* (see **Figure 5.36**).

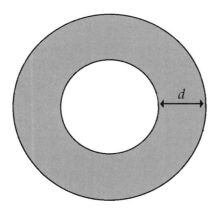

Figure 5.36

If the area of the larger circle A_R and the area of the smaller circle A_r are known, the distance between the two circles d can be found using the formula

$$d = \sqrt{\frac{A_R}{\pi}} - \sqrt{\frac{A_r}{\pi}}.$$

a) Find the distance between the two circles if $A_R = 49\pi$ square feet and $A_r = 16\pi$ square feet.

b) Find the distance between the two circles if $A_R = 300\pi$ cm^2 and $A_r = 108\pi$ cm^2.

Solution:

a) $d = \sqrt{\dfrac{A_R}{\pi}} - \sqrt{\dfrac{A_r}{\pi}} = \sqrt{\dfrac{49\pi}{\pi}} - \sqrt{\dfrac{16\pi}{\pi}}$

$\qquad\qquad\quad = \sqrt{49} - \sqrt{16}$

$\qquad\qquad\quad = 7 - 4$

$\qquad\qquad\quad = 3$ ft.

b) $d = \sqrt{\dfrac{A_R}{\pi}} - \sqrt{\dfrac{A_r}{\pi}} = \sqrt{\dfrac{300\pi}{\pi}} - \sqrt{\dfrac{108\pi}{\pi}}$

$\qquad\qquad\quad = \sqrt{300} - \sqrt{108}$

$\qquad\qquad\quad = 10\sqrt{3} - 6\sqrt{3}$

$\qquad\qquad\quad = 4\sqrt{3}$ cm.

Exercises 5.4

I. Investigations *(SE pages 62–65)*

1. Typically, the area of a triangle is found by multiplying the base of the triangle by its height and then taking one-half the product, that is, $A = \frac{1}{2}bh$.

    a) Find the area of the triangle in **Figure 5.37.**

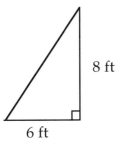

8 ft

6 ft

Figure 5.37

1. a) $A = \frac{1}{2}bh = \frac{1}{2}(6)(8) = 24\,\text{ft}^2$.

    b) A different method for finding the area of any triangle is named after the Greek mathematician Heron who lived about A.D. 50. This method is based on the *semiperimeter*, which is one-half the perimeter of the triangle.

       Find the semiperimeter of the triangle in Figure 5.37.

    b) The length of the hypotenuse is 10 ft. So the semiperimeter is $\frac{1}{2}(6 + 8 + 10) = 12\,\text{ft}$.

    c) Heron's formula states that the area of any triangle equals $\sqrt{s(s-a)(s-b)(s-c)}$, where the lengths of the sides are a, b, and c, and s is the semiperimeter. Use Heron's formula to find the area of the triangle in Figure 5.37.

    c) $A = \sqrt{s(s-a)(s-b)(s-c)} = \sqrt{12(12-6)(12-8)(12-10)}$

    $$= \sqrt{12(6)(4)(2)} = \sqrt{576} = 24 \text{ sq ft.}$$

    d) Use Heron's formula to find the exact area of the triangle whose sides are 8 in, 5 in, and 7 in.

    d) The semiperimeter of the triangle is 20/2 = 10.

    $$A = \sqrt{s(s-a)(s-b)(s-c)} = \sqrt{10(10-8)(10-5)(10-7)}$$

    $$= \sqrt{10(2)(5)(3)} = \sqrt{300} = 10\sqrt{3} \text{ sq in.}$$

2. To find the product of $\left(6+2\sqrt{3}\right)\left(3-5\sqrt{3}\right)$, we follow the same rules as when we multiply any two binomials:

$$\left(6+2\sqrt{3}\right)\left(3-5\sqrt{3}\right) = 6\left(3-5\sqrt{3}\right)+2\sqrt{3}\left(3-5\sqrt{3}\right) \quad \text{Distributive property}$$

$$= 18-30\sqrt{3}+6\sqrt{3}-10\left(\sqrt{3}\right)\left(\sqrt{3}\right) \quad \text{Distributive property}$$

$$= 18-24\sqrt{3}-30 \quad \text{Add like radicals.}$$

$$= -12-24\sqrt{3}.$$

a) Find the product: $\left(5-2\sqrt{2}\right)\left(4+\sqrt{2}\right)$.

2. a) $\left(5-2\sqrt{2}\right)\left(4+\sqrt{2}\right) = 20+5\sqrt{2}-8\sqrt{2}-2\sqrt{4} = 16-3\sqrt{2}.$

b) Find the product: $\left(4-\sqrt{7}\right)\left(4+\sqrt{7}\right)$.

b) $\left(4-\sqrt{7}\right)\left(4+\sqrt{7}\right) = 16+4\sqrt{7}-4\sqrt{7}-7 = 9.$

c) Find the product: $\left(1+3\sqrt{2}\right)^2$.

c) $\left(1+3\sqrt{2}\right)^2 = 1+3\sqrt{2}+3\sqrt{2}+\left(3\sqrt{2}\right)\left(3\sqrt{2}\right) = 1+6\sqrt{2}+18 = 19+6\sqrt{2}.$

d) What is the semiperimeter of the triangle in **Figure 5.38**? Use Heron's formula from Exercise 1 to find the exact area of the triangle.

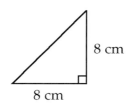

8 cm

8 cm

Figure 5.38

d) The hypotenuse of the triangle is $8\sqrt{2}$ cm, so the semiperimeter is $\frac{1}{2}\left(8\sqrt{2}+8+8\right) = 4\sqrt{2}+8$ cm. The area of the triangle is

$$A = \sqrt{s(s-a)(s-b)(s-c)} = \sqrt{\left(8+4\sqrt{2}\right)\left(4\sqrt{2}\right)\left(4\sqrt{2}\right)\left(8-4\sqrt{2}\right)} = \sqrt{32^2} = 32 \text{ cm}^2.$$

3. When asked to find the mean of two numbers such as 6 and 10, we automatically think of the **arithmetic mean,** or average of the numbers. But there is another mean, the **geometric mean.** In this investigation we will calculate the two means and compare the results.

a) To find the arithmetic mean of two numbers, we add the numbers and divide the sum by two. What is the arithmetic mean of 6 and 10? Write an expression for the arithmetic mean of a and b.

3. a) For the numbers 6 and 10, the arithmetic mean is $\dfrac{6+10}{2} = 8$. For a and b,

the arithmetic mean is $\dfrac{a+b}{2}$.

b) To find the geometric mean of two numbers, we multiply the numbers and take the square root of the product. What is the geometric mean of 6 and 10? Write an expression for the geometric mean of a and b.

b) For the numbers 6 and 10, the geometric mean is $\sqrt{6 \cdot 10} = \sqrt{60} = 2\sqrt{15}$. For a and b, the geometric mean is $\sqrt{ab}$.

c) For the numbers 6 and 10, which is greater, the arithmetic or the geometric mean?

c) The arithmetic mean is 8, the geometric mean is approximately 7.746. For these two numbers, the arithmetic mean is greater than the geometric mean.

d) Complete **Table 5.22.**

Number Pairs	Arithmetic Mean	Exact Geometric Mean	Approximate Geometric Mean
2 and 32			
3 and 18			
10 and 20			
14 and 16			
8 and 8			

Table 5.22

d)

Number Pairs	Arithmetic Mean	Exact Geometric Mean	Approximate Geometric Mean
2 and 32	$(32 + 2)/2 = 17$	$\sqrt{2 \cdot 32} = \sqrt{64} = 8$	8
3 and 18	$(3 + 18)/2 = 10.5$	$\sqrt{3 \cdot 18} = \sqrt{6 \cdot 9} = 3\sqrt{6}$	7.348
10 and 20	$(10 + 20)/2 = 15$	$\sqrt{10 \cdot 20} = \sqrt{200} = 10\sqrt{2}$	14.142
14 and 16	$(14 + 16)/2 = 15$	$\sqrt{14 \cdot 16} = 4\sqrt{14}$	14.967
8 and 8	$(8 + 8)/2 = 8$	$\sqrt{8 \cdot 8} = \sqrt{64} = 8$	8

e) For two positive numbers, which is larger, the geometric mean or the arithmetic mean? Are the geometric and arithmetic means of two numbers ever equal? If so, when?

e) The arithmetic mean is greater than the geometric mean of two positive, unequal numbers. If the two numbers are equal, the geometric and the arithmetic means are the same.

4. To summarize the discussion of roots of real numbers in this section, complete **Table 5.23** for the expression $\sqrt[n]{a}$. Indicate whether the root is positive, negative, zero, or undefined.

$\sqrt[n]{a}$	n is even	n is odd
$a > 0$		
$a = 0$		
$a < 0$		

Table 5.23

4.

$\sqrt[n]{a}$	n is even	n is odd
$a > 0$	positive	positive
$a = 0$	zero	zero
$a < 0$	undefined	negative

5. **Figure 5.39** shows a spiral that is formed by constructing right triangles whose legs are the same length as the hypotenuse of the smaller triangle.

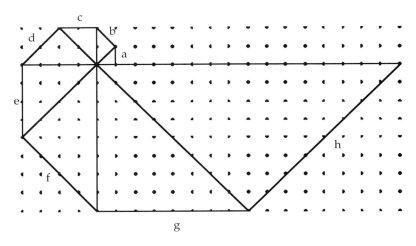

Figure 5.39

a) If the length of line segment a is 1 unit, find the lengths of the segments b, c, d, e, f, g, and h.

5. **a) $b = \sqrt{2}$, $c = 2$, $d = 2\sqrt{2}$, $e = 4$, $f = 4\sqrt{2}$, $g = 8$, $h = 8\sqrt{2}$.**

b) If this pattern were to continue, what would be the lengths of the next two segments in the spiral?

b) 16, then $16\sqrt{2}$

c) Find the total length of line segments a and b.

c) $1 + \sqrt{2}$

d) Find the total length of line segments a, b, c, and d.

d) $3 + 3\sqrt{2}$

e) Find the total length of line segments a, b, c, d, e, and f.

e) $7 + 7\sqrt{2}$

f) Find the total length of line segments a, b, c, d, e, f, g, and h.

f) $15 + 15\sqrt{2}$

g) What pattern do you see in the total lengths of the spiral if it has an even number of segments?

g) Answers will vary. Possible answers: "It's $1 + \sqrt{2}$ times a number. That number increases by twice as much each time (2, 4, 8, etc.)." or "It's $1 + \sqrt{2}$ times a number that is 1 less than a power of 2." The most precise answer would be, "If we have n pairs of segments, then it's $\left(2^n - 1\right)\left(1 + \sqrt{2}\right)$."

6. In this section, as we simplified expressions containing square roots, we examined each expression to make sure there were no perfect square factors in the radicand other than 1. Likewise, when simplifying expressions with other roots such as cube roots, fourth roots, and so forth, we need to examine the radicals for any perfect powers that occur as factors.

The following example shows how the expression $\sqrt[3]{24x^4}$ can be simplified. (Note that the perfect cube factors 8 and x^3 are factors of the radicand.)

$$\sqrt[3]{24x^4} = \sqrt[3]{8 \cdot 3 \cdot x^3 \cdot x} \qquad \text{Factor the radicand.}$$
$$= \sqrt[3]{8} \cdot \sqrt[3]{3} \cdot \sqrt[3]{x^3} \cdot \sqrt[3]{x} \qquad \text{Product property}$$
$$= 2x\sqrt[3]{3x}. \qquad \text{Cube roots of perfect cubes}$$

For (a)–(e), simplify the expression.

a) $\sqrt[4]{405}$

6. a) $\sqrt[4]{405} = \sqrt[4]{81 \cdot 5} = 3\sqrt[4]{5}.$

b) $\sqrt[5]{y^8}$

b) $\sqrt[5]{y^8} = \sqrt[5]{y^5 \cdot y^3} = y\sqrt[5]{y^3}.$

c) $\sqrt[3]{54x^8}$

c) $\sqrt[3]{54x^8} = \sqrt[3]{27 \cdot 2 \cdot x^6 \cdot x^2} = 3x^2\sqrt[3]{2x^2}.$

d) $\sqrt[3]{3} \cdot \sqrt[3]{9}$

d) $\sqrt[3]{3} \cdot \sqrt[3]{9} = \sqrt[3]{27} = 3.$

e) $\sqrt[5]{12,500}$

e) $\sqrt[5]{12,500} = \sqrt[5]{5^5 \cdot 4} = 5\sqrt[5]{4}.$

II. Projects and Group Activities (SE pages 65–67)

Materials: string (approximately 2 meters), washers or other weights, meter stick, stopwatch

7. In this activity you will investigate the factors that influence the swing of a pendulum.

Before beginning the activity, discuss with group members which of the following might cause the period of the swing to change. (Keep in mind that the period of the swing is defined as the time it takes for the pendulum to move back and forth, that is, for one complete swing.)

• The weight of the object on the end of the pendulum

• The length of the pendulum

• How far the pendulum is pulled back before releasing it

Performing the experiment:

a) Attach a weight to one end of a piece of string approximately 100 cm long. Attach the other end to a horizontal rod or hook. Measure and record the length of the pendulum (the distance from the rod to the center of the weight on the end of the string).

b) Keeping the string straight, pull the string back about 5–10 degrees. Then use a stopwatch to determine the time it takes for the pendulum to go back and forth 10 times (10 periods). (See **Figure 5.40.**)

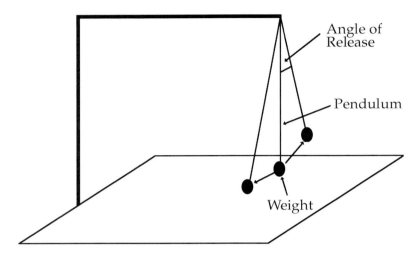

Figure 5.40

c) Calculate the time it takes for one back and forth swing (the length of one period).

d) Record the results in a table similar to **Table 5.24.**

Trial Number	Number of Weights	Length of Pendulum (centimeters)	Angle of Release (degrees)	Time of One Period (seconds)
1.				
2.				
3.				
etc.				

Table 5.24

d) Sample table:

Trial Number	Number of Weights	Length of Pendulum (centimeters)	Angle of Release (degrees)	Time of One Period (seconds)
1.	1	101	10	2.01
2.	1	101	10	2.00
3.	1	101	10	2.01
4.	1	101	10	2.02
5.	1	101	10	2.01

e) Without changing the weight, the length of the pendulum, or the angle of release, repeat the procedure four more times, and record the average period lengths.

e) Sample answer: 2.0 seconds

f) This time, change the angle of release but do not change the weight or the length of the pendulum. Perform the procedure at least five times with different angles of release, and record the times. What do you notice? Does the angle have any effect on the period? If so, what effect?

f) The period stays the same. The angle of release has no effect on the period.

g) Now change the amount of weight hanging from the pendulum. Add more weight and repeat the procedure. Be sure to keep the length and the angle of release the same. What do you notice? Does the weight of the pendulum have any effect on the period? If so, what effect?

g) The period stays the same. The weight of the pendulum has no effect on the period.

h) This time, change the length of the pendulum but keep the angle of release and the weight constant. What do you notice? Does the length of the pendulum have any effect on the period? If so, what effect?

h) The period changed. Yes, the length affects the period. As the length of the pendulum got shorter, the period got shorter.

Analyzing the Results:

i) The relationship between the time T that it takes for the pendulum to complete one period (measured in seconds) and the length of the pendulum L (measured in centimeters) is given by the equation $T = 2\pi\sqrt{\dfrac{L}{980}}$. Use this equation to predict the time of one period for the different lengths of your pendulum. How close were your actual results to the results predicted by the equation? If they were not close, how might you explain the discrepancy?

i) Sample predicted and actual times:

Length (cm)	Predicted Time (s)	Actual Time (Average) (s)
101	2.02	2.01
88	1.88	1.88
75	1.74	1.75
61	1.57	1.59
45	1.35	1.37

Discrepancies might be explained by air resistance on the pendulum and by lack of accuracy in measuring lengths and time.

III. Additional Practice *(SE pages 67–71)*

8. When the length of a pendulum L is measured in feet, its period is given by $T = 2\pi\sqrt{\dfrac{L}{32}}$, where T is measured in seconds. Find the exact period of a pendulum that is 3 feet long.

8. $T = 2\pi\sqrt{\dfrac{L}{32}} = 2\pi\sqrt{\dfrac{3}{32}} \cdot \dfrac{\sqrt{2}}{\sqrt{2}} = 2\pi\dfrac{\sqrt{6}}{\sqrt{64}} = \dfrac{2\pi}{8}\sqrt{6} = \dfrac{\pi}{4}\sqrt{6}$ seconds.

9. The formula $c = 2\sqrt{h(2r-h)}$ can be used to determine the length of a chord of a circular segment c when the radius of the circle r and the height of the segment h are known. Determine the length of the chord of the circular segment in **Figure 5.41.**

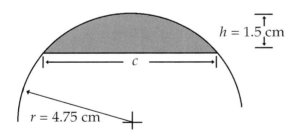

$h = 1.5$ cm

$r = 4.75$ cm

Figure 5.41

9. $c = 2\sqrt{h(2r-h)} = 2\sqrt{1.5\big(2(4.75)-1.5\big)} = 2\sqrt{12} = 4\sqrt{3}$ cm.

10. The formula $A = 6\sqrt[3]{V^2}$ relates the surface area of a cube A and its volume V. Use this formula to find the exact surface area of a cube with a volume of 18 cubic feet.

10. $A = 6\sqrt[3]{V^2} = 6\sqrt[3]{18^2} = 6\sqrt[3]{(3 \cdot 3 \cdot 2)(3 \cdot 3 \cdot 2)} = 6\sqrt[3]{3^3 \cdot 12} = 3(6)\sqrt[3]{12} = 18\sqrt[3]{12}$ sq ft.

11. Find the length of material needed to create a "Y" bracket with specifications shown in **Figure 5.42.**

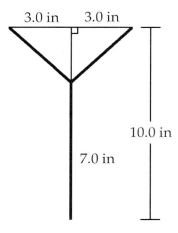

Figure 5.42

11. **Use the Pythagorean theorem to find the lengths of the "arms" of the "Y":**

   $x^2 = 3^2 + 3^2$; $x = 3\sqrt{2}$ **inches;**
   the total length is $3\sqrt{2} + 3\sqrt{2} + 7 = 7 + 6\sqrt{2}$ **inches.**

12. The formula $A = 2s\sqrt{0.25s^2 + h^2}$ can be used to find the surface area A of a square pyramid where S represents the length of one side of the square base and h represents the height of the pyramid.

   Find the surface area of the square pyramid-shaped roof in **Figure 5.43.** (Give both an exact and an approximate area.)

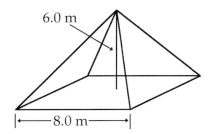

Figure 5.43

12. $A = 2s\sqrt{0.25s^2 + h^2}$; $A = 2(8)\sqrt{0.25(8)^2 + 6^2}$; $A = 16\sqrt{16 + 36}$; $A = 16 \cdot 2\sqrt{13}$;
   $A = 32\sqrt{13}$ **square meters, which is approximately 115 square meters.**

13. Clear-cutting large portions of forests often results in reduced forested areas. The border between forested and clear-cut regions exhibits what is known as an "edge effect." This edge area is often incapable of supporting the kinds of plant and animal life that thrive in the forest. Because of this, a forested area that has a significant percentage of its total area in edge may function differently as an ecosystem than an interior forest.

Figure 5.44 shows a circular Douglas fir forest that has a total area of 980,000 m². If 500,000 m² of the forest is interior forest, to what depth d does the "edge effect" penetrate? (Hint: See Example 15.)

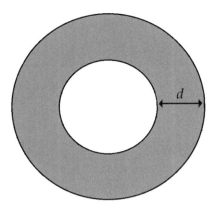

Figure 5.44

13. $d = \sqrt{\dfrac{A_R}{\pi}} - \sqrt{\dfrac{A_r}{\pi}} = \sqrt{\dfrac{980,000}{\pi}} - \sqrt{\dfrac{500,000}{\pi}}$

$= 700\sqrt{\dfrac{2}{\pi}} - 500\sqrt{\dfrac{2}{\pi}}$

$= 200\sqrt{\dfrac{2}{\pi}} \approx 160 \text{ m}^2.$

14. The formula $d = \sqrt{\dfrac{3w}{2}}$ can be used to find the diameter d (in inches) of a fiber line that is required to lift a given load w (in tons). If the load is being lifted by a wire rope, the formula used is $d = \dfrac{\sqrt{w}}{2}$.

a) Find the diameter of a fiber line that is required to lift a load of 5.0 tons. Give both an exact and an approximate length.

14. a) $d = \sqrt{\dfrac{3w}{2}} = \sqrt{\dfrac{3(5.0)}{2}} = \dfrac{\sqrt{30}}{2}$ **inches, or approximately 2.7 inches.**

b) Find the diameter of a wire rope that is required to lift a load of 5.0 tons. Give both an exact and an approximate length.

b) $d = \dfrac{\sqrt{w}}{2} = \dfrac{\sqrt{5}}{2}$ **inches, or approximately 1.1 inches.**

c) Find the circumference of a fiber line that is required to lift a 30-ton load. Give both an exact and an approximate length. (Hint: Recall that the circumference of a circle is πd.)

c) The diameter of the line is $d = \sqrt{\dfrac{3w}{2}} = \sqrt{\dfrac{3(30)}{2}} = \sqrt{45} = 3\sqrt{5}$ **inches. The circumference is** $\pi d = 3\pi\sqrt{5}$**, or approximately 21 inches.**

15. Resonance frequency f for a circuit containing an inductor L and capacitor C is found using the formula $f = \dfrac{1}{2\pi\sqrt{LC}}$. When L is measured in henrys and C is measured in farads, f is measured in hertz. Find the resonance frequency of a circuit containing a 51-microhenry (0.000051 henry) inductor and a 12-microfarad (0.000012 farad) capacitor.

15. $f = \dfrac{1}{2\pi\sqrt{LC}} = \dfrac{1}{2\pi\sqrt{(0.000051)(0.000012)}} \approx$ **6433 Hz.**

For 16–18, use the quadratic formula to solve the equation. Be sure to simplify the answers.

16. $-x^2 + 9x + 11 = 0$.

16. $\dfrac{-9 \pm \sqrt{81 - 4(-1)(11)}}{2(-1)} = \dfrac{-9 \pm \sqrt{125}}{-2} = \dfrac{-9 \pm 5\sqrt{5}}{-2}$.

17. $x^2 = 55 - 5x$.

17. $\dfrac{-5 \pm \sqrt{25 - 4(1)(-55)}}{2(1)} = \dfrac{-5 \pm \sqrt{245}}{2} = \dfrac{-5 \pm 7\sqrt{5}}{2}$.

18. $x^2 - 3x = 9$.

18. $\dfrac{-(-3) \pm \sqrt{9 - 4(1)(-9)}}{2(1)} = \dfrac{3 \pm \sqrt{45}}{2} = \dfrac{3 \pm 3\sqrt{5}}{2}$.

19. Find the edge of a cube e whose volume V is 320 cm^3 if $e = \sqrt[3]{V}$.

19. $e = \sqrt[3]{320} = \sqrt[3]{(64)(5)} = 4\sqrt[3]{5}$ **cm.**

20. Use the formula $d = \sqrt[4]{\dfrac{64I}{\pi}}$ to find the diameter d in feet of a load-bearing column whose moment of inertia I is 0.50 ft^4.

20. $d = \sqrt[4]{\dfrac{64I}{\pi}} = \sqrt[4]{\dfrac{64(0.50)}{\pi}} = \sqrt[4]{\dfrac{32}{\pi}}$ **or** $2\sqrt[4]{\dfrac{2}{\pi}}$ **feet, which is approximately 1.8 feet.**

For 21–31, rewrite the expressions in simplified radical form.

21. $\sqrt{175}$

21. $5\sqrt{7}$

22. $\sqrt{52}$

22. $2\sqrt{13}$

23. $\sqrt{567}$

23. $9\sqrt{7}$

24. $15\sqrt{8}$

24. $30\sqrt{2}$

25. $\sqrt{16x^5y^6}$

25. $4x^2y^3\sqrt{x}$

26. $\sqrt{75ab^7}$

26. $5b^3\sqrt{3ab}$

27. $\dfrac{\sqrt{2}}{\sqrt{5}}$

27. $\dfrac{\sqrt{2}}{\sqrt{5}} \cdot \dfrac{\sqrt{5}}{\sqrt{5}} = \dfrac{\sqrt{10}}{5}$.

28. $\dfrac{\sqrt{8}}{\sqrt{2}}$

28. $\dfrac{\sqrt{8}}{\sqrt{2}} \cdot \dfrac{\sqrt{2}}{\sqrt{2}} = \dfrac{\sqrt{16}}{2} = \dfrac{4}{2} = 2$.

29. $\sqrt{\dfrac{1}{7}}$

29. $\sqrt{\dfrac{1}{7}} = \dfrac{\sqrt{1}}{\sqrt{7}} \cdot \dfrac{\sqrt{7}}{\sqrt{7}} = \dfrac{\sqrt{7}}{7}$.

30. $\dfrac{r^2\sqrt{2r}}{\sqrt{3}}$

30. $\dfrac{r^2\sqrt{2r}}{\sqrt{3}} \cdot \dfrac{\sqrt{3}}{\sqrt{3}} = \dfrac{r^2\sqrt{6r}}{3}$.

31. $\dfrac{qp^3\sqrt{7q^4}}{\sqrt{13}}$

31. $\dfrac{qp^3\sqrt{7q^4}}{\sqrt{13}} \cdot \dfrac{\sqrt{13}}{\sqrt{13}} = \dfrac{q^3p^3\sqrt{91}}{13}$.

For 32–45, perform the indicated operations. Write all answers in simplified radical form.

32. $\sqrt{18} + 7\sqrt{50}$

32. $38\sqrt{2}$

33. $5\sqrt{27} - 2\sqrt{75}$

33. $5\sqrt{3}$

34. $2\sqrt{12y^4} - y^2\sqrt{27}$

34. $4y^2\sqrt{3} - 3y^2\sqrt{3} = y^2\sqrt{3}$.

35. $\sqrt{2}\left(\sqrt{2} - 5\sqrt{3}\right)$

35. $2 - 5\sqrt{6}$

36. $\sqrt{5}\left(\sqrt{10} + \sqrt{7}\right)$

36. $5\sqrt{2} + \sqrt{35}$

37. $\left(3 + \sqrt{5}\right)\left(1 + 2\sqrt{5}\right)$

37. $13 + 7\sqrt{5}$

38. $\left(2 + 4\sqrt{3}\right)\left(1 - \sqrt{3}\right)$

38. $-10 + 2\sqrt{3}$

39. $\left(4 - \sqrt{7}\right)\left(4 + \sqrt{7}\right)$

39. 9

40. $\left(2 + 3\sqrt{3}\right)\left(2 - 3\sqrt{3}\right)$

40. -23

41. $\left(1 + \sqrt{5}\right)\left(2 + \sqrt{3}\right)$

41. $2 + 2\sqrt{5} + \sqrt{3} + \sqrt{15}$

42. $\left(1 + \sqrt{5}\right)\left(2 + \sqrt{3}\right)$

42. $14 - 2\sqrt{2} + 7\sqrt{6} - 2\sqrt{3}$

43. $\left(2+\sqrt{3}\right)^2$

43. $4+4\sqrt{3}+3\ =7+4\sqrt{3}$.

44. $\left(4-\sqrt{5}\right)^2$

44. $16-8\sqrt{5}+5\ =21-8\sqrt{5}$.

45. $\left(3\sqrt{7}+1\right)^2$

45. $63+6\sqrt{7}+1\ =64+\ 6\sqrt{7}$.

Section 5.5 Fractional Exponents and Radical Equations

What You Need to Know

- How to combine fractions
- How to use the properties of exponents
- How to simplify radicals

What You Will Learn

- To express radicals using exponent notation
- To simplify expressions containing fractional exponents
- To solve equations that contain radicals

Materials

- Stand with hook
- Spring
- Hanger and weights
- Stopwatch
- Paper towel
- Oil or colored water
- Medicine dropper
- Ruler

Recall that a positive exponent is a number that indicates how many times a given number is used as a factor. For example, when we write 5^4, the exponent 4 indicates that 5 is used as a factor four times: $5^4 = 5 \cdot 5 \cdot 5 \cdot 5$.

But in the expression y^{-3}, the -3 exponent indicates the reciprocal of a product of three factors: $y^{-3} = \dfrac{1}{y^3} = \dfrac{1}{y \cdot y \cdot y}$. Recall also that the meaning of a negative exponent is derived from the properties of positive exponents. In order for the rule $\dfrac{a^m}{a^n} = a^{m-n}$ to hold for cases where m is greater than n, a negative exponent must imply a reciprocal. For instance, $\dfrac{x^3}{x^7} = x^{-4}$, but we also know that $\dfrac{x^3}{x^7} = \dfrac{x^3}{x^3 \cdot x^4} = \dfrac{1}{x^4}$; therefore, x^{-4} must be equivalent to $\dfrac{1}{x^4}$. The properties of exponents are summarized as follows.

$$a^m \cdot a^n = a^{m+n}$$

$$\frac{a^m}{a^n} = a^{m-n}, \ a \neq 0 \qquad\qquad (a \cdot b)^n = a^n b^n$$

$$a^0 = 1, \ a \neq 0 \qquad\qquad \left(\frac{a}{b}\right)^n = \frac{a^n}{b^n}$$

$$a^{-1} = \frac{1}{a}, \ a \neq 0 \qquad\qquad \left(a^m\right)^n = a^{mn}$$

$$a^{-n} = \frac{1}{a^n}, \ a \neq 0$$

In Discovery 5.4 you will investigate the meaning of a rational exponent, that is, an exponent that is a fraction.

Discovery 5.4 Fractional Exponents *(SE page 73)*

You are familiar with the use of several kinds of numbers as exponents: positive integers, negative integers, and zero. But the properties of such exponents don't show us whether fractions can have any meaning as exponents.

1. Consider the expression $9^{1/2}$. It is possible to use the properties of exponents you already know to determine the meaning of the exponent $1/2$. Write an expression for the square of $9^{1/2}$.

1. $(9^{1/2})^2$

2. Use a property of exponents to simplify and evaluate this expression.

2. $(9^{1/2})^2 = 9^{(1/2)(2)} = 9^1 = 9$.

3. The square of what number is equal to your answer to item 2?

3. 3, because $3^2 = 9$.

4. Write an equation stating the value of $9^{1/2}$.

4. $9^{1/2} = 3$.

5. Repeat items 2–4 for the expressions $25^{1/2}$ and $a^{1/2}$.

5. $(25^{1/2})^2 = 25^{(1/2)(2)} = 25^1 = 25$, and $5^2 = 25$, so $25^{1/2} = 5$.

   $(a^{1/2})^2 = a^{(1/2)(2)} = a^1 = a$, **and** $\left(\sqrt{a}\right)^2 = a$, **so** $a^{1/2} = \sqrt{a}$.

6. Use similar reasoning, with the same property of exponents you used in item 2, to determine the meaning of an exponent of 1/3. Try examining the expressions $8^{1/3}$ and $64^{1/3}$, and then finally $a^{1/3}$.

6. $(8^{1/3})^3 = 8^1 = 8$, **but also** $2^3 = 8$, **so** $8^{1/3} = 2$.

$(64^{1/3})^3 = 64^1 = 64$, **but also** $4^3 = 64$, **so** $64^{1/3} = 4$.

$(a^{1/3})^3 = a^1 = a$, **but also** $\left(\sqrt[3]{a}\right)^3 = a$, **so** $a^{1/3} = \sqrt[3]{a}$.

7. Now use the same property of exponents to write a new property that states the meaning of $a^{1/n}$.

7. $(a^{1/n})^n = a^1 = a$, **but also** $\left(\sqrt[n]{a}\right)^n = a$, **so** $a^{1/n} = \sqrt[n]{a}$.

8. Use your answer to item 7 to rewrite each of the following radicals with a fractional exponent: (a) $\sqrt{D}$; (b) $\sqrt[6]{2}$; (c) $\sqrt[4]{5x}$.

8. **a)** $D^{1/2}$

b) $2^{1/6}$

c) $(5x)^{1/4}$

In Exercise 10 of Section 5.4, we examined the formula $A = 6\sqrt[3]{V^2}$, which gives the surface area A of a cube in terms of its volume V. The expression $\sqrt[3]{V^2}$ can be written using a fractional exponent.

As seen in Discovery 5.4, the cube root can be represented by a fractional exponent:

$$\sqrt[3]{V^2} = \left(V^2\right)^{1/3} \qquad \text{Use the property } a^{1/n} = \sqrt[n]{a}.$$
$$= V^{(2)(1/3)} \qquad \text{Power of a power property}$$
$$= V^{2/3} \qquad \text{Simplify.}$$

In a similar manner, we can rewrite the expression $\left(\sqrt{b}\right)^7$ using a single fractional exponent:

$$\left(\sqrt{b}\right)^7 = \left(b^{1/2}\right)^7$$
$$= b^{(1/2)(7)}$$
$$= b^{7/2}$$

The preceding two examples can be generalized for similar expressions involving any power m with a radical of any order n:

$$\sqrt[n]{a^m} = \left(a^m\right)^{1/n} = a^{(m)(1/n)} = a^{m/n}.$$

$$\left(\sqrt[n]{a}\right)^m = \left(a^{1/n}\right)^m = a^{(1/n)(m)} = a^{m/n}.$$

Notice that these results show that there are two different possible interpretations of $a^{m/n}$. So $a^{m/n}$ is equivalent to $\sqrt[n]{a^m}$, or $a^{m/n}$ is equivalent to $\left(\sqrt[n]{a}\right)^m$.

We now have two additional properties of exponents that can be used either to rewrite radical expressions with exponent notation or to evaluate expressions that contain fractional exponents.

$$a^{1/n} = \sqrt[n]{a}.$$
$$a^{m/n} = \sqrt[n]{a^m} = \left(\sqrt[n]{a}\right)^m.$$

Example 16

Rewrite the following expressions using radical notation:

a) $x^{1/5}$ b) $c^{-1/2}$ c) $H^{2/3}$ d) $(L+W)^{1/2}$ e) $(z^2+1)^{-3/4}$

Solution:

a) $x^{1/5} = \sqrt[5]{x}$.

b) $c^{-1/2} = \dfrac{1}{c^{1/2}} = \dfrac{1}{\sqrt{c}}$.

c) $H^{2/3} = \sqrt[3]{H^2}$ or $\left(\sqrt[3]{H}\right)^2$.

d) $(L+W)^{1/2} = \sqrt{L+W}$.

e) $(z^2+1)^{-3/4} = \dfrac{1}{\left(z^2+1\right)^{3/4}} = \dfrac{1}{\sqrt[4]{\left(z^2+1\right)^3}}$ or $\dfrac{1}{\left(\sqrt[4]{z^2+1}\right)^3}$.

Example 17

Use fractional exponents to write each of the following expressions without radicals:

a) $\sqrt[4]{p}$ b) $\dfrac{1}{\sqrt[3]{L}}$ c) $\sqrt{a^2+b^2}$ d) $\left(\sqrt[5]{M-N}\right)^2$ e) $\dfrac{1}{\sqrt{x^3}}$

Solution:

a) $\sqrt[4]{p} = p^{1/4}$.

b) $\dfrac{1}{\sqrt[3]{L}} = \dfrac{1}{L^{1/3}} = L^{-1/3}$.

c) $\sqrt{a^2+b^2} = (a^2+b^2)^{1/2}$.

d) $\left(\sqrt[5]{M-N}\right)^2 = (M-N)^{2/5}$.

e) $\dfrac{1}{\sqrt{x^3}} = \dfrac{1}{x^{3/2}} = x^{-3/2}$.

Example 18

Evaluate the following expressions:

a) $49^{1/2}$ b) $(-27)^{1/3}$ c) $16^{-1/4}$ d) $32^{3/5}$ e) $64^{-2/3}$

Solution:

a) $49^{1/2} = \sqrt{49} = 7$.

b) $(-27)^{1/3} = \sqrt[3]{-27} = -3$.

c) $16^{-1/4} = \dfrac{1}{\sqrt[4]{16}} = \dfrac{1}{2}$.

d) $32^{3/5} = \left(\sqrt[5]{32}\right)^3 = 2^3 = 8$.

e) $64^{-2/3} = \dfrac{1}{64^{2/3}} = \dfrac{1}{\left(\sqrt[3]{64}\right)^2} = \dfrac{1}{4^2} = \dfrac{1}{16}$.

Because we used the properties of exponents to determine the meaning of fractional exponents, those same properties can be used to simplify expressions containing fractional exponents.

Example 19

Simplify the following expressions. Assume that all variables are nonnegative.

a) $x^{1/3}x^{2/3}$

b) $w^{1/2}w^{1/4}$

c) $\left(16a^{-4}c^2\right)^{3/2}$

d) $\left(\dfrac{8r^{3/4}}{27s^{-3}t^6}\right)^{4/3}$

Solution:

a) $x^{1/3}x^{2/3} = x^{1/3+2/3}$ Product property of exponents

   $= x^{3/3}$ Add the fractions.

   $= x^1$ Simplify.

   $= x.$

b) $w^{1/2}w^{1/4} = w^{1/2+1/4}$ Product property of exponents

   $= w^{2/4+1/4}$ Common denominator

   $= w^{3/4}.$ Add the fractions.

c) $\left(16a^{-4}c^2\right)^{3/2} = (16)^{3/2}\left(a^{-4}\right)^{3/2}\left(c^2\right)^{3/2}$ Power of a product property

   $= 16^{3/2}a^{(-4)(3/2)}c^{(2)(3/2)}$ Power of a power property

   $= 16^{3/2}a^{-6}c^3$ Simplify.

   $= \dfrac{\left(\sqrt{16}\right)^3 c^3}{a^6}$ $a^{m/n} = \left(\sqrt[n]{a}\right)^m$

   $= \dfrac{64c^3}{a^6}.$ Evaluate.

d) $\left(\dfrac{8r^{3/4}}{27s^{-3}t^6}\right)^{4/3} = \dfrac{\left(8r^{3/4}\right)^{4/3}}{\left(27s^{-3}t^6\right)^{4/3}}$ Power of a quotient property

   $= \dfrac{(8)^{4/3}\left(r^{3/4}\right)^{4/3}}{(27)^{4/3}\left(s^{-3}\right)^{4/3}\left(t^6\right)^{4/3}}$ Power of a product property

   $= \dfrac{8^{4/3}r^{(3/4)(4/3)}}{27^{4/3}s^{(-3)(4/3)}t^{(6)(4/3)}}$ Power of a power property

   $= \dfrac{8^{4/3}r^1}{27^{4/3}s^{-4}t^8}$ Simplify.

   $= \dfrac{\left(\sqrt[3]{8}\right)^4 rs^4}{\left(\sqrt[3]{27}\right)^4 t^8}$ $a^{m/n} = \left(\sqrt[n]{a}\right)^m$

   $= \dfrac{16rs^4}{81t^8}.$ Evaluate.

Solving Radical Equations

When an object falls to the ground from a height h (in meters), the impact velocity v (in meters per second) with which it strikes the ground (ignoring air resistance) is given by the formula $v = \sqrt{19.6h}$. This formula can be used directly to calculate velocity. For instance, an object dropped from a height of 10 m will strike the ground with a velocity of $\sqrt{(19.6)(10)} = 14$ m/s.

But how can we determine the height from which an object must be dropped in order for it to reach an impact velocity of 30 m/s? To do so, we must solve the radical equation $30 = \sqrt{19.6h}$ for h.

Because h occurs in the radicand, we can undo the radical by squaring both sides of the equation, as shown in the arrow diagram of **Figure 5.45:**

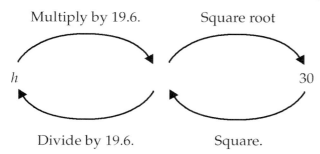

Multiply by 19.6.   Square root

h 30

Divide by 19.6.   Square.

Figure 5.45

$$30^2 = \left(\sqrt{19.6h}\right)^2$$

$$900 = 19.6h$$

$$h = 45.918... \approx 46 \text{ m.}$$

Thus, for an object to hit the ground with an impact velocity of 30 m/s, it must be dropped from a height of approximately 46 m.

A similar process can be used to solve equations containing single radicals of any order. To solve the equation $\sqrt[5]{2x} = 7$, we raise both sides to the fifth power, which is the inverse of a fifth root:

$$\left(\sqrt[5]{2x}\right)^5 = 7^5$$

$$2x = 16{,}807$$

$$x = 8403.5.$$

Example 20

Solve for y: $\sqrt[3]{4y+1} - 3 = 2$.

Solution:

Begin by isolating the radical on one side of the equation.

$$\sqrt[3]{4y+1} - 3 = 2 \qquad \text{Original equation}$$

$$\sqrt[3]{4y+1} = 5 \qquad \text{Add 3 to both sides.}$$

$$\left(\sqrt[3]{4y+1}\right)^3 = 5^3 \qquad \text{Cube both sides.}$$

$$4y + 1 = 125 \qquad \text{Simplify.}$$

$$4y = 124 \qquad \text{Subtract 1 from both sides.}$$

$$y = 31. \qquad \text{Divide both sides by 4.}$$

Check:

$$\sqrt[3]{4(31)+1} - 3 \stackrel{?}{=} 2$$

$$\sqrt[3]{125} - 3 \stackrel{?}{=} 2$$

$$5 - 3 \stackrel{?}{=} 2$$

$$2 = 2 \checkmark$$

Example 21

Solve for w: $(2x-5)^{1/3} = 4$.

Solution:

$$(2x-5)^{1/3} = 4 \qquad \text{Original equation}$$

$$\left[(2x-5)^{1/3}\right]^3 = 4^3 \qquad \text{Cube both sides.}$$

$$2x - 5 = 64 \qquad \text{Simplify.}$$

$$2x = 69 \qquad \text{Add 5 to both sides.}$$

$$x = 69/2. \qquad \text{Divide both sides by 2.}$$

Check:

$$[2(69/2) - 5]^{1/3} \stackrel{?}{=} 4$$

$$64^{1/3} \stackrel{?}{=} 4$$

$$4 = 4 \checkmark$$

Example 22

Solve for t: $2t = \sqrt{4-5t} + 1$

Solution:

$$2t = \sqrt{4-5t} + 1 \qquad \text{Original equation}$$

$$2t - 1 = \sqrt{4-5t} \qquad \text{Subtract 1 from both sides.}$$

$$(2t-1)^2 = \left(\sqrt{4-5t}\right)^2 \qquad \text{Square both sides.}$$

$$4t^2 - 4t + 1 = 4 - 5t \qquad \text{Simplify.}$$

$$4t^2 + t - 3 = 0 \qquad \text{Standard quadratic form}$$

$$t = -1 \text{ or } \frac{3}{4}. \qquad \text{Quadratic formula}$$

Check:

$$2\left(\frac{3}{4}\right) \stackrel{?}{=} \sqrt{4 - 5\left(\frac{3}{4}\right)} + 1$$

$$\frac{3}{2} \stackrel{?}{=} \sqrt{4 - \frac{15}{4}} + 1$$

$$\frac{3}{2} \stackrel{?}{=} \sqrt{\frac{1}{4}} + 1$$

$$\frac{3}{2} \stackrel{?}{=} \frac{1}{2} + 1$$

$$\frac{3}{2} = \frac{3}{2} \checkmark$$

$$2(-1) \stackrel{?}{=} \sqrt{4 - 5(-1)} + 1$$

$$-2 \stackrel{?}{=} \sqrt{9} + 1$$

$$-2 \stackrel{?}{=} 3 + 1$$

$$-2 \neq 4.$$

Notice that even though a correct procedure was used to solve the equation in Example 22, one of the solutions found did not satisfy the original equation. The number –1 is a solution of the quadratic equation $4t^2 + t - 3 = 0$, but it is not a solution of $2t = \sqrt{4-5t} + 1$. When a solution process yields a number that is not a solution of the original equation, that number is called an **extraneous solution.** Extraneous solutions are sometimes introduced by procedures that are used to solve radical equations and certain other types of equations. It is important to always check solutions in order to make sure they are not extraneous.

Exercises 5.5

I. Investigations *(SE pages 81–85)*

1. We have seen in this section that the expression $a^{m/n}$ can be interpreted in two ways—either as $(a^m)^{1/n} = \sqrt[n]{a^m}$ or as $\left(a^{1/n}\right)^m = \left(\sqrt[n]{a}\right)^m$. Both interpretations are correct and yield the same value. But there are times when the second form may be preferred.

   a) Evaluate $4^{3/2}$ using both interpretations of the exponent.

1. a) $\sqrt{4^3} = \sqrt{64} = 8;\ \left(\sqrt{4}\right)^3 = 2^3 = 8$.

   b) Now do the same for $16^{5/4}$.

b) $\sqrt[4]{16^5} = \sqrt[4]{104,8576} = 32;\ \left(\sqrt[4]{16}\right)^5 = 2^5 = 32$.

   c) Considering your answers to parts (a) and (b), which of the two forms $\sqrt[n]{a^m}$ or $\left(\sqrt[n]{a}\right)^m$ do you think would usually be easier to use to evaluate $a^{m/n}$?

c) $\left(\sqrt[n]{a}\right)^m$ would be easier in most cases because $\sqrt[n]{a^m}$ can result in very large radicands.

2. A calculator can be used to evaluate expressions with fractional exponents directly. The exponent key ("^" on most graphing calculators and computers, "y^x" on many scientific calculators) is used, followed by the fractional exponent.

   a) Evaluate $9^{3/2}$ using a calculator but without using a square root operation. That is, enter the $\dfrac{3}{2}$ directly as a fraction enclosed in parentheses. (Because you can also evaluate $9^{3/2}$ without using a calculator, you can check your answer.)

2. a) 9^(3/2) produces a result of 27. For a scientific calculator, the equivalent keystrokes are 9 y^x(3/2).

   b) Again evaluate $9^{3/2}$, but enter the exponent as a decimal.

b) 9^1.5 results in 27. For a scientific calculator, the equivalent keystrokes are 9 y^x1.5.

   c) If you try in a similar manner to evaluate $8^{2/3}$ in two ways, a difficulty arises. Explain why.

c) 8^(2/3) results in 4, the correct value, but because there is no exact decimal that equals $\dfrac{2}{3}$, no decimal exponent will yield a value of exactly 4.

3. Some decimal exponents can be interpreted as simple fractions. The expression $32^{0.2}$ is equivalent to $32^{1/5}$, which equals $\sqrt[5]{32}$, or 2. But many applications involve decimal exponents that have been determined experimentally and that do not necessarily equal simple fractions.

a) A *Parshall flume* is a type of flow meter used to measure flow rates in wastewater treatment plants and storm drainage ditches. The flow rate Q (in cubic feet per second) through one such meter depends on water depth (in feet) according to the equation $Q = 3.07H^{1.53}$. Write $H^{1.53}$ with an exponent that is an improper fraction, and interpret the expression using a radical.

3. a) $H^{1.53} = H^{\frac{153}{100}} = \left(\sqrt[100]{H}\right)^{153}$ or $\sqrt[100]{H^{153}}$.

b) Find at least two different ways of evaluating $2^{1.53}$ using a calculator, and use the result to determine the flow rate Q for a water depth of 2 feet.

b) On some calculators, any of the following keystroke sequences are possible: 2^1.53, 2^(153/100), (100$\sqrt{}$ 2)^153, 100$\sqrt{}$ (2^153). On a scientific calculator, only 2 y^x1.53 and 2 y^x(153/100) may be available. The result is $2^{1.53} \approx 2.89$, and the flow rate is approximately 8.87 ft³/s.

c) The equation $I = 3.4V^{3/2}$ gives the cathode current (in milliamperes) for a particular color picture tube as a function of voltage V (in volts). What is the cathode current when the voltage is 80 volts?

c) 2433 milliamperes

d) The surface areas of similarly shaped objects increase with size. In one study of mammals, it has been found that surface area S (in square feet) can be approximated by the function $S = 0.71W^{0.65}$, where W represents weight (in pounds). Use this formula to approximate the surface area of an 8 lb cat.

d) 2.74 ft²

e) Tools that are used in machining processes wear out faster when they are used at higher speeds. The *Taylor tool life equation* for a particular tool is $vT^{0.357} = 295$, where v represents cutting speed (in meters per minute) and T represents tool life (in minutes). What cutting speed v will result in a tool life of 20 minutes? (Hint: First solve the equation for v.)

e) $v = \dfrac{295}{T^{0.357}} = 295T^{-0.357}$; **when $T = 20$, $v = 101$ m/min.**

4. In Exercise 4 of section 5.2, you examined graphs of several power functions of the form $y = x^n$, using only positive integers for n. Each of those functions has a domain that includes all real numbers. Power functions can also contain fractional exponents.

Figures 5.46 and **5.47** show the calculator graphs of $y = x^{3/2}$ and $y = x^{5/3}$.

Graph of $y = x^{3/2}$:

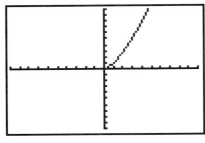

$[-10, 10] \times [-10, 10]$

Figure 5.46

Graph of $y = x^{5/3}$:

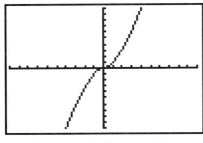

$[-10, 10] \times [-10, 10]$

Figure 5.47

By first rewriting each function in the radical form $\left(\sqrt[n]{x}\right)^m$, explain why the two graphs have such a different appearance.

4. $x^{3/2} = \left(\sqrt{x}\right)^3$, $\sqrt{x}$ **is not defined for negative values of** x. $x^{5/3} = \left(\sqrt[3]{x}\right)^5$, **and** $\sqrt[3]{x}$ **is defined for all real numbers.**

5. When an equation contains two different radicals, it may be possible to solve the equation using an extension of the method discussed in this section.

a) Consider the equation $\sqrt{x+3} = \sqrt{x-2} + 1$. Begin by squaring both sides. (Think of the right side $\sqrt{x-2} + 1$ as if it were a simple binomial like $a + b$, and recall that $(a + b)^2 = a^2 + 2ab + b^2$.)

5. a) $\left(\sqrt{x+3}\right)^2 = \left(\sqrt{x-2} + 1\right)^2$

$x + 3 = \left(\sqrt{x-2}\right)^2 + 2\left(\sqrt{x-2}\right)(1) + (1)^2$

$x + 3 = x - 2 + 2\sqrt{x-2} + 1.$

b) There is now only one radical in the equation. Isolate the term with the radical on one side of the equation, and then square both sides of the equation to complete the solution.

b)　　$4 = 2\sqrt{x-2}$

$(4)^2 = \left(2\sqrt{x-2}\right)^2$

$16 = 4(x - 2)$

$16 = 4x - 8$

$24 = 4x$

$x = 6.$

c) Check to see if your answer to (b) is a solution to the original equation.

c)　$\sqrt{6+3} \stackrel{?}{=} \sqrt{6-2} + 1$

$\sqrt{9} \stackrel{?}{=} \sqrt{4} + 1$

$3 \stackrel{?}{=} 2 + 1$

$3 = 3 \checkmark$

d) In a similar manner, solve the equation $1 - \sqrt{2x+3} = \sqrt{x+1}$ and check your solutions.

d)
$$\left(1 - \sqrt{2x+3}\right)^2 = \left(\sqrt{x+1}\right)^2$$

$$(1)^2 - 2(1)\sqrt{2x+3} + \left(\sqrt{2x+3}\right)^2 = x+1$$

$$1 - 2\sqrt{2x+3} + 2x + 3 = x + 1$$

$$-2\sqrt{2x+3} = -x - 3$$

$$\left(-2\sqrt{2x+3}\right)^2 = (-x-3)^2$$

$$4(2x + 3) = x^2 + 6x + 9$$

$$8x + 12 = x^2 + 6x + 9$$

$$0 = x^2 - 2x - 3$$

$$0 = (x + 1)(x - 3)$$

$$x = -1 \text{ or } 3.$$

Only $x = -1$ is a solution to the original equation; $x = 3$ is an extraneous solution.

6. The pitch of a musical tone is determined by its frequency. The frequency is measured in hertz (Hz), which count the number of sound vibrations produced each second by the instrument that creates the tone. For example, the A below middle C on a piano has a frequency of 220 Hz. This means that the piano strings for the A on an acoustic piano vibrate 220 times each second.

a) The tone A that is exactly one octave higher than the A below middle C has a frequency of exactly 440 Hz. What is the ratio of this tone's frequency to the frequency of the A below middle C?

> The chromatic scale that is the basis for most Western music is made up of 12 semitones. The following list shows the order of the semitones from an A to another A one octave higher:
>
> A A# B C C# D D# E F F# G G# A
>
> (Note: The symbol # stands for *sharp*, as in "A sharp.")

6. a) **440 Hz/220 Hz = $\dfrac{2}{1}$; this ratio can be written as the whole number 2.**

b) To the ears of most people, tones that harmonize well with each other often have frequency ratios that can be simplified to ratios of small whole numbers. For example, the tone E above middle C ideally has a frequency of 330 Hz. What is the ratio of this tone's frequency to the frequency of A below middle C?

b) 330 Hz/220 Hz = $\dfrac{3}{2}$; this ratio can be written as the decimal number 1.5.

In order for pianos to sound equally good in all musical keys, they are tuned to an *equally tempered scale*, which slightly changes the frequencies of the tones from their ideal values. The 12 semitones making up the chromatic scale from one A to the next are tuned so that the frequency ratios for all adjacent tones are equal. This means that the frequency ratios for adjacent pairs of tones are always equal to $2^{1/12}$. Thus, the frequency ratio between A# and A is always $2^{1/12} \approx 1.06$. The ratio between B (the next higher tone) and A# is also $2^{1/12}$.

> On an electric piano, tones are produced electronically. The A below middle C on an electronic piano has an electrical voltage signal that oscillates 220 times per second.

Now let's calculate the frequency ratio between B and A, which is an interval of 2 semitones. It is given by

$$\left(\frac{\text{frequency of B}}{\text{frequency of A\#}}\right) \cdot \left(\frac{\text{frequency of A\#}}{\text{frequency of A}}\right) = (2^{1/12})(2^{1/12}) = 2^{2/12} \approx 1.12.$$

c) Use a similar method to find the frequency ratio between middle C (the next higher semitone) and the original A below middle C.

c) $\left(\dfrac{\textbf{frequency of C}}{\textbf{frequency of B}}\right) \cdot \left(\dfrac{\textbf{frequency of B}}{\textbf{frequency of A}}\right) = (2^{1/12})(2^{2/12}) = 2^{3/12} \approx 1.19.$

d) Proceeding in a similar manner, complete **Table 5.25** to show the frequency ratios between each of the semitones in the next octave and the A below middle C.

Tone	A#	B	C	C#	D	D#	E	F	F#	G	G#	A
Ratio	$2^{1/12}$	$2^{2/12}$	$2^{3/12}$									
	1.06	1.12	1.19									

Table 5.25

d)

Tone	A#	B	C	C#	D	D#	E	F	F#	G	G#	A
Ratio	$2^{1/12}$	$2^{2/12}$	$2^{3/12}$	$2^{4/12}$	$2^{5/12}$	$2^{6/12}$	$2^{7/12}$	$2^{8/12}$	$2^{9/12}$	$2^{10/12}$	$2^{11/12}$	$2^{12/12}$
	1.06	1.12	1.19	1.26	1.33	1.41	1.5	1.59	1.68	1.78	1.89	2

e) Some of the ratios in the table are approximately equal to ratios of small whole numbers. The tones corresponding to these ratios are the tones that produce what many people consider to be the most consonant (pleasant-sounding) harmonies when combined with the A below middle C. Identify some of these tones by comparing the decimal equivalents of small whole number ratios ($\frac{3}{2}$, $\frac{4}{3}$, etc.) with the ratios in your table.

e) **E has a ratio of 1.5, which is exactly equal to $\frac{3}{2}$.**

C# has a ratio of 1.26, which is very close to $\frac{5}{4}$ = 1.25. (A-C#-E are the tones in an A major chord.)

D has a ratio of 1.33, which is almost $\frac{4}{3}$.

C has a ratio of 1.19, which is very close to $\frac{6}{5}$ = 1.2. (A-C-E are the tones in an A minor chord.)

7. A cable television company needs to run one of its cable lines from point A on one side of a 100-foot-wide expressway right-of-way to point B on the other side (see **Figure 5.48**). Local statutes require that the cable cannot be hung across the highway but must be buried underground. Underground cable installation costs $60 per foot, but above ground the cost is only $12 per foot. It is most cost effective for the company to use a straight run of buried cable under the highway to some point on the other side and another straight aboveground run to complete the installation.

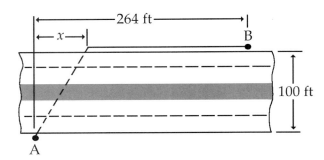

Figure 5.48

a) Write an expression in terms of the distance x for the length of the aboveground portion of the cable.

7. **a) (264 – x)**

b) Write an expression for the *cost* in dollars of laying the aboveground portion of the cable.

b) 12(264 – x)

c) Write an expression in terms of x for the length of the underground portion of cable. (Hint: Examine the right triangle whose hypotenuse is the underground portion of cable.)

c) $\sqrt{x^2 + 100^2} = \sqrt{x^2 + 10{,}000}$.

d) Write an expression in terms of x for the *cost* of the underground portion of cable.

d) $60\sqrt{x^2 + 10{,}000}$

e) Write an equation that gives the total cost C of the cable run from point A to point B as a function of x.

e) $C = 12(264 - x) + 60\sqrt{x^2 + 10{,}000}$.

f) For what value (or values) of x will the total cost be $9100?

f) When $C = 9100$, the solution process results in the quadratic equation $3456x^2 - 142{,}368x + 811{,}376 = 0$. The equation has two solutions for x: 6.8 ft and 34.4 ft. Both result in the same total cost of \$9100.

II. Projects and Group Activities *(SE pages 85–87)*

8. Materials: stand with hook, spring, hanger and weights, stopwatch

We found in Exercise 7 of Section 5.4 that the weight of a pendulum does not affect the period of the pendulum's swing. When a weight is hung on a spring, however, changing the weight does change the period of its vertical oscillations.

a) Attach the top of a spring to a hook, and then attach a hanger to the bottom of the spring on which you can place varying numbers of weights (see **Figure 5.49**). Begin with a total weight (including the hanger) of about 100 grams. Pull the weight down a few centimeters, then let it go. Use a stopwatch to measure the time it takes for 20 oscillations of the weight. Record the time.

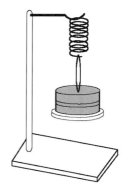

Figure 5.49

8. **a) Sample time for 20 oscillations: 7.8 s**

b) The period T is the time required for one complete oscillation of the weight. Find the period of oscillation for the weight.

b) $\dfrac{7.8 \text{ s}}{20} = 0.39$ s.

c) Complete a table similar to **Table 5.26** by measuring the time for 20 oscillations for several increasing amounts of weight.

Weight (g)	Time for 20 Oscillations (s)	Period (s)
0	0	0
100		
200		
300		
400		
500		

Table 5.26

c)

Weight (g)	Time for 20 Oscillations (s)	Period (s)
0	0	0
100	7.8	0.39
200	11.3	0.57
300	13.4	0.67
400	15.7	0.79
500	17.5	0.88

d) Construct a scatter plot of period T and weight W from the table in (c).

d)

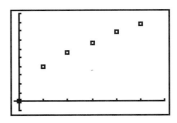

[−20, 600] **x** [−0.1, 1]

e) Find a linear regression equation for the data and make a residual plot for that equation. Is a linear model a good model for the dependence of period on weight?

e) Linear model: $T = 0.0016W + 0.14$.

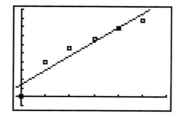

[−20, 600] x [−0.1, 1]

Residual plot:

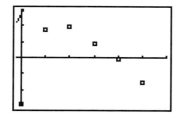

[−20, 600] x [−0.15, 0.15]

No, the residuals are not random.

f) Find a quadratic regression equation for the data and make a residual plot for that equation. Is a quadratic model a good model for the data? (Consider what you would expect to happen if you continued to increase the weight.)

f) **Quadratic model:** $T = (-3.1 \times 10^{-6})W^2 + 0.0032W + 0.039$.

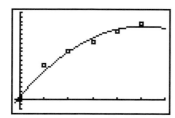

[−20, 600] x [−0.1, 1]

Residual plot:

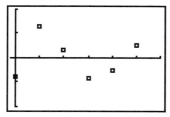

[−20, 600] x [−0.1, 0.1]

The residuals are smaller but still not random. Period should continue to increase as weight increases, but the quadratic function would decrease for higher weights.

g) Find a **power regression** model for the data and make a residual plot for that equation. (Note: Because of the way power regression is done by a calculator, you must first delete the (0, 0) data pair.) Is a power function a good model for the dependence of period on weight?

g) Power regression model: $T = 0.03916W^{0.501}$.

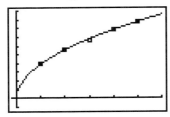

[−20, 600] x [−0.1, 1]

Residual plot:

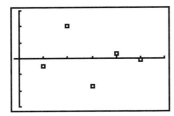

[−20, 600] x [−0.015, 0.015]

The residuals are very small and random. The power function models the data well.

h) Examine the decimal exponent in your model. Is it close to the value of a simple fractional exponent? If so, rewrite the power function model as a radical function.

h) Yes, 0.501 is very close to 0.5, or $\dfrac{1}{2}$. A possible radical model is

$$T = 0.3916\sqrt{W}, \text{ or approximately } T = 0.39\sqrt{W}.$$

9. Materials: paper towel, oil or colored water, medicine dropper, ruler

As the volume of oil in a spill increases, the diameter and area of the spill both expand. We can investigate this behavior by creating a simulation of it.

a) Let one drop of oil fall into the center of a paper towel. The oil should spread out in a circle as it soaks into the towel (see **Figure 5.50**). Find the diameter of the circle, in centimeters.

Figure 5.50

9. a) Sample diameter: 1.3 cm

b) Continue to add oil, one drop at a time, and measure the new diameter of the oil spot after each addition. Complete a table similar to **Table 5.27**.

Volume of Oil (drops)	1	2	3	4	5	6	7
Diameter (cm)							

Table 5.27

b) Sample data are shown in the table.

Volume of Oil (drops)	1	2	3	4	5	6	7
Diameter (cm)	1.3	1.9	2.2	2.5	2.6	2.9	3.1

c) Construct a scatter plot of the data from the table in (b).

c)

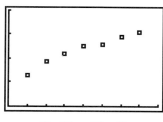

[0, 8] **x** [0, 4]

d) Find a power regression model for the dependence of oil spot diameter d on oil volume V, and make a residual plot for the model. Is a power function a good model for the data?

d) Power regression model: $d = 1.35\,V^{0.43}$.

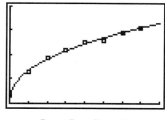

[0, 8] **x** [0, 4]

Residual plot:

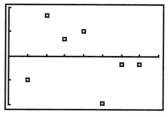

[0, 8] **x** [−0.1, 0.1]

The residuals are small and random. The power function appears to be a good model.

e) Use your model for the diameter of the oil spot to find an equation for the area A of the oil slick as a function of its volume. Recall that there are two formulas for the area of a circle, one in terms of radius and one in terms of diameter.

e) $A = \dfrac{\pi d^2}{4} = \dfrac{\pi\left(1.35V^{0.43}\right)^2}{4}$, which simplifies to $A = 1.43V^{0.86}$.

f) Use your answers to (d) and (e) to predict the diameter and area of the circle that would be formed if a total of 15 drops of oil were placed on the paper towel.

f) $d = 1.35(15)^{0.43} = 4.3$ cm; $A = 1.43(15)^{0.86} = 15$ cm^2.

III. Additional Practice (SE pages 87–89)

10. The voltage V used by a resistor can be written in terms of the power P and the resistance R as $V = \sqrt{PR}$. Rewrite this equation using a fractional exponent.

10. $V = (PR)^{1/2}$.

11. The altitude h (in kilometers) of a NOAA (National Oceanographic and Atmospheric Administration) polar-orbiting satellite is given as a function of orbital period T (in seconds) by the equation $h = 2.57\sqrt[3]{T^2}$. Rewrite this equation using a fractional exponent.

11. $h = 2.57T^{2/3}$.

For 12–14, use a fractional exponent to rewrite the expression without radicals.

12. $\sqrt[6]{x}$

12. $x^{1/6}$

13. $\sqrt[4]{A^7}$

13. $A^{7/4}$

14. $\left(\sqrt{C}\right)^5$

14. $C^{5/2}$

For 15–17, use a fractional exponent to rewrite the expression without radicals and with no variables in denominators of fractions.

15. $\dfrac{4}{\sqrt[3]{m^2 + n^2}}$

15. $4(m^2 + n^2)^{-1/3}$

16. $\dfrac{2a}{\sqrt[9]{b^4}}$

16. $2ab^{-4/9}$

17. $\dfrac{1}{5\sqrt{y}}$

17. $\dfrac{1}{5}y^{-1/2}$

18. The average velocity for open-channel fluid flow is given by the equation $\overline{v} = \dfrac{1.49r^{2/3}S^{1/2}}{n}$. Rewrite the equation using radical notation.

18. $\overline{v} = \dfrac{1.49\sqrt[3]{r^2}\sqrt{S}}{n}.$

19. A metal's surface temperature during a grinding operation is given by the equation $T = Kd^{0.75}\left(\dfrac{r_g Cv}{v_w}\right)^{0.5}D^{0.25}$. Rewrite the equation using radical notation.

19. $T = K\sqrt[4]{d^3}\sqrt{\dfrac{r_g Cv}{v_w}}\sqrt[4]{D}.$

For 20–23, rewrite the expression using radical notation.

20. $3G^{1/2}$

20. $3\sqrt{G}$

21. $(7h)^{1/5}$

21. $\sqrt[5]{7h}$

22. $(c^2 - b^2)^{-1/2}$

22. $\dfrac{1}{\sqrt{c^2 - b^2}}$

23. $L^{5/7}$

23. $\sqrt[7]{L^5}$

For 24–28, evaluate the expression.

24. $64^{1/2}$

24. $\sqrt{64} = 8.$

25. $10{,}000^{0.25}$

25. $10{,}000^{1/4} = \sqrt[4]{10{,}000} = 10.$

26. $9^{3/2}$

26. $\left(\sqrt{9}\right)^3 = 27.$

27. $16^{-5/4}$

27. $\dfrac{1}{\left(\sqrt[4]{16}\right)^5} = \dfrac{1}{32}.$

28. $\dfrac{4}{3 \cdot (-64)^{2/3}}$

28. $\dfrac{4}{3 \cdot \left(\sqrt[3]{-64}\right)^2} = \dfrac{4}{3 \cdot (-4)^2} = \dfrac{1}{12}.$

29. Simplify the following expression for the stress on soil caused by a concentrated load: $\left[\left(z^{-4/5}\right)\left(\dfrac{z^2 + r^2}{z^2}\right)\right]^{5/2}$

29. $\left(z^{-4/5}\right)^{5/2} \dfrac{\left(z^2 + r^2\right)^{5/2}}{\left(z^2\right)^{5/2}} = \dfrac{z^{-2}\left(z^2 + r^2\right)^{5/2}}{z^5} = \dfrac{\left(z^2 + r^2\right)^{5/2}}{z^7}.$

For 30–33, simplify the expression. Assume that all variables are nonnegative.

30. $r^{2/5}r^{4/5}$

30. $r^{6/5}$

31. $\dfrac{\left(27x^7\right)^{1/3}}{x^{2/3}}$

31. $2x^{5/3}$

32. $(9p^{1/2})^{1/2}$

32. $3p^{1/4}$

33. $x^6 (8x)^{1/3}(9x)^{1/2}$

33. $6x$

34. The diameter d (in inches) of a fiber line needed to lift a load weighing w tons is $d = \sqrt{1.5w}$. What load can be carried by a line with a diameter of 3 inches?

34. 6 tons

35. If a ball is dropped from a height H and bounces back to a lower height h, its *coefficient of restitution* e is given by $e = \sqrt{\dfrac{h}{H}}$. If a basketball with a coefficient of restitution of 0.84 is dropped from a height of 1.7 meters, how high will it bounce?

35. $0.84 = \sqrt{\dfrac{h}{1.7}}$; $h = 1.2$ m.

36. The period T of a pendulum (measured in seconds) is given in terms of the pendulum's length L (in feet) by $T = 2\pi\sqrt{\dfrac{L}{32}}$. What length pendulum will have a period of 1.0 second?

36. $1.0 = 2\pi\sqrt{\dfrac{L}{32}}$; L ≈ 0.81 ft.

37. A fragment of an ancient dish is found at the site of an archeological dig (see **Figure 5.51**). As discussed in Exercise 9 of Section 5.4, the formula $c = 2\sqrt{h(2r - h)}$ gives the length of a chord of a circular segment in terms of the radius r of the circle and the height h of the segment. Find the original radius of the dish.

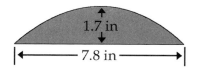

Figure 5.51

37. $7.8 = 2\sqrt{1.7(2r - 1.7)}$; r = 5.3 in.

For 38–47, solve the equation for the value of the variable.

38. $\sqrt{2x - 1} - 8 = 3$.

38. x = 61.

39. $-3 = 2\sqrt{4a + 5} - 7$.

39. $a = -\dfrac{1}{4}$.

40. $\sqrt{3 - y} + 5 = 2$.

40. No solution; y = −6 is an extraneous solution.

41. $z = \sqrt{z^2 + 3z - 5}$.

41. $z = \dfrac{5}{3}$.

42. $2\sqrt{4p^2 + p + 5} - 3p = p$.

42. p = −5.

43. $C = 2 + \sqrt{4C - 11}$.

43. C = 3 or 5.

44. $\sqrt{7R + 8} + R = 2R + 2$.

44. R = –1 or 4.

45. $x = \sqrt{2x + 2} + 3$.

45. x = 7; x = 1 is an extraneous solution.

46. $(4 - 7x)^{1/2} = 5$.

46. x = –3.

47. $(x^2 + 10)^{1/3} - 1 = 3$.

47. $x = \pm\sqrt{54} = \pm 3\sqrt{6}$.

Section 5.6 Distance in the Plane; Circles

What You Need to Know

- How to identify coordinates of points on a rectangular coordinate system
- How to use the Pythagorean theorem to find a hypotenuse
- How to simplify radicals

What You Will Learn

- To find the distance between two points in the plane
- To find the equation of a circle
- To identify a circle from its equation

Discovery 5.5 Finding the Length of a Cable *(SE pages 90–91)*

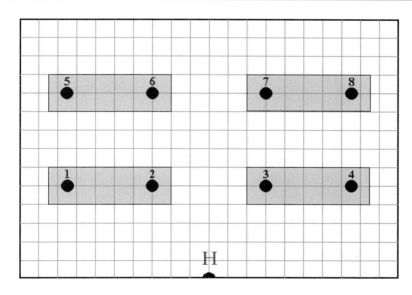

Figure 5.52

A small college classroom is being converted into a computer laboratory, as shown in **Figure 5.52**. The room will contain eight workstations on four tables. Because the computers (indicated by black dots) must all be networked to the college's main server, data cables must be run from the communications hub (H) located at the front center wall of the room. The cables will be recessed into the floor before carpeting is installed. The electrician must decide on an economical plan for connecting the workstations to the communications hub. Initially, only lengths of cable to be installed in the floor to points directly beneath each computer will be considered.

1. The room's dimensions are 20 feet by 14 feet. To aid in the wiring design, the electrician has drawn grid lines on the plan that are 1 foot apart. If the communications hub at H is chosen as the origin of a two-dimensional (x, y) coordinate system, determine the coordinates of each of the computers.

1. **#1: (–7.5, 5); #2: (–3, 5); #3: (3, 5); #4: (7.5, 5); #5: (–7.5, 10); #6: (–3, 10); #7: (3, 10); #8: (7.5, 10)**

2. Find the length of cable required to connect the hub directly to both computer #3 and computer #4; that is, find the distances from point H to #3 and from H to #4 (and denote the distances as d_3 and d_4). Round off the distances to the nearest 0.01 ft.

2. **Using the Pythagorean theorem, $d_3 = \sqrt{3^2 + 5^2} = 5.83$ feet and $d_4 = \sqrt{7.5^2 + 5^2} = 9.01$ feet.**

3. If the room layout is not yet finalized, then the exact location of computers #3 and #4 may not be known. Denote the coordinates of computer #3 as (x_3, y_3) and #4 as (x_4, y_4), and write a generalized formula for each distance.

3. **$d_3 = \sqrt{x_3^2 + y_3^2}$; $d_4 = \sqrt{x_4^2 + y_4^2}$.**

4. Another option for connecting computer #4 to the system is to run a short cable between computers #3 and #4. Write the length of this cable d_{34} in terms of the general coordinates of the points, and then find the length for the current configuration.

4. **$(x_4 - x_3) = 4.5$ ft.**

5. Now consider the connection for computer #8. It could be directly connected to the hub at H, with a cable of length $d_8 = \sqrt{x_8^2 + y_8^2} = 12.5$ ft, or connected to #7 with a 4.5 ft cable. Still another option is to use the connection at computer #3 as a "secondary hub," from which cables to all of the other three computers on the right side of the room would be connected. If that is done, write a formula for the distance from computer #3 to #8, and compute the value for the current configuration.

5. **$d_{38} = \sqrt{(x_8 - x_3)^2 + (y_8 - y_3)^2} = \sqrt{(7.5 - 3)^2 + (10 - 5)^2} = 6.73$ ft.**

6. Extend the process of computing cable lengths to include the entire room network of eight computers. Continue to consider the three types of options that you have already examined: (1) direct connection of every computer to the hub at H; (2) use of secondary hubs at computers #2 and #3 for connection of each side of the room; and (3) variations of (1) and (2) that allow for short cables connecting pairs of computers at the same desk. For each length of cable, first write a formula in terms of generalized coordinates that allow for design changes, and then compute the lengths for the configuration in Figure 5.52. What do you think would be the most economical wiring plan?

6. Wiring each computer directly to the hub uses 75.56 ft of cable. Wiring from the hub directly to computers #2 and #3 and then using each of these as a secondary hub to connect directly to the remaining three computers on each side of the room uses 44.12 ft of cable. Direct wiring of #2, #3, #6, and #7 with computers on each table connected to each other uses 50.54 ft of cable. Direct wiring of #3 with the hub and then connecting #4 and #7 with #3 and #8 to #4 (and the symmetrical arrangement on the left side) uses 40.66 ft of cable. And a variation of this in which #4 and #7 are connected to #3 with #8 connected to #7 (and the symmetrical arrangement) produces a minimum total cable usage of 39.66 ft.

When you found a formula for the distance between computers #3 and #8 in item 5 of Discovery 5.5, you discovered a particular use of the Pythagorean theorem that is important in a large variety of geometrical applications. It is worth making a special mention of it here.

The Distance Formula

The distance d between any two points having coordinates (x_1, y_1) and (x_2, y_2) may be found from the formula $d = \sqrt{(x_2 - x_1)^2 + (y_2 - y_1)^2}$.

The designations (x_1, y_1) and (x_2, y_2) in the distance formula are arbitrary, meaning that it doesn't matter which point is called the "first" point and which is the "second."

Example 23

Find the distance between the points (5, –6) and (–1, 2).

Solution: If we choose (x_1, y_1) to be (5, –6) and (x_2, y_2) to be (–1, 2), then the distance formula gives

$d = \sqrt{((-1) - 5)^2 + (2 - (-6))^2} = \sqrt{(-6)^2 + 8^2} = \sqrt{36 + 64} = \sqrt{100} = 10$. On the other hand, we could choose $(x_1, y_1) = (-1, 2)$ and $(x_2, y_2) = (5, -6)$, in which case

$d = \sqrt{(5 - (-1))^2 + ((-6) - 2)^2} = \sqrt{6^2 + (-8)^2} = \sqrt{100} = 10$.

The Equation of a Circle

Referring back to Discovery 5.5, suppose that a computer science instructor wants to be able to connect a portable computer from the back wall of the room directly to the communications hub. A 14-foot cable is needed. If the instructor then decides to move the portable computer to different locations in the back of the room, she can be no farther from the hub than 14 feet. How can we find other points that are 14 feet away from the hub? They will all be points on a circle with radius 14 feet and center at the hub (see **Figure 5.53**).

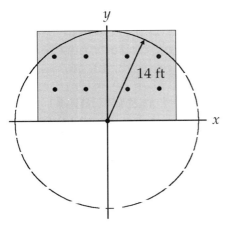

Figure 5.53

Because the distance from any point (x, y) on the circle to the center of the circle at point $(0, 0)$ is 14 feet, it must be true that $\sqrt{(x-0)^2 + (y-0)^2} = 14$. This equation can be written more simply as $\sqrt{x^2 + y^2} = 14$.

If we square both sides of this equation, the result is $x^2 + y^2 = 196$, which is usually preferable because it doesn't contain a radical. We say that the equation of the circle is $x^2 + y^2 = 196$, meaning that the sum of the squares of the coordinates of every point on the circle must be 196.

Example 24

How far back along either of the side walls of the classroom will the 14-foot cable reach?

Solution:

Point A along the right-side wall of the room (see **Figure 5.54**) is 10 feet to the right of the origin of the coordinate system. This means that the point where the circle intersects the wall has an x-coordinate of 10. Its y-coordinate must satisfy the equation $10^2 + y^2 = 196$ or $y^2 = 196 - 100 = 96$. Therefore, $y = \pm\sqrt{96} \approx \pm 9.8$ feet. But $y = -9.8$ makes no sense in this context because it would not be in the room. So the 14-foot cable will reach 9.8 feet back along the side of the room.

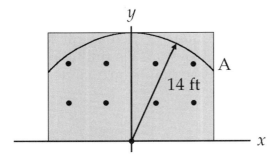

Figure 5.54

Example 25

If the instructor moves the computer in the back of the room exactly 5 feet to one side, how far forward (from the back wall) will she have to move it in order to avoid unplugging the cable?

Solution:

Point B in **Figure 5.55** has an x-coordinate of 5, so the y-coordinate must satisfy $5^2 + y^2 = 196$. Therefore, $y = \pm\sqrt{171} \approx \pm13.1$ feet. Again, only the positive result represents a point inside the room, so the computer must be moved to where $y = 13.1$ feet, which is 0.9 feet from the back wall.

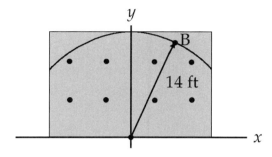

Figure 5.55

We can generalize from Example 25 and consider a circle with any radius. If the center of a circle is point $(0, 0)$, then we can replace the d in the distance formula with r to produce an equation that is valid for any such circle.

Standard Form Equation of a Circle

A circle with its center at the origin has the equation $x^2 + y^2 = r^2$, where r is the radius of the circle and (x, y) are the coordinates of any point on the circle.

Example 26

Find the standard form equation of the circle in **Figure 5.56.**

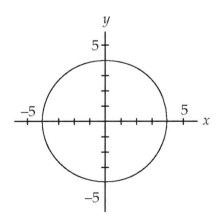

Figure 5.56

Solution:

Because the circle has its center at the origin and passes through the point (4, 0) on the x-axis, its radius is 4. Therefore, the equation of the circle is $x^2 + y^2 = 4^2$, which simplifies to $x^2 + y^2 = 16$.

Example 27

Describe and draw the circle that has an equation $x^2 + y^2 = 49$.

Solution:

Rewrite the equation as $x^2 + y^2 = 7^2$. This is the standard form equation of a circle with center at the origin and radius 7. See **Figure 5.57.**

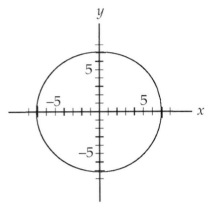

Figure 5.57

Example 28

An industrial designer has designed a wine glass as shown in **Figure 5.58.** The bowl portion is a section of a sphere with 3-inch diameter. If the opening is to have a 2.5-inch diameter, what will be the height of the bowl?

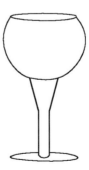

Figure 5.58

Solution:

Looking at the side of the glass, consider a vertical cross section through the widest part of the bowl to be part of a circle with a 1.5-inch radius (see **Figure 5.59**).

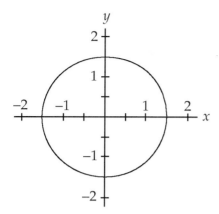

Figure 5.59

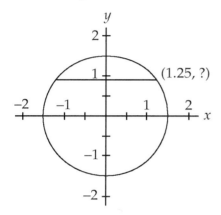

Figure 5.60

The equation of the circle is $x^2 + y^2 = (1.5)^2$ or $x^2 + y^2 = 2.25$ (x and y are both measured from the center of the circle). The x-coordinate of one of the points on the rim is half of 2.5, or 1.25 (see **Figure 5.60**). Substitute this value into the circle equation and solve for y:

$(1.25)^2 + y^2 = 2.25$

$y^2 = 0.6875$.

$y = \pm\sqrt{0.6875} \approx \pm 0.83$, but $y = -0.83$ is below the bowl, so only $y = 0.83$ is a possible solution for the rim.

The rim is 0.83 inches above the center of the bowl, so the height of the bowl from bottom to rim is (1.50 in + 0.83 in) = 2.33 in.

Exercises 5.6

I. Investigations *(SE pages 97–101)*

1. **Figure 5.61** is a drawing of a *Howe truss*, which is a type of roof truss often used in house construction. The points labeled A, B, C, and D are equally spaced, as are D, E, F, and G. Along the bottom of the truss, A, H, I, J, K, L, and G are equally spaced.

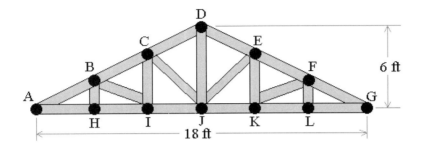

Figure 5.61

a) Choose a suitable origin for a coordinate system, and identify the coordinates of points A through L.

1. **a) If the origin is at point A, then the coordinates of the points are A(0, 0), B(3, 2), C(6, 4), D(9, 6), E(12, 4), F(15, 2), G(18, 0), H(3, 0), I(6, 0), J(9, 0), K(12, 0), and L(15, 0).**

b) Use the distance formula to help you find the total length of all the lumber needed to build this truss. Write your answer as an expression in simplified radical form.

b) AG = 18 ft; BH = FL = 2 ft; CI = EK = 4 ft; DJ = 6 ft.

$$AD = \sqrt{(9-0)^2 + (6-0)^2} = \sqrt{117} = 3\sqrt{13} \text{ ft; DG is also } 3\sqrt{13} \text{ ft.}$$

$$BI = \sqrt{(6-3)^2 + (0-2)^2} = \sqrt{13} \text{ ft; FK is also } \sqrt{13} \text{ ft.}$$

$$CJ = \sqrt{(9-6)^2 + (0-4)^2} = 5 \text{ ft; EJ is also 5 ft.}$$

The total length is

$$18 + 2(2) + 2(4) + 6 + 2(3\sqrt{13}) + 2(\sqrt{13}) + 2(5) = 46 + 8\sqrt{13} \text{ ft.}$$

c) What is the approximate total length to the nearest foot?

c) 74.84. . . . ≈ 75 ft.

2. Consider the points P and Q in **Figure 5.62**.

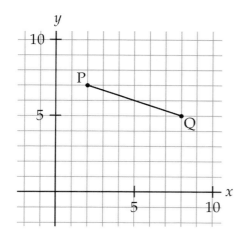

Figure 5.62

a) What are the coordinates of P and Q?

2. a) (2, 7) and (8, 5)

b) What is the length of the line segment joining points P and Q?

b) $\sqrt{(8-2)^2 + (5-7)^2} = \sqrt{40} = 2\sqrt{10}$.

c) What are the coordinates of the point that appears to be exactly in the middle of the line segment joining P and Q? (This point is called the *midpoint* of the line segment.)

c) (5, 6)

d) Use the distance formula to show that your answer to (c) is correct.

d) Distance from P to point (5, 6) = $\sqrt{(5-2)^2 + (6-7)^2} = \sqrt{10}$. **Distance from P to point (5, 6) =** $\sqrt{(8-5)^2 + (5-6)^2} = \sqrt{10}$. **The distances from point (5, 6) to each of the points P and Q are equal and half the distance from P to Q. Therefore, point (5, 6) is the midpoint of the line segment.**

Midpoint Formula

The midpoint of a line segment joining any two points (x_1, y_1) and (x_2, y_2) is $\left(\dfrac{x_1 + x_2}{2}, \dfrac{y_1 + y_2}{2} \right)$.

---

e) Use the midpoint formula to find the coordinates of the midpoints of line segments joining the following pairs of points: i. (3, 9) and (7, 4); ii. (1, –5) and (–6, 2); iii. (10, 3) and (–2, –5).

e) i. $\left(\dfrac{3+7}{2}, \dfrac{9+4}{2}\right) = \left(5, \dfrac{13}{2}\right)$ or (5, 6.5).

   ii. $\left(\dfrac{1+(-6)}{2}, \dfrac{(-5)+2}{2}\right) = \left(-\dfrac{5}{2}, -\dfrac{3}{2}\right)$ or (–2.5, –1.5).

   iii. $\left(\dfrac{10+(-2)}{2}, \dfrac{3+(-5)}{2}\right) = (4, -1)$.

3. The Global Positioning System (GPS) allows accurate determination of locations anywhere on or above Earth. The system consists of 24 satellites orbiting Earth at an altitude of 11,000 miles. Each satellite constantly broadcasts radio signals. A GPS receiver can determine its distance from a satellite, which places the receiver's location somewhere on a sphere centered on the satellite and with radius equal to the calculated distance. The intersection of spheres based on three separate satellites uniquely locates the position of the receiver.

   a) In two dimensions, only two signals would be necessary to locate an object's position, provided that the origination points of the signals are known. **Figure 5.63** shows a hypothetical simplified version of a GPS-type system. Radio signals are sent out from point A and from point B, 500 km due east of point A. A receiver at point P determines its distances from A and B to be 384 km and 267 km, respectively. Draw in a coordinate system with origin at point A and x-axis passing through point B. What are the coordinates of point B?

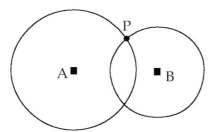

Figure 5.63

3. a) (500, 0)

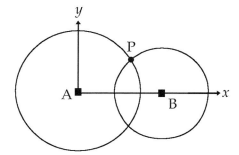

b) Write the equation of the circle with center at A and containing point P.

b) $x^2 + y^2 = (384)^2$ **or** $x^2 + y^2 = 147,456$.

c) In order to write the equation for the circle with center at B, note that the center is shifted to the right by 500 units. By replacing the x variable with the quantity $(x - 500)$ in the standard form equation for a circle, you can write the equation for the circle with center at B.

c) $(x - 500)^2 + y^2 = (267)^2$ **or** $x^2 - 1000x + 250,000 + y^2 = 71,289$.

4. **Figure 5.64** shows a top view of a drop leaf table. The leaves on the ends of the table are segments of circles. In order to lay out one of the circular arcs for cutting, it would be necessary to know the radius of the circle.

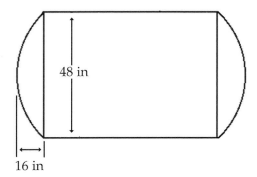

48 in

16 in

Figure 5.64

a) It is sometimes helpful to place a coordinate system on the drawing. Draw in an x-axis running horizontally through the center of the table. Consider the circular arc on the right leaf. Locate an origin at a point on the x-axis where you estimate the center of the circle to be, and draw in a y-axis. Label points A, B, and C respectively at the top, center, and bottom of the arc.

4. a)

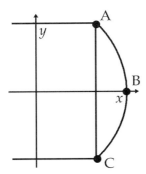

b) Let r represent the unknown radius. What are the coordinates of point B, measured with respect to the origin?

b) $(r, 0)$

c) Considering your answer to (b), what are the coordinates of point A?

c) $(r - 16, 24)$

d) Use the standard equation for a circle (with radius r) and your answer to (c) to write an equation for r.

d) $(r - 16)^2 + (24)^2 = r^2$.

e) Solve the equation to find the radius of the circular arc.

e) $r^2 - 32r + 256 + 576 = r^2$

$$832 = 32r$$

$$r = 26 \text{ in.}$$

5. Constructing the graph of a circle on a graphing calculator or computer using rectangular coordinates requires special care. Consider the circle whose equation is $x^2 + y^2 = 36$.

a) Equations to be graphed must be entered as functions. The circle equation is not a function, because there are two points on the circle for each x-value (except on the x-axis). In order to graph the circle $x^2 + y^2 = 36$, solve the equation for y to find two equations in the form of functions. Remember when taking a square root to allow for a negative, as well as a positive, result. Graph the functions using a "standard" window, which on most calculators is [–10, 10] x [–10, 10].

5. **a)** $y^2 = 36 - x^2$, **and the two functions are** $y = \sqrt{36 - x^2}$ **and** $y = -\sqrt{36 - x^2}$.

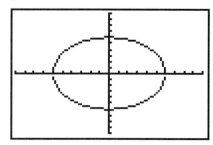

[−10, 10] **x** [−10, 10]

  b) Your graph probably does not appear circular but elliptical. This is because the standard window on most calculators compresses the vertical scale. In order to avoid distortion of geometrical shapes, it is necessary to use a "square" window for which a unit distance in each direction contains the same number of pixels. Use a "Zsquare" command on the Zoom menu, if available, to automatically compress the horizontal scale by the same amount as the vertical scale. Otherwise, if you know the dimensions of your screen in pixels, you may be able to adjust the window manually to square it off.

  b)

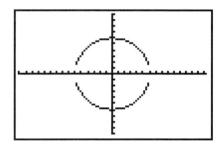

[−15.161..., 15.161...] **x** [−10, 10]

  c) Your graph should now appear to be a circle. But it may have gaps near the x-axis. This can occur if the x-intercepts of the circle don't lie exactly on the coordinates of screen pixels for the chosen window. Because the intercepts of the circle $x^2 + y^2 = 36$ are points (−6, 0) and (6, 0), only a window containing these points as pixel coordinates will show the circle as a closed curve. Find a window that remains "square" but that includes the intercepts of the circle.

c) Some calculators have a "decimal" window for which pixels are exactly 0.1 unit apart. For example, the "Zdecimal" zoom window on a TI-83 has dimensions of [–4.7, 4.7] x [–3.1, 3.1] because the screen is 95 pixels wide and 63 pixels high. However, this window is not large enough to display a circle of radius 6. Doubling the size of each window dimension will achieve the desired result.

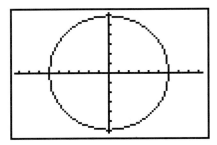

[–9.4, 9.4] x [–6.2, 6.2]

6. The standard form equation for a circle as discussed in this chapter is valid only for circles with centers at point (0, 0). But just as the distance formula was used to derive the standard equation for origin-centered circles, it can also be used to find an equation for more general circles.

a) Consider the circle in **Figure 5.65.** Its center is point (1, 2), and its radius is 5. Let P stand for any point on the circle. Use the distance formula to write an expression for the distance from the center, with coordinates (1, 2), to P, with variable coordinates (x, y).

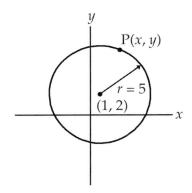

Figure 5.65

6. **a)** $\sqrt{(x-1)^2 + (y-2)^2}$

b) Because the distance from the center to any point on the circle is always equal to the radius, set your expression in (a) equal to 5. Square both sides of the equation to eliminate the radical, and you will have an equation in standard form for the circle in Figure 5.65.

b) $\sqrt{(x-1)^2 + (y-2)^2} = 5$

$(x-1)^2 + (y-2)^2 = 25.$

c) In order to generalize this result, consider the circle in **Figure 5.66**. It has an unspecified radius r, and the coordinates of its center are represented by (h, k). Following the same reasoning you used in (a) and (b), write a standard equation for a circle with radius r and center at (h, k).

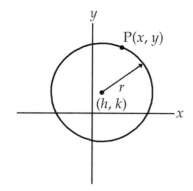

Figure 5.66

c) $\sqrt{(x-h)^2 + (y-k)^2} = r$

$(x-h)^2 + (y-k)^2 = r^2$

d) Use your answer to (c) to write an equation in standard form for the circle in **Figure 5.67**.

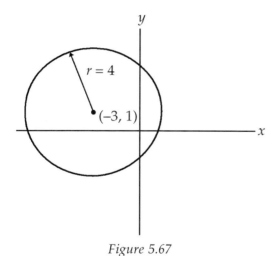

Figure 5.67

d) The center of the circle is point (–3, 1), so $h = -3$ and $k = 1$. The standard equation is $(x-(-3))^2 + (y-1)^2 = 4^2$ or $(x+3)^2 + (y-1)^2 = 16$.

e) Describe the circle whose standard equation is $(x-6)^2 + (y+8)^2 = 9$.

e) Rewrite the equation as $(x-6)^2 + (y-(-8))^2 = 3^2$. Here $h = 6$, $k = -8$, and $r = 3$. The circle has a radius of 3 and a center at point (6, –8).

II. Projects and Group Activities *(SE pages 101–102)*

7. An industrial robot on a factory floor is programmed to follow a path that is always equidistant from a wall and from a post that is 2.0 meters from the wall (see **Figure 5.68**).

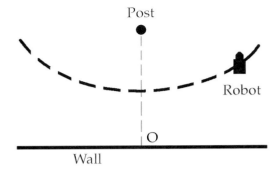

Figure 5.68

a) Let the origin of a coordinate system be located at point O on the wall, directly across from the post. Let the *x*-axis run along the wall, with the *y*-axis passing through the post. What are the coordinates of the post's location?

7. a) (0, 2.0)

b) What are the coordinates of the robot's location when it is directly between the post and the wall?

b) (0, 1.0)

c) When the robot is at a location having an *x*-coordinate of 1.5 meters, write an expression for the robot's distance from the post, using *y* for the unknown *y*-coordinate.

c) $\sqrt{(1.5)^2 + (y - 2.0)^2}$

d) What is the robot's distance from the wall at this point?

d) *y*

e) Set the two expressions from (c) and (d) equal to each other, and square both sides of the equation to solve for the unknown *y*-coordinate.

e) $\sqrt{(1.5)^2 + (y - 2.0)^2} = y$

$(1.5)^2 + (y - 2.0)^2 = y^2$

$2.25 + y^2 - 4.0y + 4.0 = y^2.$

Although at first the equation appeared to be quadratic, it is actually linear in *y* and can be solved to give *y* ≈ 1.56 m.

f) Using a spreadsheet or calculator lists, determine the coordinates of all points on the robot's path from $x = -3.0$ meters to $x = 3.0$ meters, increasing x by 0.5 meter for each calculation.

f)

x	y
-3	3.25
-2.5	2.5625
-2	2
-1.5	1.5625
-1	1.25
-0.5	1.0625
0	1
0.5	1.0625
1	1.25
1.5	1.5625
2	2
2.5	2.5625
3	3.25

g) Using regression models, determine a possible equation for the robot's path, and discuss your findings.

g) Quadratic regression results in $y = 0.25x^2 + 1$, so the robot's path is a parabola.

III. Extra Practice (SE pages 102–105)

8. A plastic bracelet is designed with an outer diameter of 8.0 cm (see **Figure 5.69a**). It can be visualized as an 8.0 cm sphere with a large hole cut out of the middle, which is shown in cross section in **Figure 5.69b.** If the diameter of the hole is 7.2 cm, how wide is the bracelet?

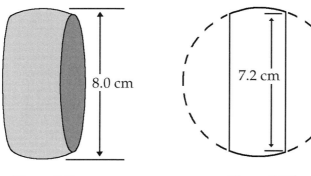

Figure 5.69a   *Figure 5.69b*

8. With an origin at the center of the circle in Figure 5.69b, the equation of the circle is $x^2 + y^2 = (4.0)^2$, or $x^2 + y^2 = 16$. The y-coordinate of a point at the upper right corner of the bracelet in that figure is 3.6. Substituting 3.6 in the circle equation gives $x \approx 1.74$, so the width of the bracelet is $2(1.74) = 3.48 \approx 3.5$ cm.

9. A Global Positioning System satellite is in orbit 11,000 miles above Earth's surface. Write an equation for the satellite's orbital path, assuming that it is circular. (The diameter of Earth is about 8000 miles.)

9. **The distance of the satellite from the center of Earth is (11,000 mi + 4000 mi) = 15,000 mi. The path is a circle with radius 15,000 mi. If the origin is placed at Earth's center, the equation of the path is $x^2 + y^2 = (15,000)^2$, or $x^2 + y^2 = 2.25 \times 10^8$.**

10. The London Eye, the world's largest observation wheel, is a circular Ferris wheel 135 meters in diameter (see **Figure 5.70**). After boarding the wheel at its bottom point, a passenger in one of the compartments moves 15 meters horizontally in the first minute. How high will the passenger rise in one minute?

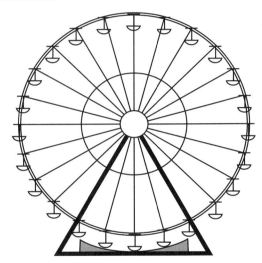

Figure 5.70

10. **The equation of the wheel is $x^2 + y^2 = 135^2$, or $x^2 + y^2 = 18{,}225$. When $x = 15$, $y \approx -134.16$ on the lower part of the circle, so the rise is only $(135 - 134.16) = 0.84$ m.**

11. Pueblo, Colorado, is 14 miles east and 33 miles south of Denver. Steamboat Springs is 71 miles west and 52 miles north of Denver. How far is it from Pueblo to Steamboat Springs? (Hint: Draw the locations of the cities on a coordinate plane.)

11.

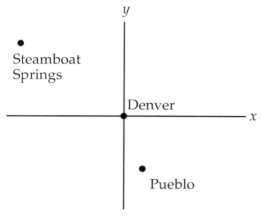

The coordinates of Pueblo are (14, –33) and of Steamboat Springs (–71, 52). The distance between them is $\sqrt{((-71) - 14)^2 + (52 - (-33))^2} \approx 120$ miles.

12. **Figure 5.71** shows a map of main roads in the vicinity of a wheat farm, with distances labeled in miles. A farmer has a choice of two grain elevators, located at points A and B, at which he can sell his wheat. There are straight dirt roads connecting the farm directly to each of the elevators. It costs 3 cents per mile traveled for the farmer to transport a bushel of wheat by truck. Which grain elevator should be used in order to minimize the costs of transportation for each bushel?

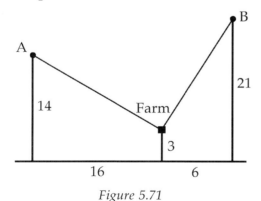

Figure 5.71

12. **If the origin is located at the intersection of two roads 3 miles south of the farm, the coordinates of the farm are (0, 3), of elevator A (–16, 14), and of elevator B (6, 21). The distance from the farm to elevator A is $\sqrt{((-16) - 0)^2 + (14 - 3)^2} \approx 19.4$ miles. The distance from the farm to elevator B is $\sqrt{(6 - 0)^2 + (21 - 3)^2} \approx 19.0$ miles. It costs (19.4 mi)(3 cents/mi) $\approx$ 58 cents to transport a bushel to elevator A and (19.0 mi)(3 cents/mi) $\approx$ 57 cents to transport a bushel to elevator B. The farmer should use elevator B.**

13. Write an equation for the circle in **Figure 5.72**.

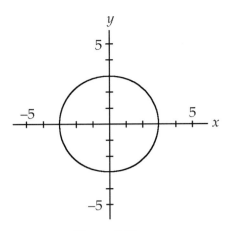

Figure 5.72

13. $x^2 + y^2 = 9$.

14. For the circle of Exercise 13, what is the y-coordinate of a point on the circle that has an x-coordinate of 1 and that is below the x-axis? (Express your answer both in simplified radical form and as a decimal rounded to the nearest thousandth.)

14. $y = -\sqrt{8} = -2\sqrt{2}$; $y \approx -2.828$.

15. For the circle of Exercise 13, what is the x-coordinate of a point in the first quadrant that has a y-coordinate of 2.1? (Express your answer as a decimal rounded to the nearest thousandth.)

15. $x = \sqrt{4.59} \approx 2.142$.

16. Write an equation for a circle with radius 8 and center at the origin.

16. $x^2 + y^2 = 64$.

17. Write an equation for a circle with radius $\sqrt{2}$ and center at the origin.

17. $x^2 + y^2 = 2$.

18. Write an equation for a circle with radius 2.7 and center at the origin.

18. $x^2 + y^2 = (2.7)^2$ or $x^2 + y^2 = 7.29$.

19. Write an equation for a circle with center at the origin that passes through the point (3, 7).

19. $r^2 = 3^2 + 7^2 = 58$, so $x^2 + y^2 = 58$.

20. Write an equation for a circle with center at the origin that passes through the point (–2, 1).

20. $r^2 = (-2)^2 + 1^2 = 5$, so $x^2 + y^2 = 5$.

21. Describe the graph of the equation $x^2 + y^2 = 81$.

21. Circle with center at the origin and radius 9

22. Describe the graph of the equation $x^2 + y^2 = 6$.

22. Circle with center at the origin and radius $\sqrt{6}$

23. Describe the graph of the equation $x^2 = 100 - y^2$.

23. Circle with center at the origin and radius 10

For 24–26, express answers both in simplified radical form and as decimals rounded to the nearest thousandth.

24. What are the y-coordinates of points on the circle $x^2 + y^2 = 25$ that have an x-coordinate of 4?

24. $x^2 = 21$, so $x = \pm3$.

25. What are the x-coordinates of points on the circle $x^2 + y^2 = 16$ that have a y-coordinate of –2?

25. $x^2 = 12$, so $x = \pm\sqrt{12} = \pm2\sqrt{3}$; $x \approx \pm3.464$.

26. What are the y-coordinates of points on the circle $x^2 + y^2 = 1$ that have an x-coordinate of $\dfrac{1}{2}$?

26. $y^2 = \dfrac{3}{4}$, so $y = \pm\sqrt{\dfrac{3}{4}} = \pm\dfrac{\sqrt{3}}{2}$; $y \approx \pm0.866$.

For 27–29, find the distance between the given points.

27. (2, 6) and (7, 18)

27. 13

28. (–10, 0) and (14, 5)

28. 25

29. (–3, 2) and (5, –4)

29. 10

For 30–32, find the distance between the given points. Express answers both in simplified radical form and as decimals rounded to the nearest thousandth.

30. (7, 2) and (0, 3)

30. $\sqrt{50} = 5\sqrt{2}$; 7.071

31. (–2, 5) and (6, 1)

31. $\sqrt{80} = 4\sqrt{5}$; 8.944

32. (4, 9) and (1, –3)

32. $\sqrt{153} = 3\sqrt{17}$; 11.402

For 33–35, find the distance between the given points. Express answers as decimals rounded to the nearest thousandth.

33. (12.7, 4.0) and (5.6, 1.5)

33. 7.527

34. (1.7, –3.2) and (6.0, 2.8)

34. 7.382

35. (–5.4, –1.9) and (–10.6, –7.3)

35. 7.497

For 36–38, determine whether the triangle formed by the given points is isosceles or scalene.

36. (–3, 2), (2, 2), and (–1, –2)

36. Two of the sides have lengths of 5 (and the third is $2\sqrt{5}$), so the triangle is isosceles.

37. (–1, 5), (5, 3), and (2, –2)

37. The lengths of the sides are $2\sqrt{10}$, $\sqrt{58}$, and $\sqrt{34}$, so the triangle is scalene.

38. (3, 2), (9, 3), and (8, –3)

38. Two of the sides have lengths of $\sqrt{37}$ (and the third is $5\sqrt{2}$), so the triangle is isosceles.

Chapter 5 Summary

Quadratic functions

Parabolas

Solving equations by factoring

Solving equations by graphing

Quadratic equations

Quadratic formula

Modeling data with quadratic functions

Roots of numbers

Simplifying radical expressions

Operations with radical expressions

Fractional exponents

Solving radical equations

Distance formula

Equation of a circle

Chapter 5 Review *(SE pages 107–111)*

1. Write a summary of the important mathematical ideas found in Chapter 5.

1. **Answers will vary. Following are some of the important ideas that should be listed:**

   Quadratic functions can be used to model many real-world applications such as suspension bridge cables.

   The standard form of a quadratic function can be written as $y = ax^2 + bx + c$.

   Graphs of quadratic functions have certain characteristics such as a U shape, a single vertex whose x-coordinate is $\dfrac{-b}{2a}$, and a vertical axis of symmetry.

   The graph of a quadratic function opens upward if a is positive and downward if a is negative. The domain of the function is all real numbers.

   Quadratic equations can be solved using several methods, including factoring, the quadratic formula, and graphically.

   The expression $b^n = a$ is equivalent to $b = \sqrt[n]{a}$ (radical notation) or $b = a^{1/n}$ (fractional exponent notation) when $a, b,$ and n are positive.

   Radical expressions can be added, subtracted, multiplied, and divided.

   When solving equations with radicals, care must be taken to check the solutions.

   The distance formula $d = \sqrt{(x_2 - x_1)^2 + (y_2 - y_1)^2}$ can be used to find the distance between two points on a coordinate plane.

   The equation $x^2 + y^2 = r^2$ represents a circle with its center at the origin.

2. A stream of wastewater exits from a horizontal pipe 8.5 meters above a concrete channel. When the stream has dropped to a height of 5.0 meters above the channel, it has gone 3.4 meters horizontally from the pipe opening.

   a) Find an equation of the parabolic path of the water stream. (Assume that the pipe opening is the vertex of the parabola.)

2. a) **The form is $y = ax^2$ if the origin is placed at the vertex. When $x = 3.4$, $y = -3.5$, so $a = \dfrac{(-3.5)}{(3.4)^2} = -0.303$, and $y = -0.303x^2$. If the origin is placed directly below the vertex at the level of the channel, then the equation is $y = 8.5 - 0.303x^2$.**

   b) How far out from the pipe opening will the stream hit the surface of the channel?

   b) **At the channel surface, $y = -8.5$, so $-0.303x^2 = -8.5$; $x = 5.3$ m.**

3. Find an equation, in the form $y = ax^2$, for the parabola with its vertex at the origin that passes through point (4, 2) (see **Figure 5.73**).

3. $a(4)^2 = 2$, **so** $a = \dfrac{1}{8}$**;** $y = \dfrac{1}{8}x^2$ **or** $y = 0.125x^2$**.**

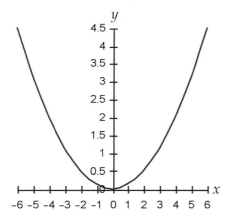

Figure 5.73

4. The main suspension cable of the Brooklyn Bridge hangs from two towers that are 142 feet high and 1596 feet apart (see **Figure 5.74**).

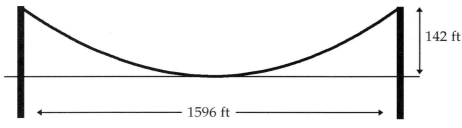

142 ft

1596 ft

Figure 5.74

a) Find an equation for the cable, assuming that the shape is a parabola with the origin of the coordinate system at the lowest point of the cable.

4. **a)** $y = ax^2$**, and the point (798, 142) is on the parabola;** $a = 142/(798)^2 =$ **0.0002230. The equation is** $y = 0.0002230x^2$**.**

b) How high is the cable above the roadway at a point that is 500 feet in from either tower? Assume that the cable is at the same height as the roadway at its center.

b) **At 500 feet from one of the towers,** $x = 798 - 500 = 298$**.**
$y = 0.0002230(298)^2 = 19.8$ **ft.**

5. A ball is thrown vertically into the air so that its height h above its point of release at any time is given by $h = 28t - 4.9t^2$, where h is measured in meters and t in seconds.

a) At what time will the ball return to its point of release?

5. a) The ball returns to the point of release when $h = 0$. So $0 = 28t - 4.9t^2$; $0 = t(28 - 4.9t)$; $t = 0$ and 5.7. Zero seconds is the time of release, and 5.7 seconds is the time of return.

b) At what time will the ball be at its highest point?

b) Maximum height will occur midway between the zeros, at 2.85 s.

c) How high will the ball rise above its point of release?

c) When $t = 2.85$ s, $h = 28(2.85) - 4.9(2.85)^2 = 40$ m.

6. The height of a projectile launched from the ground at an angle of 40° with an initial speed of 80 meters per second is given by the equation $y = 0.84x - 0.0013x^2$, where x is the horizontal distance traveled and both x and y are measured in meters. At what distance x will the projectile be 100 m above the ground?

6. $0.84x - 0.0013x^2 = 100$; standard form is $-0.0013x^2 + 0.84x - 100 = 0$, with solutions (using the quadratic formula) $x = 157$ and 489. Both are valid solutions, one (157 m) representing the ascending phase and the other (489 m) representing the descending phase of the trajectory.

7. The equation $P = 0.00256v^2$ gives the frontal pressure P (measured in pounds per square foot) on a high-rise building as a function of wind velocity v (measured in miles per hour). The New York City building code specifies that all building surfaces above 100 feet must be capable of withstanding a wind load of 20 lb/ft². What wind velocity would be necessary to produce a frontal pressure of 20 lb/ft²?

7. $0.00256v^2 = 20$, and $v^2 = 7812.5$, so $v \approx \pm 88$. A negative velocity (away from the building surface) does not make physical sense, so $v = 88$ mi/hr.

8. The 37-foot-high "Green Monster" left-field wall in Boston's Fenway Park has a 23-foot-high extension of netting above it in order to prevent many home runs from leaving the park. If a ball is hit so that its trajectory is given by $y = -0.0028x^2 + 1.05x + 4.5$, will it leave the park?

8. No, $y \approx 57$ when $x = 315$, so the ball will not clear the necessary 60-foot height.

For 9–15, solve the given equation algebraically. Give both an exact and an approximate solution. Round off approximate solutions to three decimal places.

9. $8x^2 + 3x = 0$.

9. $x(8x + 3) = 0$, $x = 0$ or $-\dfrac{3}{8}$.

10. $5d^2 + 5d - 3 = 0$.

10. Using the quadratic formula, $d = \dfrac{-5 \pm \sqrt{85}}{10}$ or $d \approx 0.422$ or -1.422.

11. $v^2 - 7v - 18 = 0$.

11. $(v - 9)(v + 2) = 0$, $v = -2$ or 9.

12. $17k = 4k^2$.

12. $17k - 4k^2 = k(17 - 4k) = 0$, $k = 0$ or $\dfrac{17}{4}$.

13. $3h^3 - 18h^2 + 15h = 0$.

13. $3h(h^2 - 6h + 5) = 3h(h - 5)(h - 1) = 0$, $h = 0$, 1, or 5.

14. $x^2 = 7x - 3$.

14. $x^2 - 7x + 3 = 0$; using the quadratic formula, $x = \dfrac{7 \pm \sqrt{37}}{2}$ or $x \approx 0.459$ or 6.541.

15. $5x^2 = 45$.

15. $5x^2 - 45 = 5(x^2 - 9) = 5(x + 3)(x - 3) = 0$, $x = -3$ or 3.

16. Solve the equation $2x^2 + 5x - 8 = 0$ by graphing. Give solutions correct to three decimal places.

16. Graph the function $y = 2x^2 + 5x - 8$.

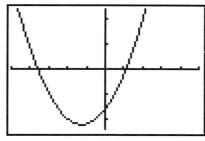

[−5, 5] x [−12, 12]

The zeros are between $x = -4$ and -3 and between 1 and 2. Zoom in on the leftmost intercept until the x-coordinate of the intercept is located to within 0.001.

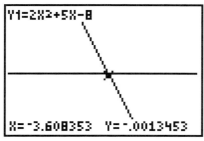

[−3.640, −3.581] x [−0.087, 0.010]

The intercept is at $x \approx -3.608$, so this is one solution. Zoom in on the other intercept.

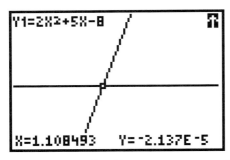

[1.108, 1.109] x [−0.0013, 0.0015]

The other intercept is at $x \approx 1.108$, so this is the other solution.

17. According to *Cigar Aficionado* magazine, the number of premium cigars imported into the United States during the early 1990s varied as shown in **Table 5.28**.

Year	Cigars Imported (millions)
1990	106
1991	102
1992	107
1993	118
1994	132
1995	176

Table 5.28

    a) Use quadratic regression to find an equation that expresses the number N of premium cigar imports, in millions of cigars, as a function of the year y.

17. a) $N = 4.9285714y^2 − 19,627.471y + 19,541,140.$

    b) Use your equation to predict cigar imports for 1996.

b) 215 million

18. Explain how the product property can be used to multiply $\sqrt{10}$ and $\sqrt{40}$.

18. The product property says that the product of two square roots is the square root of the product. That is, $\sqrt{10} \cdot \sqrt{40} = \sqrt{10 \cdot 40}$. The result $\sqrt{400}$ equals 20.

19. Find the distance d between the two circles in **Figure 5.75** if the area of the larger circle is 338π square meters and the area of the smaller circle is 200π square meters.

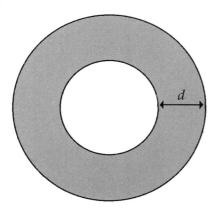

Figure 5.75

19. $d = \sqrt{\dfrac{A_R}{\pi}} - \sqrt{\dfrac{A_r}{\pi}} = \sqrt{\dfrac{338\pi}{\pi}} - \sqrt{\dfrac{200\pi}{\pi}} = 13\sqrt{2} - 10\sqrt{2} = 3\sqrt{2}$ **m.**

20. For a flexible pulley belt, the length L of one of the straight sections of the belt that is tangent to both pulleys is $L = \sqrt{(D)^2 - (R-r)^2}$. D is the center-to-center distance between the pulleys, and R and r are the radii of the pulleys (see **Figure 5.76**).

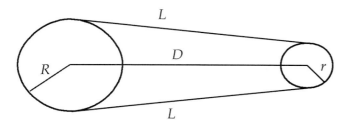

Figure 5.76

Find the length L if $D = 26$ inches, $R = 12$ inches, and $r = 4$ inches.

20. $L = \sqrt{(D)^2 - (R-r)^2} = \sqrt{(26)^2 - (12-4)^2} = \sqrt{612} = 6\sqrt{17}$ **inches, which is approximately 24.7 inches.**

21. Rewrite each expression in simplified radical form.

    a) $\sqrt{75}$

21. a) $5\sqrt{3}$

    b) $3\sqrt{80}$

  b) $12\sqrt{5}$

c) $\sqrt{\dfrac{5}{7}}$

c) $\dfrac{\sqrt{35}}{7}$

d) $\sqrt{\dfrac{1}{12}}$

d) $\dfrac{\sqrt{3}}{6}$

e) $\dfrac{3}{\sqrt{10}}$

e) $\dfrac{3\sqrt{10}}{10}$

f) $\sqrt{28a^6b^9c}$

f) $2a^3b^4\sqrt{7bc}$

22. Perform the indicated operations. Write all answers in simplified radical form.

a) $\sqrt{5}\sqrt{15}$

22. a) $5\sqrt{3}$

b) $\left(11\sqrt{5}\right)^2$

b) 605

c) $\sqrt{6}\left(\sqrt{2}-4\sqrt{15}\right)$

c) $2\sqrt{3}-12\sqrt{10}$

d) $\left(5+2\sqrt{7}\right)\left(2-\sqrt{7}\right)$

d) $-4-\sqrt{7}$

e) $5\sqrt{13}+5\sqrt{7}-9\sqrt{13}$

e) $5\sqrt{7}-4\sqrt{13}$

f) $\sqrt{45}-2\sqrt{20}+7\sqrt{5}$

f) $6\sqrt{5}$

23. Use a fractional exponent to rewrite $\sqrt[3]{b^5}$ without a radical.

23. $b^{5/3}$

24. Rewrite $M^{-1/5}$ using radical notation.

24. $\dfrac{1}{\sqrt[5]{M}}$

25. Evaluate $16^{3/4}$.

25. $\left(\sqrt[4]{16}\right)^3 = 2^3 = 8.$

26. Simplify $\left(4x^{1/3}y^{-4}\right)^{3/2}$, using only positive exponents.

26. $\dfrac{8x^{1/2}}{y^6}$

27. The line-of-sight distance d (in kilometers) to the horizon is approximately given by the formula $d = 0.5\sqrt{h}$, where h is the height (in meters) of the viewer. Solve the equation for h.

27. $h = 4d^2.$

28. Solve for t: $\sqrt{2t+1} - 3 = 2.$

28. $t = 12.$

For 29–33, find the distance between the given points. Round off approximate answers to three decimal places.

29. (3, 2) and (18, 10)

29. 17

30. (1.7, −4.5) and (−6.1, 0.8)

30. 9.430

31. Write an equation for the circle with radius 3.5 and center at the origin.

31. $x^2 + y^2 = (3.5)^2$, or $x^2 + y^2 = 12.25.$

32. Write an equation for the circle with center at the origin that passes through the point (4, −2).

32. $r^2 = (4)^2 + (-2)^2 = 20$, so $x^2 + y^2 = 20.$

33. Describe the graph of $x^2 + y^2 = 36$.

33. Circle with radius 6 and center at the origin

Chapter 6—Rational Expressions and Systems of Equations

Goals of the Chapter

- To explore basic properties of rational functions

- To perform operations with rational expressions

- To solve rational equations

- To solve systems of equations

Preparation Reading

In recent years in the United States and around the world, pollution of our lakes, rivers, and streams has been a major concern. For example, mercury is a major pollutant of our environment that causes serious diseases of the central nervous system.

A lake containing 200,000 cubic meters of water is polluted with 50 kilograms of mercury. One way to describe the concentration of this chemical is to use a **ratio.** Recall that a ratio is a comparison of two numbers using division. In this case, the concentration of mercury is equal to the ratio of the amount of mercury to the volume of the lake, or $\dfrac{50 \text{ kg}}{200,000 \text{ m}^3}$. This ratio can be rewritten as the equivalent fraction $\dfrac{1 \text{ kg}}{4000 \text{ m}^3}$ or as the decimal fraction 0.00025 kg/m^3.

What if a chemical processing plant begins to discharge additional mercury into the lake at the rate of 0.1 kilograms per day? The total amount of mercury in the lake after t days would no longer be a constant 50 kg but would increase according to the algebraic expression $(50 + 0.1t)$. If, in addition, 150 cubic meters of water is added each day to the lake during the time that mercury is being introduced, the volume of the lake would increase according to the expression $(200,000 + 150t)$. We could use these two expressions to write the ratio of the amount of mercury to the volume of the lake. This ratio, $\dfrac{50 + 0.1t}{200,000 + 150t}$, is a quotient of two polynomials and is called a **rational expression.**

In this chapter, we will examine several real-world situations that can be modeled with functions that contain rational expressions. As we investigate the behavior of these functions, we will see that rational functions have some properties that are unlike those of polynomials. We will examine how to combine rational expressions and solve equations that contain them. Methods for solving systems of equations will also be introduced.

Reflect and Discuss (SE page 113)

1. Name some quantities that can be expressed as ratios of two other quantities.

1. Velocity $= \dfrac{\text{distance}}{\text{time}}$; density $= \dfrac{\text{weight}}{\text{volume}}$; power $= \dfrac{\text{energy}}{\text{time}}$; dosage of some medicines are determined by $\dfrac{\text{dosage amount}}{\text{body weight}}$; baseball batting average $=$ $\dfrac{\text{number of hits}}{\text{number of times at bat}}$; grade on a test $= \dfrac{\text{number of points earned}}{\text{total points on test}}$; compression ratio of an engine $= \dfrac{\text{displacement plus clearance volume}}{\text{clearance volume}}$; $\pi = \dfrac{\text{circumference of a circle}}{\text{diameter}}$.

Section 6.1 Introduction to Rational Functions

What You Need to Know

- How to evaluate expressions

What You Will Learn

- To explore the behavior of a non-polynomial function

Materials

- Fettucini or similar pasta

- Ruler

- Rubber band

- Paper clip

- Small plastic bag

- Pennies

A material's strength can be measured in many ways. Material testing companies have machines designed to measure a variety of strength properties. One such machine measures the strength of a wooden beam by bending it until it breaks (see **Figure 6.1**). From tests like this, architects and structural engineers know how much weight the beam can be expected to support.

Figure 6.1

There are many applications that depend on the strength of a material. For instance, the design of a building largely depends on the strength of the wood or steel beams used in its construction. Other applications requiring strength may be less obvious. For example, the plastic used to wrap food in a supermarket and the surgical thread for medical procedures are just two of the many items for which strength is an important property.

In construction, the consequences of not paying attention to material strength can be disastrous. During a 2001 wedding in Jerusalem, a banquet hall floor collapsed killing two dozen people and injuring hundreds of others. The structural elements supporting the floor were not strong enough to withstand the weight of all the people at the party. In order to avoid such tragedies, structural engineers must have accurate information about the strength of materials used in construction.

Activity 6.1 Breaking Strength versus Length *(SE pages 115–116)*

The force required to break a beam depends on its dimensions as well as the material from which it is made. Wood or steel beams are too strong to be broken in a controlled way in the classroom, but a simulated beam strength test can be performed on a strand of pasta. In this activity, you will investigate how the breaking strength of pasta depends on its length.

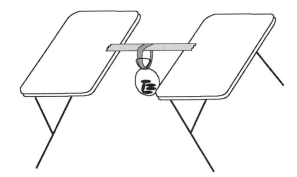

Figure 6.2

1. Use fettucini or another similar long and flat pasta. Break off a piece 7 cm long and lay it across a 5 cm gap between two desks as shown in **Figure 6.2.** Have someone press the ends of the pasta against the desks to anchor it and help prevent slipping. (The effective length of the pasta is 5 cm.) Hang a small plastic bag from the center of the pasta using a paper clip and rubber band. Add pennies one at a time until the pasta breaks, and record the force (in penny weights) required to break it in a table similar to **Table 6.1.** Repeat the measurement two more times for identical lengths of pasta and determine the average force required to break the pasta. Then complete your table by repeating the three trials using 10 cm, 15 cm, and 20 cm gaps between desks (again with an extra centimeter of overlap at each end).

Breaking Force (pennies)				
Gap Length (cm)	Trial 1	Trial 2	Trial 3	Mean Force
5				
10				
15				
20				

Table 6.1

1. **Sample table:**

Breaking Force (pennies)				
Gap Length (cm)	Trial 1	Trial 2	Trial 3	Mean Force
5				272
10				156
15				101
20				79

2. Consider the two variables in this activity: (effective) length of the pasta L and the force F that is required to break the pasta. Which variable is the independent variable, and which is the dependent variable? Explain.

2. **Length is the independent variable, and force is the dependent variable. The force that is required to break the pasta depends on the length of the pasta.**

3. What is the problem domain for this relationship?

3. **The problem domain depends on the maximum length of the pasta. So the domain must be greater than zero and less than or equal to the total length of an unbroken piece of pasta.**

4. Using your completed Table 6.1, plot points corresponding to your length and average force values on a graph.

4. **Sample graph:**

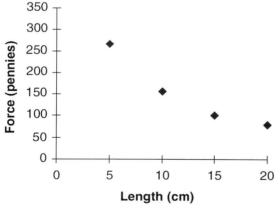

5. As the length of the pasta L increases, what happens to the force F that is required to break it?

5. **It decreases.**

6. As the length of the pasta L decreases, what happens to the force F that is required to break it?

6. **It increases.**

7. What would happen to the force required to break the pasta if your piece of pasta could get longer and longer without a limit to its length?

7. It would get closer and closer to zero.

8. What would happen to the force required to break the pasta if your piece of pasta could get smaller and smaller until it was very close to 0 cm long?

8. It would take more and more force to break the pasta.

9. Use your answers to items 5–8 to help you complete the graph that you began in item 4, including the entire problem domain.

9. Sample graph:

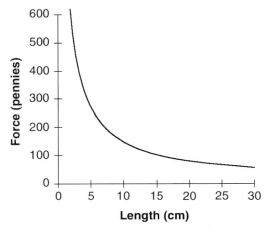

10. Would a linear function make sense to describe the relationship between force and length for the pasta? Explain.

10. No, the graph should be curved.

11. Would a quadratic function make sense to describe the relationship between force and length for the pasta? Explain.

11. No, the graph of a quadratic function would curve upward for greater values of length.

Extend the Activity *(SE page 116)*

12. By holding down the ends of the pasta, you simulated a beam that is anchored at its ends. Another type of beam is called "simply supported" because its ends merely rest on their supports without being anchored. Try repeating Activity 6.1 without holding down the ends of the pasta and see if your results change. Or use a "cantilevered" arrangement, in which the pasta sticks out horizontally and is anchored at one end only.

12. **For simple supports at the ends, the pasta will bend more and will usually support a larger load without breaking. The cantilevered configuration will require attaching loads to the end of the beam and should result in lowered breaking strength.**

13. Consider your graph in item 9 of Activity 6.1. Could the graph ever touch the vertical axis? What characteristic of a mathematical function could account for this behavior?

13. **No, it couldn't. The amount of force required to break a beam of length 0 makes no sense. The function would be undefined when $L = 0$.**

14. Describe how the steepness of the graph in item 9 changes as the length of the pasta beam increases.

14. **The graph is very steep for small length values, but it gets less and less steep as length increases.**

Section 6.2 Modeling with Rational Functions

What You Need to Know

- How to use a table of values to construct the graph of a function
- How to find the problem domain for an applied function

What You Will Learn

- To examine real-world situations that can be modeled with rational functions
- To evaluate a rational function for given values of the independent variable
- To identify asymptotes on the graph of a rational function

Materials

- Spaghetti
- Small kitchen scale
- Ruler

In Chapter 5, we examined functions whose domains were restricted because only certain input values made sense in the context of the problem. For example, in Example 2 of Section 5.2, we examined the elevation of a road passing over a hill. The mathematical domain of the function $y = 0.082x - 0.00025x^2$ consisted of all real numbers. But because the ends of the curved portion of the road were at $x = 0$ feet and $x = 328$ feet, the problem domain included only values from 0 to 328 inclusive.

In Discovery 6.1, we will examine a function that has a restriction on its domain that provides meaningful insight into a problem situation.

Discovery 6.1 A Model for Beam Strength (SE pages 118–119)

In Activity 6.1, you found that the vertical force required to break a horizontal beam decreases with the length of the beam in a way that is not explained adequately by either a linear or a quadratic function. What is needed is a function that increases in value as input values approach zero. The function must also decrease in value as the input values get larger and larger. But even though the function decreases, the value of the function must remain positive no matter how large the input value becomes.

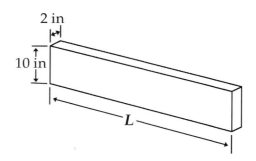

Figure 6.3

Consider the wooden beam shown in **Figure 6.3.** This type of beam is used to support upper floors of a house, or a first floor over a full basement. When used in this way, it is called a *joist*. A typical joist is made of 2" x 10" lumber, with a variable length L that depends on the length of the floor it supports. The vertical force that is required to break such a beam at its center can be measured directly with a testing machine like the one shown in Figure 6.1. For 2" x 10" beams made of yellow pine, it has been found that breaking strength can be modeled by the formula $F = \dfrac{40,000}{L}$, where L is measured in feet and F in pounds.

1. Use the function $F = \dfrac{40,000}{L}$ to complete **Table 6.2** and show the force required to break beams of various lengths.

Length of Beam (ft)	Breaking Force (lb)
0	
1	
2	
5	
10	
20	
100	

Table 6.2

1.

Length of Beam (ft)	Breaking Force (lb)
0	undefined
1	40,000
2	20,000
5	8000
10	4000
20	2000
100	400

2. Recall the characteristics that you considered when building the model for the force required to break pasta in items 5–8 of Activity 6.1. Use those characteristics and the information from your table in item 1 of this Discovery to explain why the function $F = \dfrac{40,000}{L}$ is a reasonable model for the force required to break a beam.

2. The force F increases as L decreases, but L cannot equal 0. F increases as L increases, but F cannot be reduced to 0 for any value of L. The function is a good model for explaining the observed properties.

The function $F = \dfrac{40,000}{L}$ is an example of a **rational function.** A rational function is a function that can be written as a ratio of two polynomial functions.

3. Rational functions frequently exhibit the kind of behavior you observed in Table 6.2. That is, they may be undefined for some value or values of the independent variable. Explain what would cause a rational function to be undefined.

3. The function will be undefined whenever a denominator is equal to zero.

The value or values of the independent variable that cause a function to be undefined are called **excluded values** because they are values that must be excluded from the domain of the function.

4. What is the excluded value for the function $F = \dfrac{40,000}{L}$?

4. $L = 0$.

5. For this particular situation, how is this excluded value related to the problem domain?

5. The excluded value $L = 0$ is the lower limit of the problem domain.

6. Using a graphing calculator or computer graphing program, construct a graph of the function $F = \dfrac{40,000}{L}$. Use a window that includes beam lengths from 0 to 20 feet and forces from 0 to 40,000 pounds.

6.

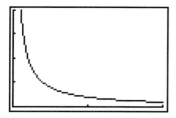

[0, 20] x [0, 40,000]

7. What are the coordinates of the leftmost point on the graph that appears in this window?

7. **Sample answer: (1.0638298, 37,600)**

8. If you double the vertical maximum of the graph window, what are the coordinates of the leftmost point on the graph that appears in the new window?

8. **Sample answer: (0.63829787, 62,666.667)**

9. What will happen to the coordinates of the leftmost point displayed on the graph if the vertical maximum becomes increasingly larger and larger?

9. **The x-coordinate will become closer and closer to zero, and the y-coordinate will increase.**

Vertical Asymptotes

As we have seen in Discovery 6.1, the graph of the function $F = \dfrac{40,000}{L}$ gets closer and closer to the vertical axis as the length of the beam gets close to zero, but it can never touch the vertical axis. This type of behavior is said to be **asymptotic,** and the line that the graph approaches is called an **asymptote.** In this case, the vertical axis is called a **vertical asymptote.**

A vertical asymptote may occur on the graph of a rational function when a value of the independent variable is excluded from the domain of the function (see Exercise 6). For a function that models a real-world situation, a vertical asymptote often coincides with a boundary of the problem domain.

Example 1

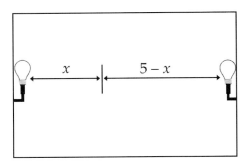

Figure 6.4

The amount of light reaching an object from a light source is called *illumination*. It depends on the intensity of the light source and the distance from the light source to the object. **Figure 6.4** shows two 100-watt light bulbs placed at opposite ends of a 5-meter-long room. The illumination (in *lux*) at any point that is a distance x from the bulb on the left along a line joining the bulbs is given by the function $E = \dfrac{125}{x^2} + \dfrac{125}{(5-x)^2}$.

a) Find the illumination when $x = 2$ m and when $x = 0.25$ m.

b) At what points is the illumination undefined? How are these points related to the problem domain?

c) Use a table of values to construct a graph of the function. Identify all vertical asymptotes.

Solution:

a) When x is 2 m, E is $\dfrac{125}{2^2} + \dfrac{125}{(5-2)^2} = 45$ lux. When x is 0.25 m, E is $\dfrac{125}{(0.25)^2} + \dfrac{125}{(5-0.25)^2} = 2006$ lux.

b) E is undefined for $x = 0$ meters and for $x = 5$ meters because each number makes one of the denominators equal zero. These are the locations of the light bulbs. The problem domain includes only values of x between 0 and 5.

c) See **Table 6.3** and **Figure 6.5**.

x (meters)	E (lux)
0.25	2006
0.5	506
1	133
2	45
3	45
4	133
4.5	506
4.75	2006

Table 6.3

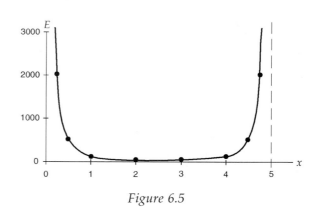

Figure 6.5

The vertical asymptotes are at $x = 0$ and $x = 5$. The dashed line indicates the location of the $x = 5$ asymptote.

Example 2

The function $P = \dfrac{400}{R + 100}$ represents the power P (in watts) used by an electrical circuit that contains a variable resistor R (in ohms) in series with another resistor of 100 ohms. Power is measured by nonnegative numbers.

a) What is the power used by the circuit when R is 0, 50, and 100 ohms?

b) For what value of R is the power undefined? How is this value of R related to the problem domain?

c) Use a graphing calculator or computer to graph the function. Choose a window that includes the values from (a) and (b). Identify any vertical asymptotes.

Solution:

a) For $R = 0$, P is $\dfrac{400}{0 + 100} = 4$ watts.

For $R = 50$, P is $\dfrac{400}{50 + 100} = \dfrac{8}{3}$, or about 2.7 watts.

For $R = 100$, P is $\dfrac{400}{100 + 100} = 2$ watts.

b) P is undefined for $R = -100$ because this value makes the denominator equal zero. But the problem domain doesn't include negative numbers, so $R = -100$ is not part of the problem domain.

c) See **Figure 6.6**.

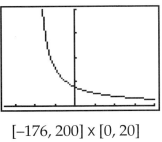

[−176, 200] × [0, 20]

Figure 6.6

There is a vertical asymptote at $R = -100$.

> The window of [−176, 200] × [0, 20] was chosen to prevent a vertical line being drawn at $R = -100$. Without a careful choice of window, the calculator will incorrectly connect points across a vertical asymptote. The connection will appear as a vertical or nearly vertical line. The error can be avoided by using "dot" mode instead of "connected" mode, but then the smoothness of the graph is lost (see Exercise 4).

Asymptotes are not always vertical lines. Depending on the function, asymptotes can be vertical, horizontal, or oblique. In Discovery 6.2, we will examine a function that has a **horizontal asymptote**.

Discovery 6.2 Alcohol Elimination and Limiting Values *(SE pp. 122–123)*

Our bodies eliminate chemicals that we ingest in ways that can be modeled mathematically. For some chemicals, like caffeine, the fraction eliminated each hour is independent of the amount of caffeine in the blood. But for alcohol, the fraction eliminated depends on how much alcohol is present in the blood.

For a typical male adult, the actual amount E of alcohol that is eliminated from the blood in one hour is approximated by the function $E = \dfrac{10a}{a+4}$, where a is the initial amount of alcohol in the blood (in grams). For example, for an alcohol content of 16 grams, the amount eliminated in one hour is $E = \dfrac{10(16)}{16+4} = 8$ grams. So half the alcohol is eliminated in one hour.

1. If a male's blood contains 12 grams of alcohol, find the amount and the fraction of alcohol eliminated in one hour.

1. **The amount eliminated is** $E = \dfrac{10(12)}{12+4} = 7.5$ **g; the fraction is** $\dfrac{7.5}{12} = 0.625$, **or** $\dfrac{5}{8}$ **of the initial amount.**

2. What is the problem domain for the alcohol elimination function?

2. **The lowest value of a is 0. There is no obvious upper limit except for the fact that consumption of very high levels of alcohol will result in unconsciousness or death, which would put a practical limit on the amount.**

3. Complete **Table 6.4** by evaluating the function $E = \dfrac{10a}{a+4}$.

a (grams)	E
0	
10	
20	
30	
40	
50	

Table 6.4

3.

a (grams)	E
0	**0**
10	**7.14**
20	**8.33**
30	**8.82**
40	**9.09**
50	**9.26**

4. Use a graphing calculator or computer graphing program to graph the function for initial alcohol amounts up to 50 grams. What is the maximum amount eliminated, according to this graph?

4.

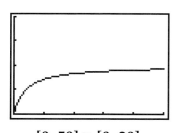

[0, 50] x [0, 20]

About 9.26 grams are eliminated when *a* **= 50 g.**

5. Change the window of your graph so it shows *a* values of up to 100 g. What is the maximum amount eliminated according to this graph?

5.

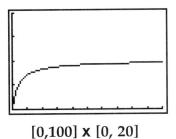

[0,100] x [0, 20]

About 9.62 grams are eliminated when *a* = 100 g.

6. Continue to increase the maximum *a* value shown on the graph. What can you conclude about the maximum amount of alcohol that can be eliminated by the body in one hour?

6. As *a* increases, the value of *E* approaches 10 but never reaches it. Ten grams is the limiting value for the amount of alcohol that can be eliminated from a typical male adult's body in one hour, no matter how much is ingested.

Horizontal Asymptotes

In Discovery 6.2, we saw that the value of the function approached a limiting value as the amount of alcohol increased. This function exhibits asymptotic behavior. Its graph approaches a horizontal line as the amount of alcohol becomes increasingly large.

Whereas a vertical asymptote results when a function is undefined for some value of the independent variable, a horizontal asymptote is an indication of an unreachable value of the dependent variable as the absolute value of the independent variable becomes increasingly large.

Recall that as you move to the left on a number line, numbers become smaller. For example, $7 < 10$ and $-5 < -3$. But if you examine the absolute values of the numbers less than 0, as you move to the left, the absolute values of the numbers increase.

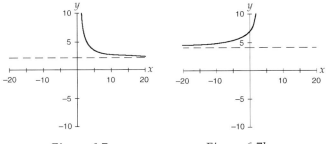

Figure 6.7a   *Figure 6.7b*

For the function in **Figure 6.7a,** as the absolute value of *x* increases through positive values of *x*, the graph approaches the horizontal asymptote $y = 2$.

In **Figure 6.7b,** as the absolute value of *x* increases through negative values of *x*, the graph approaches the horizontal asymptote $y = 4$.

Example 3

When a business orders raw materials, the total cost is often a multiple of the cost of a single unit (called a *raw material unit*, or RMU) plus a fixed cost for shipping.

A photography supply store orders frames at $12 each from a supplier that charges a fixed price of $135 for shipping. The total cost (in dollars) of n frames is $12n + 135$. The *average* cost C of n frames is $C = \dfrac{12n + 135}{n}$.

a) Find the total cost and the average cost of 10 frames.

b) Construct a graph that shows the average cost of ordering any number of frames between 5 and 50.

c) What does the shape of the graph indicate about average frame costs?

d) Extend the scale of the graph to include the average cost of up to 500 frames.

e) Identify the horizontal asymptote and interpret it in the context of ordering frames.

Solution:

a) The total cost of 10 frames is $12(10) + 135 = \$255$. The average cost is $\dfrac{12(10) + 135}{10} = \$25.50.$

b) See **Figure 6.8.**

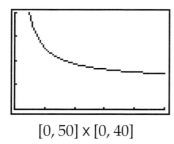

[0, 50] × [0, 40]

Figure 6.8

c) As the number of frames ordered increases, the average cost of a frame decreases.

d) See **Figure 6.9.**

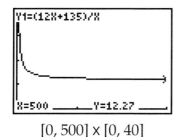

[0, 500] × [0, 40]

Figure 6.9

e) The horizontal asymptote is $n = 12$. As the order size increases, the average cost of a frame approaches $12. The average cost can never equal $12 due to the shipping cost, but for very large orders, the shipping cost is spread out over a large number of frames. The average shipping cost per frame therefore becomes very small.

Example 4

Use a graphing calculator or computer to graph the rational function
$y = \dfrac{5x^2}{(x+1)(x-2)}$. Identify the vertical and horizontal asymptotes.

Solution:

The function is undefined for $x = -1$ or 2, so the lines $x = -1$ and $x = 2$ are vertical asymptotes (see **Figure 6.10**).

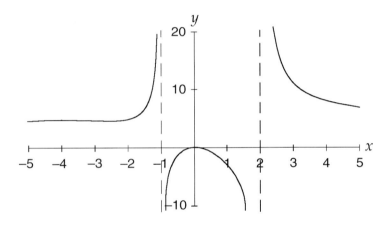

Figure 6.10

By continuing to enlarge the scale of the graph, horizontal asymptotes become evident (see **Figure 6.11**). On this scale, the details of the graph may be lost, but the asymptotic behavior can be seen as the absolute value of x becomes larger. There is a horizontal asymptote at $y = 5$.

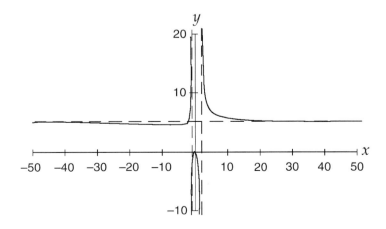

Figure 6.11

Exercises 6.2

I. Investigations *(SE pages 126–130)*

1. Some medications work best when the amount of medicine in the body is maintained at a constant level, also called the equilibrium level. The relationship between the equilibrium level and the dosage depends on a number of factors that include the rate at which the body eliminates the medication. If this relationship is known, it may be possible to determine what equilibrium level will result for a given dosage.

    The equilibrium level L for a particular medication is related to the dose d (given at regular intervals) by the function $L = \dfrac{10d}{15-d}$. Both d and L are measured in milligrams.

    a) What equilibrium levels will be reached for doses of 0, 6, or 10 mg?

1. **a)** $\dfrac{10(0)}{15-0} = 0$ **mg,** $\dfrac{10(6)}{15-6} = 6\dfrac{2}{3}$ **mg,** $\dfrac{10(10)}{15-10} = 20$ **mg.**

    b) For what doses is the function undefined?

    b) 15 mg, because d = 15 is an excluded value.

    c) As the dose increases toward 15 mg, what happens to the equilibrium value?

    c) It increases without limit.

    d) Graph the function for doses from 0 to 30 mg, and identify any vertical asymptotes. Explain the connection between the graph and your answer to (c). Interpret the result for this context.

    d)

    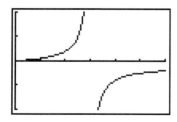

    [0, 30] x [–100, 100]

    A vertical asymptote occurs at the excluded value d = 15. Doses of 15 mg or greater cannot result in equilibrium levels for the medication. The body would not be able to eliminate the medication fast enough to allow an equilibrium level to be maintained.

    e) What is the problem domain?

    e) d is greater than or equal to 0 and less than 15.

2. A backyard gardener wants to fence off a rectangular area for a vegetable garden with a total area of 600 ft². The number of linear feet of fencing needed is equal to the perimeter of the garden, which depends on its dimensions. Recall that the perimeter P of a rectangle is given by the formula $P = 2L + 2W$, where L and W are the length and width of the rectangle.

a) Complete a table similar to **Table 6.5** that lists some possible dimensions for the garden that will result in an area of 600 ft². Also list the perimeter of the garden.

Length L (ft)	Width W (ft)	Perimeter P (ft)
60		
50		
40		
30		

Table 6.5

2. a)

Length L (ft)	Width W (ft)	Perimeter P (ft)
60	10	140
50	12	124
40	15	110
30	20	100

b) How did you determine the values for the width? Write a formula that gives the width W of the 600 ft² garden for any length L.

b) The width equals the area divided by the length, or $W = \dfrac{600}{L}$.

c) Substitute your formula for W into the perimeter formula to find perimeter as a function of length only.

c) $P = 2L + 2\left(\dfrac{600}{L}\right)$ or $P = 2L + \dfrac{1200}{L}$.

d) Where on the graph of this function would you expect to find a vertical asymptote?

d) A vertical asymptote will occur at $L = 0$.

e) Interpret the meaning of the vertical asymptote in this context.

e) The length of a rectangle cannot be 0, but it can be very close to 0. (Actually, this dimension would be called the width if it were the smaller dimension.)

f) What is the domain of the function?

f) L can be any number but 0.

g) What is the problem domain?

g) *L* must be greater than 0.

h) Graph the perimeter function over the problem domain. Determine the minimum length of fence needed for the vegetable garden. What dimensions should the garden have in order to minimize the perimeter?

h)

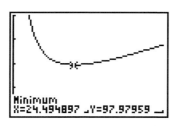

[0, 60] × [0, 200]

The minimum perimeter is about 98 feet, for a square garden 24.5 feet on each side.

3. In Discovery 6.1, we examined the function $F = \dfrac{40{,}000}{L}$ that gives breaking strength as a function of beam length. The value of 40,000 in the numerator of the rational expression is only valid for a 2" x 10" beam made out of yellow pine. In general, we can write $F = \dfrac{k}{L}$ to express the relationship between breaking strength and length for all similar beams. Here *k* represents a constant that depends on the cross-sectional dimensions of the beam and the type of wood from which it is made.

There are many pairs of quantities that are related to each other in the same way as breaking strength and beam length. For example, the pressure *P* of a gas in a container is related to the volume *V* of the gas by the function $P = \dfrac{k}{V}$, and the electrical resistance *R* of a wire is related to its cross-sectional area *A* by $R = \dfrac{k}{A}$. The type of relationship exhibited by these functions is called **inverse variation**. In general, if a variable *y* is related to *x* by the function $y = \dfrac{k}{x}$, then *y* is said to **vary inversely** with *x*.

a) Use a window of [–4.7, 4.7] × [–3.1, 3.1] to graph the function $y = \dfrac{1}{x}$.

3. a)

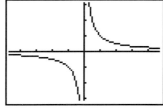

[–4.7. 4.7] × [–3.1, 3.1]

b) On the same screen, graph the function $y = \dfrac{k}{x}$, using $k = 2, 3$, and 4. Describe what you see.

b)

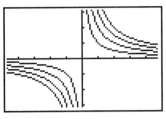

[–4.7, 4.7] x [–3.1, 3.1]

All four graphs occur only in quadrants 1 and 3. They have the same general shape, and as k increases, the graph moves further from the origin.

c) Graph the function $y = \dfrac{k}{x}$, using $k = -1, -2, -3$, and -4. Describe what you see.

c) Sample graph:

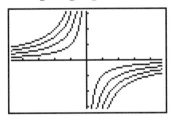

[–4.7, 4.7] x [–3.1, 3.1]

The graphs occur only in quadrants 2 and 4. They have the same general shape, which is the same shape as the graphs for positive k values reflected across the y-axis. As the absolute value of k increases, the graph moves further from the origin.

d) Repeat (a) and (b) using a window of [0, 4.7] x [0, 3.1]. Describe what you see with the viewing area restricted to quadrant 1.

d)

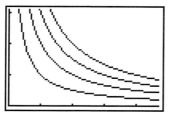

[0, 4.7] x [0, 3.1]

On each graph, as x increases y decreases, and as x decreases y increases.

4. In Example 4, we saw that rational functions can be analyzed graphically to determine their behavior for large values of the independent variable. Asymptotes can often be identified in this way. But although graphing calculators are convenient for much graphical analysis, they can sometimes provide misleading information when used to graph rational functions.

a) Figure 6.10 shows a graph of the function $y = \dfrac{5x^2}{(x+1)(x-2)}$, drawn using a computer program. Graph this function on a graphing calculator in connected mode, using a window of [–5, 5] x [–10, 20]. Compare the calculator graph to the graph in Figure 6.10.

4. a)

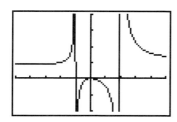

[–5, 5] x [10, 20]

The calculator graph shows two lines that are not present in Figure 6.10.

b) Now graph the function using the same window, but put the calculator in dot mode.

b)

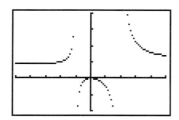

[–5, 5] x [–10, 20]

c) In order to understand the behavior of the calculator in graphing this function, it will be helpful to trace the graph near one of the asymptotes. We know that there is an asymptote for $x = -1$ because the function is undefined for $x = -1$. Trace the graph as x increases from -2 to 0, and describe what you see.

c) The value of the function increases as x nears -1 but then jumps to a negative value as the x-coordinate moves from -1.06383 to -0.9574468.

For the window $[-5, 5] \times [-10, 20]$, the excluded value $x = -1$ does not appear on a pixel of the calculator screen. In connected mode, the calculator joins the points $(-1.06383, 28.935185)$ and $(-0.9574468, -36.42086)$ with a line as if the graph were actually connected across those points. However, we know that there cannot be a connection here, because the function is undefined at $x = -1$. The extra lines drawn by the calculator when in connected mode are a result of the limited resolution of the calculator screen.

d) Again, graph the same function in connected mode as in (a), but this time use a window of $[-4.7, 4.7] \times [-10, 20]$. Trace the graph as you did in (c), and explain why the appearance of the graph differs from your result in (a).

d)

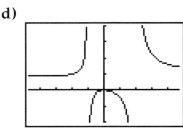

$[-4.7, 4.7] \times [-10, 20]$

The value $x = -1$ is now located on one of the screen pixels. The calculator is forced to attempt to evaluate the function for $x = -1$ and cannot do so. Therefore, it recognizes that there is a gap in the graph for $x = -1$ and does not try to connect the points to the immediate left and right of $x = -1$.

e) Now use the same window, $[-50, 50] \times [-10, 20]$, as shown in Figure 6.11. Recall that the purpose of that graph was to help identify the horizontal asymptote. Describe what you see.

e)

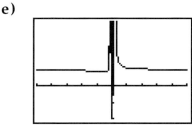

$[-50, 50] \times [-10, 20]$

By tracing to the right and left edges of the screen, the graph can be seen to approach the line $y = 5$. This suggests that there is a horizontal asymptote at $y = 5$. But the details of the graph near the vertical asymptotes are lost.

5. The graph of the function $P = 2L + \dfrac{1200}{L}$ from Exercise 2 contains a vertical asymptote for $L = 0$. But there is also another asymptote that can be found by examining the graph of the function using a large window.

    a) Use the window [−200, 200] × [−1000, 1000] to graph the function
$P = 2L + \dfrac{1200}{L}$.

5. a)

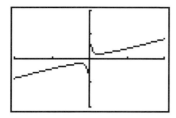

[−200, 200] x [−1000, 1000]

At this scale, you can see the graph approach a straight-line asymptote that is neither vertical nor horizontal. This kind of asymptote is called an **oblique asymptote.**

    b) Assuming that the oblique asymptote passes through the origin, identify a point near the right-hand edge of the calculator screen and approximate its slope.

    b) One point on the graph has approximate coordinates (195.7, 397.6), and the asymptote would pass very close to this point. Therefore, an approximate slope for the asymptote is $\dfrac{397.6 - 0}{195.7 - 0} \approx 2.$

    c) Write an equation for the oblique asymptote, and graph it along with the graph from (a).

    c) $P = 2L.$

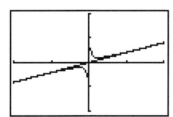

[−200, 200] x [−1000, 1000]

d) The length of a belt joining two pulleys of radii R and r that are a distance D apart (**Figure 6.12**) can be approximated using the formula

$L = \pi(R + r) + 2D + \dfrac{2(R - r)^2}{D}$. If R and r equal 8 cm and 5 cm, respectively, write a formula giving L as a function of D.

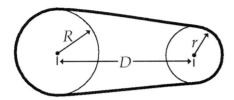

Figure 6.12

d) $L = 13\pi + 2D + \dfrac{18}{D}$.

e) Identify the domain of the function and also the problem domain.

e) The domain of the function includes all real numbers except 0. The minimum value of D occurs where the pulleys touch, so D must be greater than (8 cm + 5 cm) = 13 cm.

f) Use a graphing calculator to discuss the asymptotic behavior of the graph of the function. How many vertical, horizontal, and oblique asymptotes are found on the graph?

f)

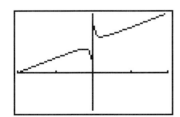

[–20, 20] x [–50, 80]

There is one vertical asymptote at $D = 0$. There are no horizontal asymptotes. There is one oblique asymptote.

6. Consider the rational function $y = \dfrac{2x^2 - 2x}{x - 1}$.

a) Use the equation of the function to complete **Table 6.6.**

x	y
−4	
−3	
−2	
−1	
0	
0.5	
0.9	
0.99	
1	
1.01	
1.1	
1.5	
2	
3	
4	

Table 6.6

6. a)

x	y
−4	−8
−3	−6
−2	−4
−1	−2
0	0
0.5	1
0.9	1.8
0.99	1.98
1	undefined
1.01	2.02
1.1	2.2
1.5	3
2	4
3	6
4	8

b) From your table, would you expect the graph of the function to have a vertical asymptote?

b) No, the *y*-values should increase or decrease by much larger amounts near a vertical asymptote.

c) Use a graphing calculator to graph the function $y = \dfrac{2x^2 - 2x}{x - 1}$. Take care to use a window that locates the excluded value exactly on a pixel on the graph screen. (A window of [−4.7, 4.7] × [−3.1, 3.1] will suffice on many calculators.) Do you see any asymptotic behavior?

c)

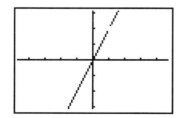

[−4.7, 4.7] × [−3.1, 3.1]

There is no vertical asymptote. The graph is a straight line, but there is a gap in the graph where *x* = 1, at a height corresponding to *y* = 2.

d) Factor the numerator of $y = \dfrac{2x^2 - 2x}{x-1}$ and then compare the numerator and denominator.

d) $\dfrac{2x^2 - 2x}{x-1} = \dfrac{2x(x-1)}{x-1}$. **The numerator and denominator contain a common factor of $(x-1)$.**

When a rational function has factors common to the numerator and denominator, the excluded value or values related to the common factors may not result in asymptotic behavior. Instead, the graph of the function has a "hole" in it and just appears to be missing a single point. If there are no common factors in the numerator and denominator, an excluded value always results in a vertical asymptote.

II. Projects and Group Activities *(SE pages 131–132)*

7. Materials: spaghetti, small kitchen scale, ruler

   Columns of various kinds are built to support various kinds of loads. Thick pillars help support buildings and bridges, whereas thinner columns suspend flags, wires and cables, and lights. If the vertical "point" load P on a thin column is large enough, it can cause the column to "buckle," or bend sideways (see **Figure 6.13**). If such a load is not removed, the column will fail.

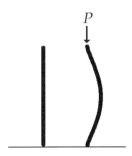

Figure 6.13

   a) Measure the length L of a strand of spaghetti. Position it vertically using one finger on a small scale that is sensitive enough to detect weights as small as 20 grams, or about an ounce. If you increase the downward force slightly, the spaghetti will buckle; that is, it will bend slightly to the side. Record the force required to buckle the spaghetti.

7. **a) It takes about 20 g of force to buckle a 25 cm strand of regular spaghetti.**

b) Complete a table like **Table 6.7** by snapping off shorter pieces of spaghetti and observing the forces required to buckle them.

Length L (cm)	Force P (g)
25	
20	
15	
10	
8	
6	

Table 6.7

b)

Length L (cm)	Force P (g)
25	20
20	40
15	90
10	250
8	350
6	550

c) Make a scatter plot of the data.

c)

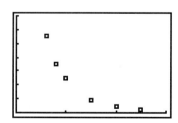

[0, 30] x [0, 700]

d) Consider the shape of the graph as well as the buckling loads you would expect for extremely short and extremely long strands of spaghetti. Would a rational function be an appropriate model for the relationship between length and buckling force? Include a discussion of asymptotes.

d) A very long strand should buckle with very little force, which would be consistent with a horizontal asymptote of $L = 0$. Shorter and shorter strands should require increasing amounts of force, which would be consistent with a vertical asymptote of $P = 0$. A rational function should also have a curved shape and is an appropriate model for the data and graph.

e) One possible model for the buckling data is similar to the one for breaking of horizontal beams, as discussed in Discovery 6.1; that is, $P = \dfrac{k}{L}$, where k is some constant. To investigate this inverse variation model (see Exercise 3), add a third column to Table 6.7 labeled "k for $P = \dfrac{k}{L}$." Then calculate the value of k that would fit this model for each pair of values in the table. For example, if a 25 cm strand buckles with a load of 20 g, then the relation $20 = \dfrac{k}{25}$ would give a value of $k = 500$.

e) Sample table:

Length L (cm)	Force P (g)	k for $P = \dfrac{k}{L}$
25	20	500
20	40	800
15	90	1350
10	250	2500
8	350	2800
6	550	3300

f) To see whether the $P = \dfrac{k}{L}$ model is a good one, find the mean of the k-values in your table and graph the resulting function along with your scatter plot. Is $P = \dfrac{k}{L}$ a good model for the data?

f) The mean value of k is 1875, so the model is $P = \dfrac{1875}{L}$.

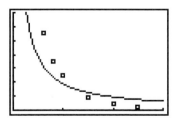

[0, 30] x [0, 700]

The model is not a good one. For shorter lengths, the graph underestimates the buckling load, and for longer lengths it overestimates.

g) Add a fourth column to your table labeled "k for $P = \dfrac{k}{L^2}$" and test this "inverse square" model in a similar manner to the way you evaluated the simple inverse variation model in (e) and (f).

g) Sample table:

Length L (cm)	Force P (g)	k for $P = \dfrac{k}{L}$	k for $P = \dfrac{k}{L^2}$
25	20	500	12,500
20	40	800	16,000
15	90	1350	20,250
10	250	2500	25,000
8	350	2800	22,400
6	550	3300	19,800

The mean value of k is 19,325, so the model is $P = \dfrac{19,325}{L^2}$.

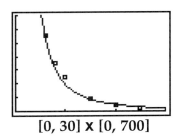

[0, 30] x [0, 700]

The function models the data fairly well.

III. Additional Practice (SE pages 132–136)

8. The rotational speed S (in rpm) of a grinder that is necessary for tool grinding at a surface speed of 5500 feet per minute depends on the diameter D (in inches) of the grinding wheel, according to the function $S = \dfrac{5500}{0.2618D}$. Find the required rotational speeds for grinding wheels with diameters of (a) 3 inches, (b) 5 inches, and (c) 10 inches.

8. a) 7003 rpm

   b) 4202 rpm

   c) 2101 rpm

9. The time t (in hours) required for a particular combine to cut 15 acres of grain depends on the traveling speed v (in miles per hour) of the combine, according to the formula $t = \dfrac{10.3}{v}$. Find the time required to cut 15 acres of grain at speeds of (a) 2.0 mph and (b) 3.5 mph.

9. **a) 5.15 hr**

    b) 2.9 hr

10. The electrical reactance X (in ohms) of a capacitor is related to its capacitance C (in farads) and the frequency f (in hertz) of the circuit according to the function $X = \dfrac{1}{2\pi f C}$. Find the electrical reactance of a 200 microfarad (200×10^{-6} farad) capacitor for frequencies of (a) 100 hertz, and (b) 500 hertz.

10. **a) 7.96 ohms**

    b) 1.59 ohms

11. The low E string on a particular acoustic guitar has a length of 63 cm and produces a musical tone (a "low E") with a frequency of about 82 hertz when plucked. The frequency (and therefore the pitch of the tone) produced by the string can be changed by shortening the effective length L, according to the function $f = \dfrac{5170}{L}$.

    a) What is the problem domain for this function?

11. **a) L is greater than 0 and less than or equal to 63: $0 < L \le 63$.**

b) Complete a table similar to **Table 6.8** showing the frequencies *f* of tones produced by the low E string for various effective lengths *L*. Use about six evenly spaced values for *L* in your table.

String Effective Length *L* (cm)	Frequency of Tone *f* (hertz)
63	82

Table 6.8

b) Sample table:

String Effective Length *L* (cm)	Frequency of Tone *f* (hertz)
63	82
54	96
45	115
36	144
27	191
18	287
9	574

c) Use your table from (b) to construct a graph of $f = \dfrac{5170}{L}$.

c) Sample graph:

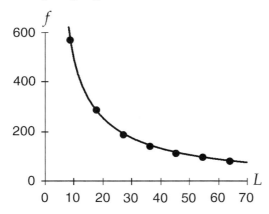

12. Consider the function $y = \dfrac{x}{x+2}$.

   a) Complete **Table 6.9** by evaluating the function for the indicated values of x.

x	y
−5	
−4	
−3	
−2.5	
−2	
−1.5	
−1	
0	
1	
2	
3	

Table 6.9

12. a)

x	y
−5	**5/3**
−4	**2**
−3	**3**
−2.5	**5**
−2	**undefined**
−1.5	**−3**
−1	**−1**
0	**0**
1	**1/3**
2	**1/2**
3	**3/5**

b) Use the table in (a) to construct a graph of the function.

b)

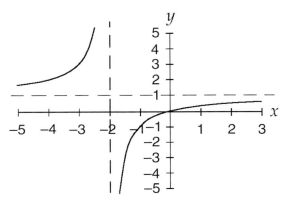

13. Consider the function $y = \dfrac{4}{(x-1)^2}$.

   a) Complete **Table 6.10** by evaluating the function for the indicated values of x.

x	y
−4	
−3	
−2	
−1	
0	
0.5	
1	
1.5	
2	
3	
4	

Table 6.10

a)

x	y
−4	$\dfrac{4}{25} = 0.16$
−3	$\dfrac{1}{4} = 0.25$
−2	$\dfrac{4}{11} \approx 0.44$
−1	1
0	4
0.5	16
1	**undefined**
1.5	16
2	4
3	1
4	$\dfrac{4}{11} \approx 0.44$

b) Use the table in (a) to construct a graph of the function.

b)

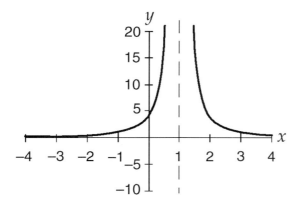

14. Consider the function $y = 3 - \dfrac{2}{x^2 + 1}$.

   a) Complete **Table 6.11** by evaluating the function for the indicated values of x.

x	y
–5	
–4	
–3	
–2	
–1	
–0.5	
0	
0.5	
1	
2	
3	
4	
5	

Table 6.11

a)

x	y
–5	$\dfrac{38}{13} \approx 2.92$
–4	$\dfrac{49}{17} \approx 2.88$
–3	$\dfrac{14}{5} = 2.8$
–2	$\dfrac{13}{5} = 2.6$
–1	2
–0.5	$\dfrac{7}{5} = 1.4$
0	1

x	y
0.5	$\dfrac{7}{5} = 1.4$
1	2
2	$\dfrac{13}{5} = 2.6$
3	$\dfrac{14}{5} = 2.8$
4	$\dfrac{49}{17} \approx 2.88$
5	$\dfrac{38}{13} \approx 2.92$

b) Use the table in (a) to construct a graph of the function.

b)

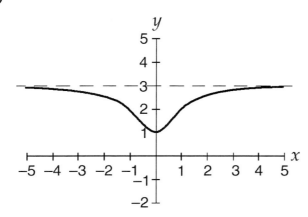

For 15–21, identify excluded values for the independent variable in each given function.

15. $y = \dfrac{x^2 - 1}{x + 6}$.

15. –6

16. $P = \dfrac{1700}{h(h - 2)}$.

16. 0 and 2

17. $L = 32t - \dfrac{15}{t^2}$.

17. 0

18. $y = \dfrac{x}{x^2 + 3x - 10}$.

18. –5 and 3

19. $F = \dfrac{r^2}{r - 4}$.

19. 4

20. $w = \dfrac{1}{d^2 + 5d}$.

20. –5 and 0

21. $s = \dfrac{3}{v + 2} + \dfrac{5}{v - 2}$.

21. –2 and 2

For 22–27, (a) identify vertical asymptotes on the graph of the function by inspecting the function, and (b) identify horizontal asymptotes by examining the graph with a graphing calculator.

22. $y = \dfrac{x}{x - 3}$.

22. a) $x = 3$ is a vertical asymptote.

   b) Sample graph:

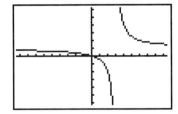

   [–9.4, 9.4] x [–6.2, 6.2]

   $x = 1$ is a horizontal asymptote.

23. $y = \dfrac{x + 5}{(x + 1)(2x - 1)}$.

23. a) $x = -1$ and $x = \dfrac{1}{2}$ are vertical asymptotes.

   b) Sample graph:

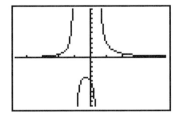

   [–4.7, 4.7] x [–10, 10]

   $x = 0$ is a horizontal asymptote.

24. $y = \dfrac{x-1}{x^2+4}$.

24. a) There are no vertical asymptotes.

   b) Sample graph:

   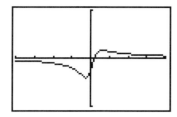

   [–20, 20] **x** [–1, 1]

   $x = 0$ is a horizontal asymptote.

25. $y = \dfrac{2x^2+3}{1-x^2}$.

25. a) $x = 1$ and $x = -1$ are vertical asymptotes.

   b) Sample graph:

   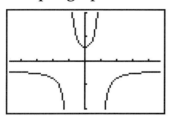

   [–4.7, 4.7] **x** [–10, 10]

   $x = -2$ is a horizontal asymptote.

26. $E = \dfrac{25a}{a+7}$.

26. a) $a = -7$ is a vertical asymptote.

   b) Sample graph:

   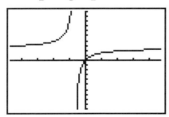

   [–47, 47] **x** [–100, 100]

   $E = 25$ is a horizontal asymptote.

27. $P = \dfrac{500R}{10R + 75}.$

27. a) $R = -7.5$ is a vertical asymptote.

b) Sample graph:

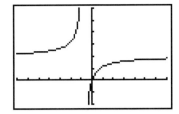

$[-70.5, 70.5] \times [-50, 150]$

$R = 50$ is a horizontal asymptote.

Section 6.3 Multiplying and Dividing Rational Expressions

What You Need to Know

- How to evaluate algebraic expressions

- How to reduce fractions to lowest terms

- How to factor quadratic expressions

- How to find the volumes and surface areas of spheres and rectangular solids

What You Will Learn

- To reduce rational expressions to lowest terms

- To multiply and divide rational expressions

- To find values for which a rational expression is undefined

- To use rational expressions to model real-world situations

Materials

- None

Packaging plays an important part in each of our lives. It is difficult to imagine life without it. Boxes and bags are used to hold everything from cereal to electronic equipment to baby diapers. Containers shaped like cylinders are used to store such things as soft drinks, medications, and motor oil.

Because packages are three-dimensional objects with two-dimensional sides, both volume and area play an important part in packaging efficiency. The design of efficient packages requires a knowledge of geometry and algebra. Discovery 6.3 uses algebraic expressions to describe the efficiency of package design.

Discovery 6.3 Using Expressions to Compare Surface Area and Volume (SE pages 137–139)

The surface area of a container helps determine the amount of material needed to make the container, and the volume of the container helps determine how much the container holds. One way to explore the efficiency of the container is to examine the ratio of its surface area to its volume.

When comparing two containers of equal volume, the one with the smaller ratio of surface area to volume would require less material to construct and could be considered more efficient. This, in turn, might mean that it is less costly to manufacture and possibly cheaper to ship.

1. Consider the container in **Figure 6.14.** Find the surface area of the container, its volume, and the ratio of its surface area to its volume. Recall that the surface area of a geometric solid is the sum of all the areas of the faces of the object.

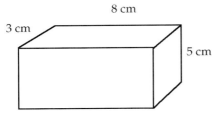

Figure 6.14

1. **Surface area: 2(5)(3) + 2(3)(8) + 2(5)(8) = 158 cm².**

   Volume: 3(8)(5) = 120 cm³.

   Ratio of surface area to volume: $\dfrac{158 \text{ cm}^2}{120 \text{ cm}^3} = \dfrac{79 \text{ cm}^2}{60 \text{ cm}^3}$.

2. The container in **Figure 6.15** has a depth of x cm, a width of $3x$ cm, and a height of 15 cm. Write an expression for the surface area of the container. (Hint: Don't forget to include all six of the sides.)

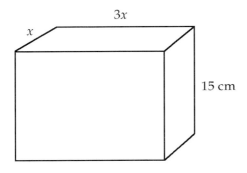

Figure 6.15

2. **$3x^2 + 3x^2 + 15x + 15x + 45x + 45x = 6x^2 + 120x$ cm².**

3. Write an expression for the volume of the solid in Figure 6.15.

3. **$x(3x)15 = 45x^2$ cm³.**

4. Use your answers from items 1 and 2 to write an expression for the ratio of the surface area of the container to its volume.

4. **$\dfrac{6x^2 + 120x}{45x^2}$**

5. Are there any values of x for which the expression is undefined? If so, what are they?

5. **Yes, the expression is undefined when $x = 0$.**

Recall how to reduce a fraction to lowest terms. For example, $\dfrac{35}{42}$ can be reduced by factoring 35 and 42 and dividing the numerator and denominator by the common factor 7.

$$\frac{35}{42} = \frac{5 \cdot \cancel{7}}{6 \cdot \cancel{7}}$$
$$= \frac{5}{6}.$$

6. Use your expression from item 4. Factor the numerator and denominator of the expression. Then divide out any factors common to the numerator and denominator. Write your simplified expression.

6. $\dfrac{6x^2 + 120x}{45x^2} = \dfrac{6x(x+20)}{45x^2} = \dfrac{2x+40}{15x}.$

7. Use your expression for the ratio of surface area to volume from item 4 and your simplified ratio from item 6 to complete **Table 6.12.**

x	Ratio of Surface Area to Volume	Simplified Ratio
1.0		
2.0		
3.0		
4.0		
5.0		

Table 6.12

7. **Answer**

x	Ratio of Surface Area to Volume	Simplified Ratio
1.0	126/45 = 14/5	42/15 = 14/5
2.0	264/180 = 22/15	44/30 = 22/15
3.0	414/405 = 46/45	46/45
4.0	576/720 = 4/5	48/60 = 4/5
5.0	750/1125 = 2/3	50/75 = 2/3

8. What do you notice about the ratios in the second and third column?

8. **The values in the second and third columns are the same.**

9. What do you think would have happened if you examined additional values of x in your table?

9. The values in the second and third columns would be the same.

10. How are the expressions used to complete the second and third columns of your table related?

10. The expressions are equivalent.

Exploring Rational Expressions

In Discovery 6.3, we explored a ratio that was created by dividing an expression that represented surface area by a second expression that represented volume. Any expression such as the algebraic fraction $\dfrac{6x^2 + 120x}{45x^2}$ that is a quotient of two polynomials is called a **rational expression.** Other examples of rational expressions are

$$\frac{2x^2 - 3x + 5}{5x + 3}, \quad \frac{t^3 + 1}{1 - t^3}, \quad \frac{3}{n + 5}, \quad \frac{1}{r^3 + 2r - 6}, \text{ and } \frac{y - 5}{2y}.$$

In the Discovery, you also simplified your original rational expression and noticed that the new expression was equivalent to the original. We say that a rational expression is **reduced to lowest terms** or **simplified** if the numerator and denominator have no factors in common other than 1. To reduce a rational expression to lowest terms, we use the following property:

Reducing Rational Expressions to Lowest Terms

If a, b, and c are polynomial expressions and $b \neq 0$ and $c \neq 0$, then $\dfrac{ac}{bc} = \dfrac{a}{b}$.

We use this property when reducing both numeric fractions and algebraic fractions to lowest terms. For example, when reducing the fraction $\dfrac{15}{33}$, we use the following procedure:

$$\frac{15}{33} = \frac{3 \cdot 5}{3 \cdot 11} \quad \text{Factor numerator and denominator.}$$

$$= \frac{5}{11} \quad \text{Divide numerator and denominator by 3.}$$

The same method is applied when reducing algebraic fractions to lowest terms. For example, in the Preparation Reading of this chapter, the expression $\dfrac{50 + 0.1t}{200{,}000 + 150t}$ was used to model the concentration of pollutants in a lake after t days. Suppose the discharge of the pollutant was 2 kg per day. If this were the case, the expression $\dfrac{50 + 2t}{200{,}000 + 150t}$ could be used to model the concentration. This expression can be reduced to lowest terms.

$$\frac{50+2t}{200{,}000+150t} = \frac{2(25+t)}{50(4000+3t)} \quad \text{Factor numerator and denominator.}$$

$$= \frac{25+t}{25(4000+3t)} \quad \text{Divide numerator and denominator by 2.}$$

Simplifying rational expressions (reducing them to lowest terms), usually requires two steps.

Simplifying Rational Expressions

- Factor the numerator and denominator.
- Divide the numerator and denominator by any common factors.

As a word of caution, make sure you only divide the numerator and denominator by common factors and *not* common terms.

Example 5

Simplify each rational expression if possible. (Assume all variables are non-zero.)

a) $\dfrac{r^2-3r}{2r-6}$

b) $\dfrac{a^2-9a+20}{a^2-16}$

c) $\dfrac{s-5}{s}$

Solution:

a) $\dfrac{r^2-3r}{2r-6} = \dfrac{r(r-3)}{2(r-3)}$ Factor the numerator and denominator.

$\qquad\qquad = \dfrac{r}{2}$ Divide the numerator and denominator by the common factor.

b) $\dfrac{a^2-9a+20}{a^2-16} = \dfrac{(a-5)(a-4)}{(a+4)(a-4)}$ Factor the numerator and denominator.

$\qquad\qquad = \dfrac{a-5}{a+4}$ Divide the numerator and denominator by the common factor.

c) There are no factors common to the numerator and denominator, and we cannot divide the common term *s*. So the expression cannot be reduced.

When working with rational expressions, it is important to check for values for which the expression is not defined. For example, the expression $\dfrac{z-4}{z+4}$ is not defined when $z = -4$, because -4 makes the denominator of the rational expression equal to zero. When a rational expression contains factors common to the numerator and denominator, it is important to check for these values prior to reducing the expression.

Example 6

State the value(s) of the variable for which the expression is undefined.

a) $\dfrac{a^2 + 2a + 6}{a^2}$

b) $\dfrac{2r + 7}{(r + 2)(2r - 5)}$

c) $\dfrac{4y^3 + 1}{y^2 - 2y - 3}$

Solution:

a) The expression is not defined when $a = 0$.

b) The expression is not defined when $r = -2$ or when $r = \dfrac{5}{2}$.

c) To find when the denominator of this expression is equal to zero, first find its factors: $y^2 - 2y - 3 = (y - 3)(y + 1)$. When $y = 3$ or $y = -1$, the expression is undefined.

Multiplying and Dividing Rational Expressions

The procedure for multiplying algebraic fractions is the same as the procedure for multiplying and dividing numeric fractions.

When we multiply algebraic fractions, we multiply the numerators, multiply the denominators, and simplify the resulting expression.

Multiplying Rational Expressions

If a, b, c, and d are polynomial expressions and $b \neq 0$ and $d \neq 0$, then $\dfrac{a}{b} \cdot \dfrac{c}{d} = \dfrac{ac}{bd}$.

Example 7

Multiply the expressions and simplify the results.

a) $\dfrac{3x}{5y} \cdot \dfrac{-30y^2}{21x^3}$

b) $\dfrac{4x+8y}{3xy} \cdot \dfrac{2y}{x^2+2xy}$

c) $\dfrac{3t}{t^2-5t+6} \cdot (t-3)$

Solution:

a) $\dfrac{3x}{5y} \cdot \dfrac{-30y^2}{21x^3} = \dfrac{-(3 \cdot 30)xy^2}{(5 \cdot 21)x^3y}$ Indicate the multiplication of the numerators and denominators.

$= -\dfrac{6y}{7x^2}$ Divide the numerator and denominator by the common factor $15xy$.

b) $\dfrac{4x+8y}{3xy} \cdot \dfrac{2y}{x^2+2xy} = \dfrac{(4x+8y)2y}{3xy(x^2+2xy)}$ Indicate the multiplication of the numerators and denominators.

$= \dfrac{4(x+2y)2y}{(3xy)x(x+2y)}$ Factor the numerator and denominator.

$= \dfrac{8}{3x^2}$ Divide the numerator and denominator by $y(x + 2y)$.

c) $\dfrac{3t}{t^2-5t+6} \cdot (t-3) = \dfrac{3t}{t^2-5t+6} \cdot \dfrac{t-3}{1}$

$= \dfrac{3t(t-3)}{t^2-5t+6}$ Indicate the multiplication of the numerators and denominators.

$= \dfrac{3t(t-3)}{(t-2)(t-3)}$ Factor the numerator and denominator.

$= \dfrac{3t}{t-2}$ Divide the numerator and denominator by $(t - 3)$.

When we divide one algebraic fraction by another, we multiply the first fraction by the reciprocal of the second. The resulting product should then be written in simplified form.

Dividing Rational Expressions

If a, b, c, and d are polynomial expressions, and $b \neq 0$, $c \neq 0$, and $d \neq 0$, then

$$\frac{a}{b} \div \frac{c}{d} = \frac{a}{b} \cdot \frac{d}{c} = \frac{ad}{bc}.$$

As a word of caution, remember that only the expression that you are dividing by is inverted.

Example 8

Divide the expressions and simplify the results.

a) $\dfrac{4m}{5n^2} \div \dfrac{20m^5}{30n^3}$

b) $\dfrac{y^2 + y}{y^2} \div (y^2 - 1)$

c) $\dfrac{\dfrac{x}{x-3}}{\dfrac{2x^3}{x^2-9}}$

Solution:

a) $\dfrac{4m}{5n^2} \div \dfrac{20m^5}{30n^3} = \dfrac{4m}{5n^2} \cdot \dfrac{30n^3}{20m^5}$ Multiply by the reciprocal.

$\qquad = \dfrac{120n^3 m}{100n^2 m^5}$ Multiply the numerators and denominators.

$\qquad = \dfrac{6n}{5m^4}$ Divide the numerator and denominator by $20n^2 m$.

b) $\dfrac{y^2 + y}{y^2} \div (y^2 - 1) = \dfrac{y^2 + y}{y^2} \div \dfrac{y^2 - 1}{1}$

$\qquad = \dfrac{y^2 + y}{y^2} \cdot \dfrac{1}{y^2 - 1}$ Multiply by the reciprocal.

$\qquad = \dfrac{y(y + 1)}{y^2(y - 1)(y + 1)}$ Multiply the numerators and denominators, then factor.

$\qquad = \dfrac{1}{y(y - 1)}$ Divide the numerator and denominator by $y(y + 1)$.

c) $\dfrac{\dfrac{x}{x-3}}{\dfrac{2x^3}{x^2-9}} = \dfrac{x}{x-3} \div \dfrac{2x^3}{x^2-9}$ Rewrite using the $\div$ symbol.

$= \dfrac{x}{x-3} \cdot \dfrac{x^2-9}{2x^3}$ Multiply by the reciprocal.

$= \dfrac{x(x-3)(x+3)}{(x-3)2x^3}$ Indicate the multiplication of the numerators and

denominators, then factor.

$= \dfrac{x+3}{2x}$ Divide the numerator and denominator by $x(x-3)$.

Exercises 6.3

I. Investigations *(SE pages 145–149)*

1. Companies often pay their employees a salary for a 40-hour workweek and then pay overtime for additional hours worked.

   a) Suppose an employee makes $800 per week. What is the employee's average hourly wage?

   1. a) $\dfrac{\$800}{40} = \$20$ **per hour.**

   b) Suppose that in addition to the weekly salary, the employee is paid for overtime at a rate of $30 per hour. What are the weekly earnings if the employee works 42 hours? For a 42-hour week, what is the average hourly wage?

   b) Weekly earnings: $800 + 2(30) = $860; Average: $\dfrac{\$860}{42} = \$20.48$ **per hour.**

   c) Let t represent the number of overtime hours worked. Write an expression for the weekly earnings for the employee. Write an expression for the average hourly wage for a week.

   c) Weekly earnings: $800 + 30$t; **Average:** $\dfrac{800 + 30t}{40 + t}$.

   d) Use your expression from (c) to complete **Table 6.13**. What happens to the average hourly wage as the number of overtime hours worked increases?

Number of Hours Worked	Weekly Earnings	Average Hourly Wage
40		
45		
50		
55		
60		

   Table 6.13

d) The average hourly wage increases as the number of hours of overtime worked increases.

Number of Hours Worked	Weekly Earnings	Average Hourly Wage
40	$800	$20
45	$950	$21.11
50	$1100	$22
55	$1250	$22.73
60	$1400	$23.33

2. **a)** For what value(s) of y is the expression $\dfrac{y^2 + 2y - 3}{y - 1}$ undefined?

2. **a)** For $y = 1$.

b) Reduce the expression $\dfrac{y^2 + 2y - 3}{y - 1}$ to lowest terms.

b) $\dfrac{y^2 + 2y - 3}{y - 1} = \dfrac{(y + 3)(y - 1)}{y - 1} = y + 3$.

c) Complete **Table 6.14** for the expressions $\dfrac{y^2 + 2y - 3}{y - 1}$ and $y + 3$.

	–2	–1	0	1	2	3
$\dfrac{y^2 + 2y - 3}{y - 1}$						
$y + 3$						

Table 6.14

c)

	–2	–1	0	1	2	3
$\dfrac{y^2 + 2y - 3}{y - 1}$	1	2	3	undefined	5	6
$y + 3$	1	2	3	4	5	6

d) Use the information from (a)–(c) to discuss the equivalence of $\dfrac{y^2+2y-3}{y-1}$ and $y+3$.

d) Sample answer: The two expressions are equivalent for all values for which y is defined for both expressions.

3. A packaging engineer wants to design a package so that it fits into a cube x cm on each edge (see **Figure 6.16**).

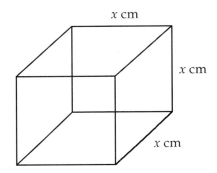

x cm

x cm

x cm

Figure 6.16

Possible designs are shown in **Figures 6.17a** and **6.17b.** The possibilities considered are a cube with edges x cm and a sphere with a diameter of x cm.

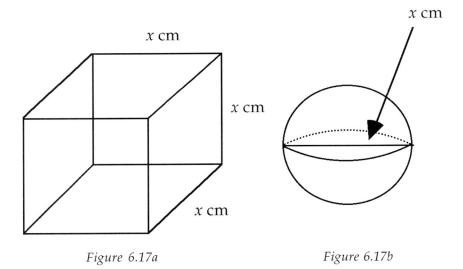

x cm

x cm

x cm

x cm

Figure 6.17a   *Figure 6.17b*

a) Suppose the only criterion for efficiency being considered is the surface area to volume ratio. Using this criterion, the most efficient design would be the figure with the smallest ratio. Without calculating the ratio, which design do you think is the most efficient?

3. a) Either the cube or the sphere could be chosen.

To test your conjecture in (a), answer (b)–(c).

b) Write an expression for the ratio of the surface area of the cube in Figure 6.17a to its volume.

b) $SA = 6x^2$; $V = x^3$; $\dfrac{SA}{V} = \dfrac{6x^2}{x^3} = \dfrac{6 \text{ cm}^2}{x \text{ cm}^2}$.

c) Write an expression for the ratio of the surface area of the sphere in Figure 6.17b to its volume.

c) $SA = 4\pi r^2$; $SA = 4\pi\left(\dfrac{x}{2}\right)^2 = \pi x^2$; $V = \dfrac{4}{3}\pi r^3$; $V = \dfrac{4}{3}\pi\left(\dfrac{x}{2}\right)^3 x = \dfrac{\pi x^3}{6}$;

$\dfrac{SA}{V} = \dfrac{\pi x^2}{\dfrac{\pi x^3}{6}} = \dfrac{6 \text{ cm}^2}{x \text{ cm}^2}$.

d) According to the criterion of the surface area to volume ratio, which design is the most efficient? Explain.

d) According to the criterion, the efficiency is the same for both models.

4. For a rational expression containing only one variable, a graphing calculator or computer graphing software can be used to support algebraic calculations.

a) Examine the expression $\dfrac{x^2 - 2x}{x^2 + 3x - 10}$. Graph the numerator and the denominator of the expression on the same screen. Notice that the two graphs have a common x-intercept. Write the coordinates of that point.

4. a) (2, 0)

b) Now simplify the expression by factoring the numerator and denominator. Write the expression in lowest terms. (Hint: Don't forget the common factor x in the numerator.)

b) $\dfrac{x^2 - 2x}{x^2 + 3x - 10} = \dfrac{x(x - 2)}{(x + 5)(x - 2)} = \dfrac{x}{(x + 5)}$.

c) If you weren't sure of the factors in (b), how could you have used the graphs of the numerator and denominator from (a) to help you simplify the expression algebraically?

c) Sample answer: The x-intercept of 2 gives a hint that $x - 2$ is a factor in both the numerator and the denominator.

d) Simplify the expression $\dfrac{x^2 + 12x - 85}{x^2 - 16x + 55}$. Graph the numerator and denominator if you need help with the factoring.

d) There is a common x-intercept of (0, 5). Hence, $(x - 5)$ must be a factor common to both the numerator and the denominator.

$\dfrac{x^2 + 12x - 85}{x^2 - 16x + 55} = \dfrac{(x - 5)(x + 17)}{(x - 5)(x - 11)} = \dfrac{x + 17}{x - 11}$.

5. In Discovery 6.2, you found that if a is the initial amount of alcohol in the blood of a typical male, the expression $\dfrac{10a}{a+4}$ can be used to model the amount of alcohol that is eliminated from his blood in one hour. The alcohol elimination rate for a typical female is lower than the elimination rate for a typical male.

    a) Suppose the alcohol elimination rate for a particular woman is 2/3 of the elimination rate for the typical male. Write an expression for her alcohol elimination rate.

5. a) $\dfrac{2}{3} \cdot \dfrac{10a}{a+4} = \dfrac{20a}{3a+12}$.

    b) Use your expression from (a) to complete **Table 6.15**.

Initial Amount of Alcohol (g)	0	5	10	50	100	250	500
Amount Eliminated in One Hour (g)							

Table 6.15

b)

Initial Amount of Alcohol (g)	0	5	10	50	100	250	500
Amount Eliminated in One Hour (g)	0	3.7	4.8	6.2	6.4	6.6	6.6

    c) From Table 6.15, for this particular female, what appears to be the maximum number of grams of alcohol that can be eliminated in an hour no matter how much alcohol is initially in the blood?

c) Approximately 6.6 grams.

6. In Chapter 5, we found that the quadratic formula $x = \dfrac{-b \pm \sqrt{b^2 - 4ac}}{2a}$ can be used to solve any quadratic equation. Often the resulting solution can be simplified. For example, the solution $x = \dfrac{-8 \pm 2\sqrt{5}}{6}$ can be reduced to lowest terms by factoring the numerator and denominator:

$$x = \frac{-8 \pm 2\sqrt{5}}{6} = \frac{2(-4 \pm \sqrt{5})}{2 \cdot 3}$$
$$= \frac{-4 \pm \sqrt{5}}{3}.$$

Solve the given quadratic equations using the quadratic formula. Be sure to simplify your solutions.

    a) $4t^2 + 2t - 1 = 0$.

6. a) $t = \dfrac{-2 \pm \sqrt{2^2 - 4(4)(-1)}}{2 \cdot 4}$; $t = \dfrac{-2 \pm \sqrt{20}}{8}$; $t = \dfrac{-2 \pm 2\sqrt{5}}{8}$; $t = \dfrac{-1 \pm \sqrt{5}}{4}$.

b) $3x^2 + 18x - 2 = 0$.

   b) $x = \dfrac{-18 \pm \sqrt{18^2 - 4(3)(-2)}}{2 \cdot 3}$; $x = \dfrac{-18 \pm 2\sqrt{87}}{6}$; $x = \dfrac{-9 \pm \sqrt{87}}{3}$.

c) $2m^2 + 4m = -1$.

   c) $m = \dfrac{-4 \pm \sqrt{4^2 - 4(2)(1)}}{2 \cdot 2}$; $m = \dfrac{-4 \pm 2\sqrt{2}}{4}$; $m = \dfrac{-2 \pm \sqrt{2}}{2}$.

7. In Chapter 5, we simplified rational expressions that contained radicals in their denominators. For example, to simplify $\dfrac{3}{\sqrt{2}}$, we multiplied the numerator and denominator by $\sqrt{2}$. This resulted in the expression $\dfrac{3\sqrt{2}}{2}$. But what if we have an expression such as $\dfrac{5}{7 - \sqrt{2}}$? Multiplying the numerator and denominator by $\sqrt{2}$ does not eliminate the radical in the denominator. To simplify this expression, first consider the following:

a) Find the product: $(3 + \sqrt{5})(3 - \sqrt{5})$.

7. a) $(3 + \sqrt{5})(3 - \sqrt{5}) = 9 + 3\sqrt{5} - 3\sqrt{5} - 5 = 9 - 5 = 4$.

Note that when we multiply $(3 + \sqrt{5})(3 - \sqrt{5})$, the product is a number that does not contain a radical. The pair of expressions $(3 + \sqrt{5})$ and $(3 - \sqrt{5})$ is called a **conjugate pair**. In general, $a + b$ and $a - b$ are called conjugates and their product is $a^2 - b^2$.

This gives us a hint on how to simplify the expression $\dfrac{5}{7 - \sqrt{2}}$. To simplify it, multiply the numerator and denominator of the expression by the conjugate of $7 - \sqrt{2}$.

$$
\begin{aligned}
\frac{5}{7 - \sqrt{2}} &= \frac{5}{7 - \sqrt{2}} \left(\frac{7 + \sqrt{2}}{7 + \sqrt{2}} \right) \\
&= \frac{5(7 + \sqrt{2})}{49 - 7\sqrt{2} + 7\sqrt{2} - 2} \\
&= \frac{35 + 5\sqrt{2}}{49 - 2} \\
&= \frac{35 + 5\sqrt{2}}{47}.
\end{aligned}
$$

b) Simplify: $\dfrac{3}{4+\sqrt{5}}$

b) $\dfrac{3}{4+\sqrt{5}} = \dfrac{3\left(4-\sqrt{5}\right)}{16-5} = \dfrac{12-3\sqrt{5}}{11}.$

c) Simplify: $\dfrac{5}{\sqrt{3}-3}$

c) $\dfrac{5}{\sqrt{3}-3} = \dfrac{5\left(\sqrt{3}+3\right)}{3-9} = \dfrac{5\sqrt{3}+15}{-6}.$

d) Simplify: $\dfrac{-1}{8+\sqrt{y}}$

d) $\dfrac{-1}{8+\sqrt{y}} = \dfrac{-1\left(8-\sqrt{y}\right)}{64-y} = \dfrac{-8+\sqrt{y}}{64-y}.$

II. Projects and Group Activities *(SE page 149)*

8. Mathematics is often used to help make fairness decisions such as how to divide an estate fairly among its heirs. Sometimes it is impossible to find a perfect solution to the problem, and often in those instances, solving equations or inequalities can help optimize fairness. In the case of apportioning the seats in the United States House of Representatives, rational equations are helpful.

There are several methods that can be used to apportion the seats in the House of Representatives. The Jefferson method, the Adams method, and the Webster method are three of many methods known as *divisor methods*.

Research these three methods of apportionment. Write a short paper comparing and contrasting the methods.

III. Additional Practice *(SE pages 150–152)*

9. Suppose a copy machine produces 10,000 copies in 130 minutes. Write an expression for the number of copies it can make in 1 minute, in 5 minutes, and in t minutes.

9. 1 minute: $\dfrac{10{,}000}{130}$ or approximately 77 copies; 5 minutes: $\dfrac{10{,}000(5)}{130}$ or approximately 385 copies; t minutes: $\dfrac{10{,}000t}{130}$ or $\dfrac{1000t}{13}$ copies.

10. **Figure 6.18** shows a cylinder with a base of radius r and a height of $2r$.

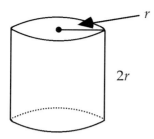

Figure 6.18

   a) Write expressions for the volume and the surface area of the cylinder in terms of r.

10. a) Volume: $\pi r^2 (2r) = 2\pi r^3$; **surface area:** $2\pi r^2 + 2\pi r(2r) = 6\pi r^2$.

   b) Write an expression for the ratio of the volume to the surface area. (Be sure to simplify the expression.)

b) $\dfrac{2\pi r^3}{6\pi r^2} = \dfrac{r}{3}$.

11. a) Write an expression for the average cost of n units of materials if materials cost \$25 per unit and there is a fixed delivery charge of \$75.

11. a) $\dfrac{25n + 75}{n}$

   b) Find the average cost of 10 units, 100 units, and 1000 units.

b) $\dfrac{25(10) + 75}{10} = \$32.50;\ \dfrac{25(100) + 75}{100} = \$25.75;\ \dfrac{25(1000) + 75}{1000} = \$25.08.$

12. Suppose one of your college courses has six 100-point tests scheduled during the semester. You've taken five of them, and your scores are 87, 98, 83, 91, and 79. Write an expression for your average test score if your score on the final test is x.

12. $\dfrac{87 + 98 + 83 + 91 + 79 + x}{6}$

13. The expression $\left(\dfrac{r}{h + r}\right)^2 w_0$ gives the weight of an object at a distance h above Earth's surface if the radius of Earth is r and the weight of the object at sea level is w_0. How much will a 150-pound person weigh if the person is 250 miles above the surface of Earth? (Recall that the radius of Earth is approximately 3950 miles.)

13. $\left(\dfrac{3950}{250 + 3950}\right)^2 150 \approx 133$ lb.

For 14–23, reduce each rational expression to lowest terms. (Assume all variables are nonzero.)

14. $\dfrac{6ax}{2bx}$

14. $\dfrac{3a}{b}$

15. $\dfrac{9a^2w}{6a}$

15. $\dfrac{3aw}{2}$

16. $\dfrac{(b+3)(k-5)}{(k-5)(b+1)}$

16. $\dfrac{b+3}{b+1}$

17. $\dfrac{2x+6}{3x+9}$

17. $\dfrac{2}{3}$

18. $\dfrac{2x^2+12x}{6x^3}$

18. $\dfrac{x+6}{3x^2}$

19. $\dfrac{5ac+bc}{2bc}$

19. $\dfrac{5a+b}{2b}$

20. $\dfrac{a^2-2a-3}{a^2-3a}$

20. $\dfrac{a+1}{a}$

21. $\dfrac{m^3+2m^2}{m^2+7m+10}$

21. $\dfrac{m^2}{m+5}$

22. $\dfrac{9s^2 - 4}{6s + 4}$

22. $\dfrac{3s - 2}{2}$

23. $\dfrac{t^2 - 3t - 10}{t^2 - 25}$

23. $\dfrac{t + 2}{t + 5}$

For 24–29, state the value(s), if any, for which the expression is undefined.

24. $\dfrac{r + 3}{r + 1}$

24. The expression is undefined when $r = -1$.

25. $\dfrac{3t - 5}{t^2 - 4}$

25. The expression is undefined when $t = 2$ or when $t = -2$.

26. $\dfrac{x^2 + 2x}{2}$

26. The expression is defined for all numbers.

27. $\dfrac{q - 7}{q^2 - 5q + 6}$

27. The expression is undefined when $q = 2$ or when $q = 3$.

28. $\dfrac{a(b + 1)}{b(a - 1)}$

28. The expression is undefined when $b = 0$ or when $a = 1$.

29. $\dfrac{r + s}{rs}$

29. The expression is undefined when $r = 0$ or $s = 0$.

30. The expression $\dfrac{\dfrac{A_1}{L_1}}{\dfrac{A_2}{L_2}}$ shows the conductance ratio for two different-sized wires

of the same material. Rewrite the expression as a single simple fraction.

30. $\dfrac{\dfrac{A_1}{L_1}}{\dfrac{A_2}{L_2}} = \dfrac{A_1}{L_1} \div \dfrac{A_2}{L_2} = \dfrac{A_1}{L_1} \cdot \dfrac{L_2}{A_2} = \dfrac{A_1 L_2}{L_1 A_2}.$

31. Find two different pairs of rational expressions whose product is $\dfrac{x^2-25}{x^2-x-12}$.

31. $\dfrac{x-5}{x-4}$ and $\dfrac{x+5}{x+3}$ or $\dfrac{x+5}{x-4}$ and $\dfrac{x-5}{x+3}$.

For 32–43, perform the indicated operations. Reduce answers to lowest terms. (Assume all variables are nonzero.)

32. $\dfrac{8x^2y^3}{3ab^3} \cdot \dfrac{9a^2x}{10aby^5}$

32. $\dfrac{12x^3}{5b^4y^2}$

33. $\dfrac{a(b+1)}{6(b-1)} \cdot \dfrac{9(b-1)}{a(2b+1)}$

33. $\dfrac{3(b+1)}{2(2b+1)}$

34. $\dfrac{5a^2b^3}{a^2-b^2} \cdot \dfrac{3a+3b}{30a^2b}$

34. $\dfrac{b^2}{2(a-b)}$

35. $\dfrac{x^2-5x+6}{x^2+2x} \cdot \dfrac{x^2-4}{x^2-x-2}$

35. $\dfrac{(x-3)(x-2)(x-2)(x+2)}{x(x+2)(x-2)(x+1)} = \dfrac{(x-3)(x-2)}{x(x+1)}$.

36. $\dfrac{x^2+2x}{x^2-3x+2} \cdot \left(x^2-4\right)$

36. $\dfrac{x(x+2)^2}{x-1}$

37. $\dfrac{5a^2b}{9c^2d^3} \div \dfrac{15a^5c^2}{6bd^4}$

37. $\dfrac{5a^2b}{9c^2d^3} \div \dfrac{15a^5c^2}{6bd^4} = \dfrac{5a^2b}{9c^2d^3} \cdot \dfrac{6bd^4}{15a^5c^2} = \dfrac{2b^2d}{9a^3c^4}$.

38. $\dfrac{12x^2+6x}{6xy} \div \dfrac{4x^2-1}{3x^2y^3}$

38. $\dfrac{12x^2+6x}{6xy} \div \dfrac{4x^2-1}{3x^2y^3} = \dfrac{12x^2+6x}{6xy} \cdot \dfrac{3x^2y^3}{4x^2-1} = \dfrac{3x^2y^2}{2x-1}$.

39. $\dfrac{x^2-4}{x} \div (x-2)$

39. $\dfrac{x^2-4}{x} \div (x-2) = \dfrac{x^2-4}{x} \cdot \dfrac{1}{x-2} = \dfrac{x+2}{x}.$

40. $\dfrac{\dfrac{2x}{3y^2}}{\dfrac{6x^3}{15y^5}}$

40. $\dfrac{\dfrac{2x}{3y^2}}{\dfrac{6x^3}{15y^5}} = \dfrac{2x}{3y^2} \div \dfrac{6x^3}{15y^5} = \dfrac{2x}{3y^2} \cdot \dfrac{15y^5}{6x^3} = \dfrac{5y^3}{3x^2}.$

41. $\dfrac{\dfrac{8r}{3r+9}}{\dfrac{4r^2}{r+3}}$

41. $\dfrac{\dfrac{8r}{3r+9}}{\dfrac{4r^2}{r+3}} = \dfrac{8r}{3r+9} \div \dfrac{4r^2}{r+3} = \dfrac{8r}{3r+9} \cdot \dfrac{r+3}{4r^2} = \dfrac{2}{3r}.$

42. $\dfrac{\dfrac{x+4}{x-3}}{\dfrac{1}{x^2+x-12}}$

42. $\dfrac{\dfrac{x+4}{x-3}}{\dfrac{1}{x^2+x-12}} = \dfrac{x+4}{x-3} \div \dfrac{1}{x^2+x-12} = \dfrac{x+4}{x-3} \cdot \dfrac{x^2+x-12}{1} = (x+4)^2.$

43. $\dfrac{\dfrac{x^2-8x+15}{x^2-2x-3}}{\dfrac{6x-30}{x+1}}$

43. $\dfrac{\dfrac{x^2-8x+15}{x^2-2x-3}}{\dfrac{6x-30}{x+1}} = \dfrac{(x-5)(x-3)}{(x-3)(x+1)} \div \dfrac{6(x-5)}{(x+1)} = \dfrac{(x-5)(x-3)}{(x-3)(x+1)} \cdot \dfrac{(x+1)}{6(x-5)} = \dfrac{1}{6}.$

Section 6.4 Solving Rational Equations; Adding and Subtracting Rational Expressions

What You Need to Know

- How to evaluate rational expressions
- How to simplify rational expressions
- How to multiply rational expressions
- How to add and subtract fractions
- How to find the volume and surface area of a cylinder

What You Will Learn

- To solve rational equations algebraically
- To find common denominators for two or more rational expressions
- To add and subtract rational expressions
- To use rational expressions to model real-world situations

Materials

- None

A rational expression may contain one or more variables in its denominator. If an equation contains a rational expression, it may be necessary to solve for a variable that occurs in a denominator. In this section, we will examine how such equations can be solved.

Discovery 6.4 Finding the Length of a Beam *(SE pages 153–156)*

In Activity 6.1, you modeled a structural beam using pasta. You found that long pieces of pasta were easier to break than short ones. In this Discovery, you will explore the behavior of wooden beams in three different situations.

The behavior of beams is well known. The force required to break a 2" x 10" wooden beam such as might be used for a joist supporting a floor in a house (see **Figure 6.19**) is given by the rational function $F = \dfrac{40,000}{L}$. In this function, the length of the beam L is measured in feet and the force F that is applied to the center of the beam is measured in pounds.

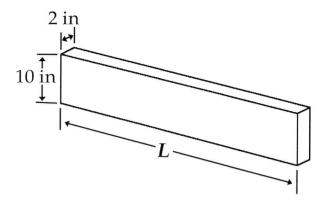

Figure 6.19

Suppose that the design of a house requires the use of a 2" x 10" beam with a breaking strength of 3600 pounds. How can the maximum length of the beam be found?

1. Rewrite the function $F = \dfrac{40{,}000}{L}$, substituting the value of 3600 for F.

1. $3600 = \dfrac{40{,}000}{L}$.

2. Because the right side of the equation $3600 = \dfrac{40{,}000}{L}$ involves division by L, multiply both sides of the equation by L to undo the division. Now that you've transformed the rational equation into an equivalent linear equation in L, solve for the length of the beam.

2. $3600(L) = \dfrac{40{,}000}{L}(L)$

$\qquad 3600L = 40{,}000$

$\qquad\qquad L = 11.1111\dots \text{ or } L \approx 11.1 \text{ ft.}$

Now suppose that a kitchen design has a partially open pantry at one end that is $3\frac{1}{2}$ feet wide (see **Figure 6.20**). If l is the length of the kitchen area, the total length of the room (kitchen and pantry) can be written ($l + 3.5$). To help support the floor above the kitchen, a beam spanning the entire kitchen (including the pantry) must have a breaking strength of 2600 pounds.

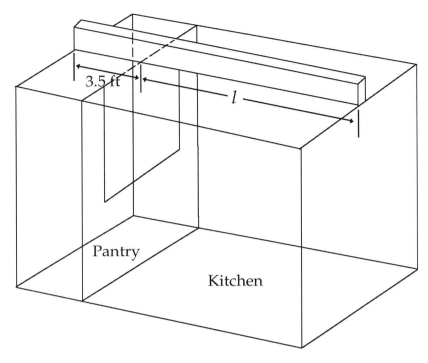

3.5 ft

l

Pantry

Kitchen

Figure 6.20

3. Write an equation for the maximum length l of the kitchen area.

3. $2600 = \dfrac{40{,}000}{l + 3.5}$.

4. To solve this equation for l, begin by multiplying both sides of the equation by $l + 3.5$. Now that you have transformed the equation from a rational equation into an equivalent linear equation, solve for l to find the maximum length of the kitchen area.

4.
$$2600 = \frac{40{,}000}{l + 3.5}$$
$$2600(l + 3.5) = \frac{40{,}000}{l + 3.5}(l + 3.5)$$
$$2600l + 9100 = 40{,}000$$
$$2600l = 30{,}900$$

$l = 11.8846\ldots$ **or** $l \approx 11.89$ **ft, which is about 11 ft 11 in.**

A tall, thin column is not as likely to fail by breaking as it is to "buckle," or bend, when too large a load is applied to it. The force that causes elastic buckling is called the *Euler Load*. Suppose that a 4-inch-square wooden column in the center of a large lobby supports a vaulted ceiling, as in **Figure 6.21.** The vertical force F (in pounds) that will cause such a column of length L feet to buckle is given by the formula $F = \dfrac{1{,}250{,}000\pi^2}{L^2}$.

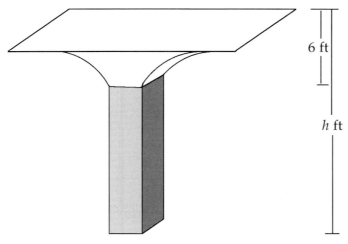

6 ft

h ft

Figure 6.21

5. The maximum height h of the vaulted ceiling is 6 feet greater than the length of the wooden column. The column must have a Euler Load of 100,000 pounds. Write an equation for h.

5. $100{,}000 = \dfrac{1{,}250{,}000\pi^2}{(h-6)^2}$.

6. To find the maximum height h of the ceiling, solve this equation for h. Begin by using an appropriate inverse operation to eliminate the denominator.

6. $100{,}000(h-6)^2 = \dfrac{1{,}250{,}000\pi^2}{(h-6)^2}(h-6)^2$

$(h-6)^2 = 12.5\pi^2$

$\sqrt{(h-6)^2} = \sqrt{12.5\pi^2}$

$h - 6 \approx 11.1$

$h \approx 17.1 \text{ ft.}$

Solving Rational Equations

In Discovery 6.4, we examined three situations that could be modeled by rational equations. In each case, we solved the equation by using an inverse operation (multiplication) to undo the indicated division in the equation. This same process works for other rational equations as well. For example, consider the equation $\dfrac{x}{6} = \dfrac{3x+8}{6}$. We can undo the division by 6 by multiplying both sides of the equation by 6.

$$\frac{x}{6} = \frac{3x+8}{6}$$

$$(6)\frac{x}{6} = (6)\frac{3x+8}{6} \qquad \text{Multiply both sides of the equation by 6.}$$

$$x = 3x + 8 \qquad \text{Simplify.}$$

$$-2x = 8 \qquad \text{Subtract 3x from both sides.}$$

$$x = -4 \qquad \text{Divide both sides by } -2.$$

To check the solution, substitute -4 in the original equation.

$$\frac{-4}{6} \overset{?}{=} \frac{3(-4)+8}{6}$$

$$-\frac{2}{3} \overset{?}{=} \frac{-4}{6}$$

$$-\frac{2}{3} = -\frac{2}{3} \quad \checkmark$$

Undoing the division in equations that contain algebraic fractions by multiplying both sides of an equation by some number or expression produces a polynomial equation that can then be solved by standard methods.

Example 9

a) Solve for m: $\dfrac{1}{4} + \dfrac{1}{2m} = \dfrac{7}{3m}$.

b) Solve for x: $\dfrac{x}{x+4} = \dfrac{4}{x-4} - \dfrac{28}{x^2-16}$.

Solution:

a) $\dfrac{1}{4} + \dfrac{1}{2m} = \dfrac{7}{3m}$.

$$12m\left(\dfrac{1}{4} + \dfrac{1}{2m}\right) = 12m\left(\dfrac{7}{3m}\right)$$ Find a common denominator; multiply both sides of the equation by it.

$$3m + 6 = 28$$ Simplify.

$$3m = 22$$ Subtract 6 from both sides.

$$m = \dfrac{22}{3}$$ Divide both sides by 3.

Be sure to check your solution in the original equation.

b) $\dfrac{x}{x+4} = \dfrac{4}{x-4} - \dfrac{28}{x^2 - 16}$

$$\dfrac{x}{x+4} = \dfrac{4}{x-4} - \dfrac{28}{(x-4)(x+4)}$$ Factor the denominators.

$$(x-4)(x+4)\left(\dfrac{x}{x+4}\right) = (x-4)(x+4)\left(\dfrac{4}{x-4} - \dfrac{28}{(x-4)(x+4)}\right)$$ Find a common denominator; multiply both sides of the equation by it.

$$(x-4)(x+4)\left(\dfrac{x}{x+4}\right) = (x-4)(x+4)\left(\dfrac{4}{x-4}\right) - (x-4)(x+4)\left(\dfrac{28}{(x-4)(x+4)}\right)$$ Distributive property.

$$x(x-4) = 4(x+4) - 28$$ Simplify.

$$x^2 - 4x = 4x + 16 - 28$$ Distributive property.

$$x^2 - 8x + 12 = 0$$ Set equal to 0.

$$(x-6)(x-2) = 0$$ Factor.

$$x = 6; \ x = 2$$

Be sure to check your solution in the original equation.

The steps for solving equations such as those shown in Example 9 can be summarized as follows:

1. Factor the denominators of the rational expressions in the equation.

2. Find an expression that is a multiple of each of the denominators of the fractions in the equation. This multiple is often called a **common denominator.**

3. Multiply both sides of the equation by the common denominator.

4. Solve the resulting polynomial equation.

5. Check the solution(s).

Note that it is important to check the solutions to the equation because multiplying both sides of the equation by a common denominator may introduce extraneous (or false) solutions (see Exercise 6).

There are times that it is advantageous to solve a formula for one variable in terms of the others. Often formulas contain rational expressions. For example, the relationship among the focal length of a lens f, the distance from the lens to the object d, and the distance from the lens to the image i is given by the formula $\frac{1}{f} = \frac{1}{d} + \frac{1}{i}$. In the case of a camera lens, the film must be placed at specific distance i from the lens in order to produce a focused photo. To find the distance from the lens to the image, we can solve the formula for i in terms of the other variables:

$$\frac{1}{f} = \frac{1}{d} + \frac{1}{i}$$

$(fdi)\dfrac{1}{f} = (fdi)\left(\dfrac{1}{d} + \dfrac{1}{i}\right)$ Multiply both sides by fdi.

$\qquad di = fi + fd$ Simplify.

$\quad di - fi = fd$ Subtract fi from both sides.

$i(d - f) = fd$ Factor.

$\qquad i = \dfrac{fd}{d - f}$ Divide both sides by $d - f$.

Adding and Subtracting Fractions with Like Denominators

In the remaining portion of this section, we will extend our ability to work with rational expressions by learning to add and subtract them. Adding and subtracting algebraic fractions is very similar to adding and subtracting numeric fractions. As long as the denominators are the same, the numerators can be added. Just remember to simplify the sum or difference.

Example 10

a) Add: $\dfrac{5}{12} + \dfrac{3}{12}$

b) Add: $\dfrac{2y}{6x} + \dfrac{y}{6x}$

c) Add: $\dfrac{3t}{t+2} + \dfrac{6}{t+2}$

d) Subtract: $\dfrac{4m+2}{m+1} - \dfrac{m+7}{m+1}$

Solution:

a) $\dfrac{5}{12} + \dfrac{3}{12} = \dfrac{8}{12}$ Add numerators.

$\qquad\qquad = \dfrac{2}{3}$ Simplify.

b) $\dfrac{2y}{6x} + \dfrac{y}{6x} = \dfrac{3y}{6x}$ Add numerators.

$\qquad\qquad = \dfrac{y}{2x}$ Simplify.

c) $\dfrac{3t}{t+2} + \dfrac{6}{t+2} = \dfrac{3t+6}{t+2}$ Add numerators.

$\qquad\qquad\quad = \dfrac{3(t+2)}{t+2}$ Factor.

$\qquad\qquad\quad = 3$ Simplify.

d) $\dfrac{4m+2}{m+1} - \dfrac{m+7}{m+1} = \dfrac{4m+2-(m+7)}{m+1}$ Subtract numerators.

$\qquad\qquad\qquad = \dfrac{4m+2-m-7}{m+1}$ Remember to subtract every term of the polynomial being subtracted.

$\qquad\qquad\qquad = \dfrac{3m-5}{m+1}$ Simplify.

Adding and Subtracting Fractions with Unlike Denominators

As when adding and subtracting rational expressions with like denominators, the procedure for adding and subtracting rational expressions with unlike denominators is similar to adding and subtracting numeric fractions with unlike denominators.

Recall the process of adding and subtracting fractions with unlike denominators.

Example 11

a) Add: $\dfrac{1}{8} + \dfrac{3}{4}$

b) Subtract: $\dfrac{5}{6} - \dfrac{2}{15}$

Solution:

a) $\dfrac{1}{8} + \dfrac{3}{4} = \dfrac{1}{8} + \dfrac{3}{4} \cdot \dfrac{2}{2}$ Multiply the second term by 1 $\left(\text{in the form } \dfrac{2}{2} \right)$

to change its denominator to the common denominator.

$= \dfrac{1}{8} + \dfrac{6}{8}$ Simplify.

$= \dfrac{7}{8}$ Add numerators.

b) $\dfrac{5}{6} - \dfrac{2}{15} = \dfrac{5}{6} \cdot \dfrac{5}{5} - \dfrac{2}{15} \cdot \dfrac{2}{2}$ Multiply each term by 1 $\left(\text{in the forms } \dfrac{5}{5} \text{ and } \dfrac{2}{2} \right)$

to change their denominators to the common denominator.

$= \dfrac{25}{30} - \dfrac{4}{30}$ Simplify.

$= \dfrac{21}{30}$ Subtract numerators.

$= \dfrac{7}{10}$ Simplify.

When adding and subtracting rational expressions with different denominators, we must first find a common denominator. Then each algebraic fraction must be written as an equivalent expression that has the same (common) denominator.

Example 12

a) Add: $\dfrac{7}{3x} + \dfrac{2}{12x}$

b) Add: $\dfrac{4}{y-1} + \dfrac{6}{y-5}$

c) Subtract: $\dfrac{x}{x^2 - 3x + 2} - \dfrac{3}{x^2 - 1}$

d) Add: $\dfrac{t}{t^2 + 6t + 9} + \dfrac{6}{t+3}$

Solution:

a) $\dfrac{7}{3x} + \dfrac{2}{12x} = \dfrac{7}{3x} + \dfrac{2}{3 \cdot 4 \cdot x}$ 　　Factor denominators.

$\qquad = \dfrac{7}{3x}\left(\dfrac{4}{4}\right) + \dfrac{2}{12x}$ 　　Multiply the first term by 1 $\left(\text{in the form } \dfrac{4}{4}\right)$.

$\qquad$ to change its denominator to the common denominator.

$\qquad = \dfrac{28}{12x} + \dfrac{2}{12x}$ 　　Simplify.

$\qquad = \dfrac{30}{12x}$ 　　Add numerators.

$\qquad = \dfrac{5}{2x}$ 　　Simplify.

b) $\dfrac{4}{y-1}+\dfrac{6}{y-5}=\dfrac{4}{y-1}\left(\dfrac{y-5}{y-5}\right)+\dfrac{6}{y-5}\left(\dfrac{y-1}{y-1}\right)$

Multiply each term by 1 $\left(\text{in the forms }\dfrac{y-5}{y-5}\text{ and }\dfrac{y-1}{y-1}\right)$ to change their denominators to the common denominator.

$=\dfrac{4y-20}{(y-1)(y-5)}+\dfrac{6y-6}{(y-5)(y-1)}$

Simplify.

$=\dfrac{10y-26}{(y-1)(y-5)}$

Add numerators.

$=\dfrac{2(5y-13)}{(y-1)(y-5)}$

Factor to see if the fraction can be simplified.

c) $\dfrac{x}{x^2-3x+2}-\dfrac{3}{x^2-1}=\dfrac{x}{(x-1)(x-2)}-\dfrac{3}{(x-1)(x+1)}$

Factor.

$=\dfrac{x}{(x-1)(x-2)}\left(\dfrac{x+1}{x+1}\right)-\dfrac{3}{(x-1)(x+1)}\left(\dfrac{x-2}{x-2}\right)$

Change each denominator to the common denominator.

$=\dfrac{x^2+x}{(x-1)(x-2)(x+1)}-\dfrac{3x-6}{(x-1)(x+1)(x-2)}$

Simplify.

$=\dfrac{x^2-2x+6}{(x-1)(x-2)(x+1)}$

Subtract numerators.

d) $\dfrac{t}{t^2+6t+9}+\dfrac{6}{t+3}=\dfrac{t}{(t+3)^2}+\dfrac{6}{t+3}$ Factor denominators.

$\qquad\qquad = \dfrac{t}{(t+3)^2}+\dfrac{6}{t+3}\left(\dfrac{t+3}{t+3}\right)$ Multiply the second term by 1 $\left(\text{in the form } \dfrac{t+3}{t+3}\right)$ to change its denominator to the common denominator.

$\qquad\qquad = \dfrac{t}{(t+3)^2}+\dfrac{6(t+3)}{(t+3)^2}$ Simplify.

$\qquad\qquad = \dfrac{t+6(t+3)}{(t+3)^2}$ Add numerators.

$\qquad\qquad = \dfrac{7t+18}{(t+3)^2}$ Simplify.

The flow chart in **Figure 6.22** provides a summary of the procedures for adding and subtracting algebraic fractions.

Adding and Subtracting Rational Expressions

Do the rational expressions have a common denominator?

↓   ↓

Yes:
- Add (or subtract) the numerators and place the result over the common denominator.
- Simplify the result if possible.

No:
- Find a common denominator. To find the smallest (or least) common denominator, it is helpful to factor the denominators.
- Rewrite each rational expression as an equivalent expression that has the common denominator. Do this by multiplying each expression by another rational expression that is equal to 1.
- Add (or subtract) the numerators and place the result over the common denominator.
- Simplify the result if possible.

Figure 6.22

Exercises 6.4

I. Investigations *(SE pages 164–168)*

1. A company pays $4 per unit of raw material plus shipping of $500 regardless of the size of the order.

    a) What is the cost of ordering 200 units of raw material?

1. a) 4(200) + 500 = $1300.

    b) What is the average cost per unit of ordering 200 units of raw materials?

b) 1300/200 = $6.50.

    c) Write an expression for the total cost of ordering *n* units of raw materials.

c) 4*n* + 500

    d) Express the average cost per unit of raw materials *A* as a function of the number of units *n* of raw materials ordered.

d) $A = \dfrac{4n + 500}{n}.$

    e) Use your equation from (d) to complete **Table 6.16.**

Number of Units	10	50	100	200	500	1000	5000
Average Cost (dollars)							

Table 6.16

e)

Number of Units	10	50	100	200	500	1000	5000
Average Cost (dollars)	54.00	14.00	9.00	6.50	5.00	4.50	4.10

f) Use your completed table from (e) to sketch a graph of the average cost function.

f)

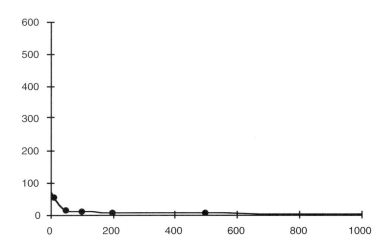

g) What is the horizontal asymptote for the function? What does this horizontal asymptote mean in the context of this problem?

g) The horizontal line is $y = 4$. It means that as the number of units of raw materials placed in each order gets larger and larger, the average cost per unit gets closer to $4.00.

h) Suppose the manufacturer wants the average cost per unit of raw materials to equal $4.25. How many units of raw materials would have to be ordered?

h) $4.25 = \dfrac{4n + 500}{n}$

$4.25n = 4n + 500$

$0.25n = 500$

$n = 2000$ **units.**

i) Suppose the manufacturer wants the average cost per unit of raw materials to be less than $4.02. How many units of raw materials would have to be ordered?

i) To find the number of units of raw materials that would have to be ordered for the average cost per unit A to equal $4.02, solve

$4.02 = \dfrac{4n + 500}{n}$ **for n; n = 25,000. For the cost per unit to be less than $4.02, more than 25,000 units would have to be ordered.**

2. For each hour of a course in which a student earns an A, a college awards 4 grade points. For example, a student would earn 12 grade points for an A in a three-hour course and 16 grade points for an A in a four-hour course. Similarly, students earn points for other grades: 3 points for a B, 2 for a C, 1 for a D, and 0 points for an F. A student's grade point average (GPA) is the number of grade points earned by the student divided by the number of hours taken.

a) If a student has a hours with a grade of A, b hours of B, c hours of C, d hours of D, and f hours of F, write an expression for the student's GPA in terms of a, b, c, d, and f.

2. a) $\dfrac{4a + 3b + 2c + 1d + 0f}{a + b + c + d + f} = \dfrac{4a + 3b + 2c + d}{a + b + c + d + f}$.

b) Suppose a student has completed the following: 15 hours of A, 21 hours of B, 12 hours of C, 0 hours of D, and 0 hours of F. What is this student's current GPA?

b) $\dfrac{15(4) + 21(3) + 12(2)}{15 + 21 + 12} = 3.06$.

c) Suppose the student in (b) plans to earn A's in all remaining courses. If h represents the number of hours of the remaining courses, write an expression for the student's future GPA.

c) $\dfrac{15(4) + 21(3) + 12(2) + 4h}{15 + 21 + 12 + h} = \dfrac{147 + 4h}{48 + h}$.

d) How many hours of courses will this student need to take in order to have a GPA of 3.4?

d) $\dfrac{147 + 4h}{48 + h} = 3.4$; $147 + 4h = 3.4(48 + h)$; $h = 27$ **hours of A.**

3. In Section 6.3, Exercise 3, we explored a packaging problem that involved designing a package that would fit into a cube x cm on each edge (see **Figure 6.23**).

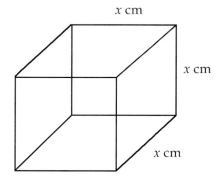

x cm

x cm

x cm

Figure 6.23

In that exercise, we considered two designs, a cube and a sphere, and found that if efficiency were defined as the ratio of the surface area of the solid to its volume, then the two solids were equally efficient with ratios of $\dfrac{6 \text{ cm}^2}{x \text{ cm}^3}$.

a) **Figure 6.24** shows a cylinder that would fit into the given cube. If its diameter and height are x cm, find the ratio of its surface area to its volume.

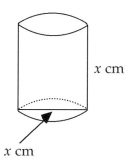

x cm

x cm

Figure 6.24

3. a) $SA = \pi dh + 2\pi r^2$; $SA = \pi x^2 + 2\pi \left(\dfrac{x}{2}\right)^2 = \dfrac{6\pi x^2}{4}$

$$V = \pi r^2 h; \; V = \pi\left(\dfrac{x}{2}\right)^2 x = \dfrac{\pi x^3}{4}; \; \dfrac{SA}{V} = \dfrac{\dfrac{6\pi x^2}{4}}{\dfrac{\pi x^3}{4}} = \dfrac{6 \text{ cm}^2}{x \text{ cm}^3}.$$

b) If efficiency is based on the ratio of the surface area of the solid to its volume, which design (cube, sphere, or cylinder) is the most efficient? Explain.

b) According to the criterion, the efficiency is the same for all three models.

4. In Exercise 3, you revisited an investigation from Section 6.3 and examined the efficiency of three solids that could all fit into a given space, in this case a cube x cm on each side.

To have a better understanding of the efficiency of a design, you need to compare differently shaped containers that have the same volume. When comparing two solids whose volumes are equal, the ratio of surface area to volume could be interpreted as the amount of material needed per unit volume of storage. In this exercise, you will examine the efficiencies of a cube and a sphere when their volumes are equal.

a) Assume the volume of the cube in **Figure 6.25** is 64 cm³. Calculate the length of an edge e.

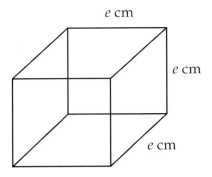

e cm

e cm

e cm

Figure 6.25

4. a) edge $= \sqrt[3]{64} = 4$ **cm.**

b) Write an expression for the ratio of the surface area of a cube to its volume in terms of e. Then use your expression and to find the efficiency of a cube whose volume is 64 cm³.

b) $\dfrac{SA}{V} = \dfrac{6e^2}{e^3} = \dfrac{6}{e}.$

When $e = 4$ cm, $SA/V = 1.5$ cm²/cm³.

c) Assume the volume of the sphere in **Figure 6.26** is 64 cm³. Find an approximate value for the radius r.

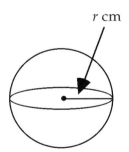

r cm

Figure 6.26

c) $64 = \dfrac{4\pi r^3}{3}; \quad r^3 = \dfrac{192}{4\pi}; \quad r = \sqrt[3]{\dfrac{192}{4\pi}}; \quad r \approx 5.32$ **cm.**

d) Write an expression for the ratio of the surface area of a sphere to its volume in terms of r. Then use your expression to find the efficiency of a sphere whose volume is 64 cm³.

d) $\dfrac{SA}{V} = \dfrac{4\pi r^2}{\dfrac{4\pi r^3}{3}} = \dfrac{3}{r}$ **cm² / cm³.**

When $r \approx 5.32$, $SA/V \approx 3/5.32$ or 0.56 cm²/cm³.

e) Find the edge of a cube and the radius of a sphere whose volume is 100 cm³. Use those values and your expressions from (b) and (d) to determine the efficiencies of the two solid figures.

e) Cube:

$$e = \sqrt[3]{100} \approx 4.64 \text{ cm.}$$

$$\frac{SA}{V} = \frac{6}{e} \approx \frac{6}{4.64} \text{ or } 1.29 \text{ cm}^2 / \text{ cm}^3.$$

Sphere:

$$r = \sqrt[3]{\frac{300}{4\pi}} \approx 6.18 \text{ cm.}$$

$$\frac{SA}{V} = \frac{3}{r} \approx \frac{3}{6.18} \text{ or } 0.49 \text{ cm}^2 / \text{ cm}^3.$$

f) If the only criterion being considered for efficiency of an object is its ratio of surface area to volume, and assuming that the smaller the ratio, the more desirable the design is to the packager, which of these two solids is the most efficient when both solids have a volume of 64 cm³? When both solids have a volume of 100 cm³?

f) The sphere; the sphere

5. Electrical current I, voltage V, and resistance R are related according to Ohm's law, $V = IR$. (Current is measured in amperes, voltage in volts, and resistance in ohms.) In a home with typical voltage (120V), the amount of resistance caused by a 60-watt light bulb is 240 ohms. The diagram in **Figure 6.27** illustrates this circuit.

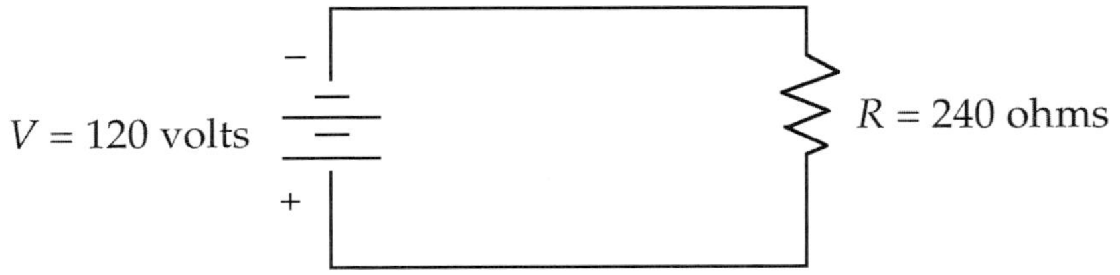

Figure 6.27

a) Solve Ohm's law for I, then use your formula to determine the current in amperes passing through the circuit in Figure 6.27.

5. a) $I = \dfrac{V}{R}; \ I = \dfrac{120 \text{ volts}}{240 \text{ ohms}} = \dfrac{1}{2}$ or 0.5 amperes.

b) Suppose the resistance in Figure 6.27 were caused by a 100-watt bulb that has a resistance of 144 ohms. Determine the current passing through the circuit.

b) $I = \dfrac{120 \text{ volts}}{144 \text{ ohms}} = \dfrac{5}{6}$ amperes.

c) **Figure 6.28** shows a 60-watt bulb with a resistance $R_1 = 240$ ohms, wired in parallel with a 100-watt bulb with a resistance $R_2 = 144$ ohms. If the total voltage is 120 volts and the total current is the sum of the currents through each bulb (found in (a) and (b)), use Ohm's law to find the total resistance of the circuit.

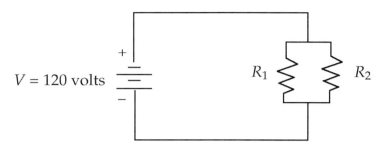

Figure 6.28

c) $R = \dfrac{V}{I}$; $R = \dfrac{120}{\dfrac{1}{2} + \dfrac{5}{6}}$; $R = \dfrac{120}{\dfrac{8}{6}} = \textbf{90 ohms.}$

d) Find an expression for the total resistance R of the circuit represented in Figure 6.28. Write your expression for R in terms of R_1 and R_2.

d) $R = \dfrac{V}{I}$; $R = \dfrac{120}{\dfrac{120}{R_1} + \dfrac{120}{R_2}}$; $R = \dfrac{120}{\dfrac{120R_2 + 120R_1}{R_1 R_2}} = \dfrac{120 R_1 R_2}{120(R_2 + R_1)} = \dfrac{R_1 R_2}{R_2 + R_1}$ **ohms.**

6. As mentioned in this section, multiplying both sides of an equation containing algebraic fractions may introduce extraneous solutions. To determine whether a solution is extraneous, all possible solutions should be checked in the original equation.

   a) Solve for m:

   $$\dfrac{m}{m-1} - \dfrac{2}{m^2 - 1} = \dfrac{7}{m+1}.$$

6. a)
$$\dfrac{m}{m-1} - \dfrac{2}{m^2 - 1} = \dfrac{7}{m+1}$$

$$\dfrac{m}{m-1} - \dfrac{2}{(m-1)(m+1)} = \dfrac{7}{m+1}$$

$$(m-1)(m+1)\left(\dfrac{m}{m-1} - \dfrac{2}{(m-1)(m+1)}\right) = (m-1)(m+1)\left(\dfrac{7}{m+1}\right)$$

$$m(m+1) - 2 = 7(m-1)$$

$$m^2 - 6m + 5 = 0$$

$$(m-5)(m-1) = 0$$

$$m = 5 \text{ or } m = 1.$$

b) Check your possible solutions. Are any of the solutions extraneous? If so, explain why.

b) **The possible solution $m = 1$ is not a solution to the equation, because the rational expressions $\dfrac{m}{m-1}$ and $\dfrac{2}{m^2-1}$ are not defined when $m = 1$.**

II. Projects and Group Activities *(SE pages 168–169)*

7. Materials: magnifying glass, light bulb, rigid paper or cardboard

    A glass lens can focus light in such a way that an image of an object can be seen when viewed through the lens at a particular distance.

    Figure 6.29 shows a magnifying glass placed a distance d in front of a light bulb. The lens in the magnifying glass will produce a focused image of the bulb at a distance i on the other side of the lens. For a lens with focal length f, the relationship among d, i, and f is given by the equation $\dfrac{1}{f} = \dfrac{1}{d} + \dfrac{1}{i}$.

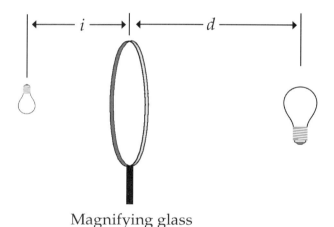

Magnifying glass

Figure 6.29

a) Hold the magnifying glass at a distance of 30 cm from a light bulb. If you look through the glass at the bulb, the image will probably appear out of focus. However, by changing the distance of your eye from the glass, you should be able to find a viewing distance from which the bulb appears in focus. If you place a sheet of heavy paper or cardboard at that distance, you will see a focused image of the bulb on the paper, although it will appear upside down. Record this distance.

7. **a) Sample answer: 60 cm**

b) Complete a table similar to **Table 6.17** by varying the distance d from the magnifying glass to the light bulb and by observing the distance i from the magnifying glass to the point where the image of the bulb is produced.

d (cm)	i (cm)

Table 6.17

b) **Sample table:**

d (cm)	i (cm)
30	60
40	39
50	34
60	31
70	27

c) Add a third column to your table and label it "f." Use the formula $\dfrac{1}{f} = \dfrac{1}{d} + \dfrac{1}{i}$ to compute the experimental value of the magnifying glass's focal length f (in cm) for each of the pairs of d and i values in your table.

c) **Sample calculation:**

$$\frac{1}{f} = \frac{1}{30} + \frac{1}{60}$$

$$\frac{1}{f} = \frac{1}{20}$$

$$f = 20.$$

Sample table:

d (cm)	i (cm)	f (cm)
30	60	20
40	39	19.7
50	34	20.2
60	31	20.4
70	27	19.5

d) Find the average of the values in the f column of your table, which is an experimental value for the focal length of the magnifying glass.

d) 19.96 cm ≈ 20 cm.

e) Solve the equation $\dfrac{1}{f} = \dfrac{1}{d} + \dfrac{1}{i}$ algebraically for f to find a formula for f in terms of d and i.

e)

$\dfrac{1}{f} = \dfrac{1}{d} + \dfrac{1}{i}$	**Original equation**
$di = fi + fd$	**Multiply both sides by fdi.**
$di = f(i + d)$	**Factor the right side.**
$f = \dfrac{di}{i+d}$	**Divide both sides by $(i + d)$.**

f) If the sun is shining, you can determine the focal length of the magnifying glass directly by shining the light from the sun through the lens and placing a sheet of paper at a point where you can see a small focused image of the sun on the paper. The distance between the lens and paper is the focal length of the lens. How does this measurement compare with your calculated focal length from (d)?

f) The result should be very close to the calculated value of f, in this case 20 cm.

III. Additional Practice *(SE pages 169–172)*

8. Recall from Discovery 6.2 that the equation $E = \dfrac{10a}{a+4}$ can be used to model the amount of alcohol E that is eliminated from the blood in one hour, where a is the initial amount of alcohol in the blood.

Suppose that a urine test indicates that a typical male eliminated 8 g of alcohol over the past hour. How much alcohol did he have in his body at the beginning of the hour?

8. $8 = \dfrac{10a}{a+4}$ **;** $8(a+4) = 10a$ **;** $a = 16$ **g.**

9. The relationship among the illumination of a surface E, the luminous intensity of the light source I, and the distance of the light source from the surface R is given by the equation $E = \dfrac{I}{R^2}$. The intensity I of the light source is measured in candela (cd), and the illumination E of a surface is measured in lumens per square meter (lux) or in lumens per square foot.

a) What is the illumination produced by a 200 cd light source on a surface 4.5 m away?

9. a) $E = \dfrac{200}{(4.5)^2}$ **;** $E \approx 9.9$ **lux.**

b) At what distance from a wall will a 125 cd light source produce an illumination of 13.9 lux?

b) $13.9 = \dfrac{125}{R^2}$; $R^2 = 9$; $R = 3$ m.

10. In this section, you solved the equation $\dfrac{1}{f} = \dfrac{1}{p} + \dfrac{1}{i}$ for i and found that

$i = \dfrac{fp}{p-f}$.

a) Find the distance from the lens to a focused image i if the focal length of the lens $f = 12$ cm and the distance from the object to the lens $p = 40$ cm.

10. a) $i = \dfrac{fp}{p-f}$; $i = \dfrac{(12)(40)}{40-12}$; $i \approx 17$ cm.

b) Find the focal length of a lens f if a lens is held 4 cm from an object and produces a focused image 8 cm from the lens.

b) $8 = \dfrac{f(4)}{4-f}$; $8(4-f) = 4f$; $12f = 32$; $f = \dfrac{8}{3}$ or $2\dfrac{2}{3}$ cm.

11. One copy machine produces 10,000 copies in 120 minutes, whereas a second copy machine produces 10,000 copies in 80 minutes.

a) Write an expression for the number of copies each machine can produce in t minutes.

11. a) 1st machine: $\dfrac{10,000t}{120}$ **or** $\dfrac{250t}{3}$ **copies in t minutes;**

2nd machine: $\dfrac{10,000t}{80}$ **or $125t$ copies in t minutes**

b) Write an expression for the total number of copies the two machines working together at the same time can produce in t minutes.

b) $\dfrac{250t}{3} + 125t = \dfrac{625t}{3}$.

c) Suppose the two machines are running simultaneously. How long will it take to produce 10,000 copies? How many copies will each machine make?

c) $10,000 = \dfrac{625t}{3}$; $t = 48$ **minutes; 1st machine makes 4000 copies; 2nd machine makes 6000 copies.**

For 12–18, solve for the unknown. Be sure to check for extraneous solutions.

12. Solve for a: $\dfrac{a+2}{a-1} = 2$.

12. $a = 4$.

13. Solve for x: $\dfrac{3}{2x-3} = \dfrac{5}{3x+7}$.

13. $x = 36$.

14. Solve for t: $\dfrac{t}{t-3} + t = \dfrac{35}{t-3}$.

14. $t = 7$ or $t = -5$.

15. Solve for x: $\dfrac{6}{x-2} + x = \dfrac{3x}{x-2}$.

15. $x = 3$ is the only solution; $x = 2$ is an extraneous solution.

16. Solve for y: $\dfrac{2y}{2y-3} = \dfrac{y+2}{y-2}$.

16. $y = 6/5$ or 1.2.

17. Solve for r: $\dfrac{r}{r^2-r-2} + \dfrac{2}{r-2} = \dfrac{4}{r+1}$.

17. $r = 10$.

18. Solve for x: $\dfrac{6}{x^2-9} = \dfrac{3}{2x-6}$.

18. $x = 1$.

19. Solve for L: $F = \dfrac{40{,}000}{L}$.

19. $L = \dfrac{40{,}000}{F}$.

20. Solve for L: $F = \dfrac{1{,}250{,}000\pi^2}{L^2}$.

20. $F = \dfrac{1{,}250{,}000\pi^2}{L^2}$; $L^2 = \dfrac{1{,}250{,}000\pi^2}{F}$; $L = \sqrt{\dfrac{1{,}250{,}000\pi^2}{F}}$ or $\dfrac{500\pi}{F}\sqrt{5F}$.

21. Solve for f: $\dfrac{1}{f} = \dfrac{1}{d} + \dfrac{1}{i}$.

21. $\dfrac{1}{f} = \dfrac{1}{d} + \dfrac{1}{i}$; $di = fi + fd$; $di = f(i + d)$; $f = \dfrac{di}{i + d}$.

22. The relation $\dfrac{P_1 V_1}{T_1} = \dfrac{P_2 V_2}{T_2}$ combines Boyle's and Charles's laws and is used to describe the thermal behavior of gases. P_1, V_1, and T_1 are the initial pressure, volume, and temperature, respectively, of a given gas, whereas P_2, V_2, and T_2 are the final pressure, volume, and the temperature of the gas. Solve this equation for the final pressure of the gas P_2.

22. $\dfrac{P_1 V_1}{T_1} = \dfrac{P_2 V_2}{T_2}$; $P_1 V_1 T_2 = P_2 V_2 T_1$; $P_2 = \dfrac{P_1 V_1 T_2}{V_2 T_1}$.

23. Consider the fractions $1/10$ and $1/12$. Which of the following are common denominators of these two fractions: 10, 12, 30, 60, 120? Explain.

23. Only 60 and 120 are common denominators because they are the only numbers in the list that are multiples of both 10 and 12.

24. Consider the rational expressions $\dfrac{1}{x(x-2)}$ and $\dfrac{1}{(x-2)(x+2)}$. Which of the following are common denominators of these two fractions: $x(x-2)$, $(x-2)(x+2)$, $x(x-2)(x+2)$, $x(x-2)(x-2)(x+2)$? Explain.

24. Only $x(x-2)(x+2)$, and $x(x-2)(x-2)(x+2)$ are common denominators because they are the only expressions in the list that are multiples of both $x(x-2)$ and $(x-2)(x+2)$.

25. **Figure 6.30** shows two light sources of equal intensity placed at opposite ends of a 5-meter-long room. The expression $\dfrac{I}{x^2} + \dfrac{I}{(5-x)^2}$ gives the illumination of an object where I is the intensity of the light source in candela (cd) and x represents the distance from the bulb on the left along a line joining the two.

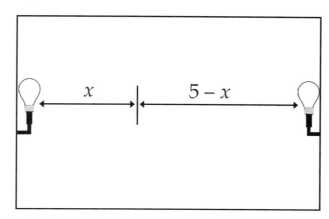

Figure 6.30

a) Rewrite the expression $\dfrac{I}{x^2} + \dfrac{I}{(5-x)^2}$ as a single fraction in simplest terms.

25. a) $\dfrac{I}{x^2} + \dfrac{I}{(5-x)^2} = \dfrac{I(5-x)^2 + Ix^2}{x^2(5-x)^2} = \dfrac{I[(5-x)^2 + x^2]}{x^2(5-x)^2} = \dfrac{I(25 - 10x + 2x^2)}{x^2(5-x)^2}.$

b) Use your expression to find the illumination of a point when $x = 3.5$ m and $I = 250$ cd.

b) $\dfrac{I(25 - 10x + 2x^2)}{x^2(5-x)^2} = \dfrac{250(25 - 10(3.5) + 2(3.5)^2)}{(3.5)^2(1.5)^2} \approx 132 \,\text{lux}.$

26. The expression $1 - \dfrac{g_d}{g_s} - \dfrac{wg_d}{g_w}$ gives the fraction of air in a unit volume of soil. Rewrite the expression as a single simplified fraction.

26. $1 - \dfrac{g_d}{g_s} - \dfrac{wg_d}{g_w} = \dfrac{g_s g_w - g_d g_w - wg_d g_s}{g_s g_w}.$

27. The expression $\dfrac{\dfrac{E_1}{R_1} + \dfrac{E_2}{R_2}}{\dfrac{1}{R_1} + \dfrac{1}{R_2} + \dfrac{1}{R_i}}$ gives the output voltage of a computer gating circuit. Rewrite the expression as a single simplified fraction.

27. $\dfrac{\dfrac{E_1}{R_1} + \dfrac{E_2}{R_2}}{\dfrac{1}{R_1} + \dfrac{1}{R_2} + \dfrac{1}{R_i}} = \dfrac{\dfrac{E_1 R_2 + E_2 R_1}{R_1 R_2}}{\dfrac{R_2 R_i + R_1 R_i + R_1 R_2}{R_1 R_2 R_i}} = \dfrac{E_1 R_2 + E_2 R_1}{R_1 R_2} \div \dfrac{R_2 R_i + R_1 R_i + R_1 R_2}{R_1 R_2 R_i}$

$= \dfrac{E_1 R_2 + E_2 R_1}{R_1 R_2} \cdot \dfrac{R_1 R_2 R_i}{R_2 R_i + R_1 R_i + R_1 R_2} = \dfrac{R_i(E_1 R_2 + E_2 R_1)}{R_2 R_i + R_1 R_i + R_1 R_2}.$

For 28–35, perform the indicated operations. Reduce answers to lowest terms. (Assume all variables are nonzero.)

28. $\dfrac{a}{b} + \dfrac{2a}{b^2}$

28. $\dfrac{ab + 2a}{b^2}$ or $\dfrac{a(b + 2)}{b^2}$

29. $\dfrac{2x}{3ab^4} - \dfrac{5y}{6a^3b^3}$

29. $\dfrac{4a^2x - 5by}{6a^3b^4}$

30. $\dfrac{4r^3}{9s^2t^4} + \dfrac{s}{6t^3r}$

30. $\dfrac{8r^4 + 3s^3t}{18rs^2t^4}$

31. $\dfrac{1}{x-3} + \dfrac{5}{x+7}$

31. $\dfrac{1}{x-3} + \dfrac{5}{x+7} = \dfrac{x+7+5(x-3)}{(x-3)(x+7)} = \dfrac{6x-8}{(x-3)(x+7)}$ or $\dfrac{2(3x-4)}{(x-3)(x+7)}.$

32. $\dfrac{r-7}{r^2+5r} - \dfrac{r-1}{r^2-25}$

32. $\dfrac{r-7}{r^2+5r} - \dfrac{r-1}{r^2-25} = \dfrac{r-7}{r(r+5)} - \dfrac{r-1}{(r+5)(r-5)} = \dfrac{(r-7)(r-5)-(r-1)r}{r(r+5)(r-5)} = \dfrac{-11r+35}{r(r+5)(r-5)}.$

33. $\dfrac{x}{x^2-16} + \dfrac{5}{x-4}$

33. $\dfrac{x}{x^2-16} + \dfrac{5}{x-4} = \dfrac{x}{(x-4)(x+4)} + \dfrac{5}{x-4} = \dfrac{x+5(x+4)}{(x-4)(x+4)} = \dfrac{6x+20}{(x-4)(x+4)}$ or $\dfrac{3(2x+10)}{(x-4)(x+4)}.$

34. $\dfrac{4}{y^2-6y-7} + \dfrac{y}{y-7}$

34. $\dfrac{4}{y^2-6y-7} + \dfrac{y}{y-7} = \dfrac{4}{(y-7)(y+1)} + \dfrac{y}{y-7} = \dfrac{4+y(y+1)}{(y-7)(y+1)} = \dfrac{y^2+y+4}{(y-7)(y+1)}.$

35. $\dfrac{1}{x^2-64} + \dfrac{2x}{x-8} - 2$

35. $\dfrac{1}{x^2-64} + \dfrac{2x}{x-8} - 2 = \dfrac{1}{(x-8)(x+8)} + \dfrac{2x}{x-8} - \dfrac{2}{1} = \dfrac{1+2x(x+8)-2(x-8)(x+8)}{(x-8)(x+8)}$

$= \dfrac{1+2x^2+16x-2x^2+128}{(x-8)(x+8)} = \dfrac{16x+129}{(x-8)(x+8)}.$

Section 6.5 Systems of Equations

What You Need to Know

- How to graph linear, quadratic, and rational functions
- How to solve linear, quadratic, and rational equations

What You Will Learn

- To solve linear systems of equations by substitution
- To solve linear systems of equations graphically
- To solve systems containing both a linear and a nonlinear equation by substitution
- To solve systems containing both a linear and a nonlinear equation graphically

Materials

- Small straws
- Scissors
- Ruler

We have seen many problem situations in which a single function or equation is sufficient to model the problem. In such cases, we often find that we need to solve an equation in order to accomplish a particular task, such as finding lengths of bridge cables, determining the distance from a camera lens to the film, or predicting Internet usage.

There are other situations that require us to consider multiple aspects of a problem. These often involve combining information that is contained in more than one equation in order to reach our objective. In this section, we will examine methods of solving **systems of equations** and some of the applications in which they may occur.

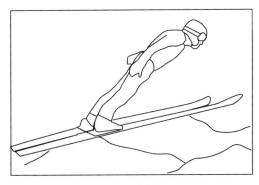

Figure 6.31

In Olympic ski-jumping competition on what is called a "normal hill," an athlete takes off into the air at a speed of about 55 miles per hour (25 m/s) and lands about 90 meters down the hill (see **Figure 6.31**).

Assume that a particular jumper has a parabolic trajectory given by the function $y_1 = 86 - 0.0079x^2$, where x represents the horizontal distance traveled (measured from the takeoff point) at any point of the trajectory and y_1 represents the height above the lowest point of the final runout. Further assume that the landing area on the hill is a straight slope with equation $y_2 = -0.60x + 80$, where x is again the horizontal distance and y_2 is the height of the hill. If you want to predict the exact location of the jumper's landing spot, you must find the point with coordinates (x, y) that make both equations true.

The solution to this system of two equations in two unknowns can be found by several methods.

Using a Table

1. Use the equations for y_1 and y_2 to complete a table similar to **Table 6.18**. You may want to use a spreadsheet or graphing calculator to do so. Continue to increase the x-values in increments of 10 until you see the height of the jumper (y_1) go below the height of the hill (y_2). The jumper will land at a point where the height of the jump (y_1) equals the height of the hill (y_2).

x	y_1 (height of jumper)	y_2 (height of hill)
0		
10		
20		
30		
etc.		

Table 6.18

1.

x	y_1 (height of jumper)	y_2 (height of hill)
0	86	80
10	85.21	74
20	82.84	68
30	78.89	62
40	73.36	56
50	66.25	50
60	57.56	44
70	47.29	38
80	35.44	32
90	22.01	26

The height of the jumper (y_1) equals the height of the hill (y_2) for some x-value between 80 m and 90 m.

2. You now know the horizontal distance of the jump, at least to the nearest 10 meters. Refine your estimate by examining a smaller section of the table more closely. Beginning with the last x-coordinate for which y_1 is greater than y_2, change the increment of x to 1 and again continue until y_1 is less than y_2. To the nearest meter, what is the horizontal distance of the jump?

2.

x	y_1	y_2
80	35.44	32
81	34.168	31.4
82	32.88	30.8
83	31.577	30.2
84	30.258	29.6
85	28.923	29

The values of y_1 and y_2 are equal at some point between $x = 84$ m and $x = 85$ m. Also, 28.923 and 29 are closer to each other than 30.258 and 29.6, so it is likely that y_1 and y_2 are equal near 85 m. Therefore, 85 m is the distance of the jump to the nearest meter.

3. If you continue refining the table, you can determine the horizontal distance of the jump as accurately as desired. Find the distance accurate to the nearest meter.

3. $x \approx 84.90$ m.

Using a Graph

4. A graph can also be used to estimate the landing point. Create a graph that shows both the "jump" function and the "hill" function.

4.

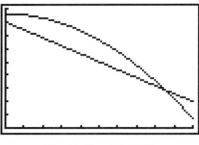

[0, 100] x [0, 90]

5. The point where the graphs of the two functions cross is the point at which the height of the jump and the height of the hill are equal. To get an accurate estimate of both coordinates of the intersection point, zoom in to magnify the graph in the area near the point. By zooming in repeatedly, you should be able to again estimate distances to the nearest centimeter.

5. $x \approx 84.90$ m.

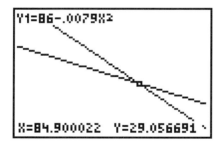

[83.97, 85.40] x [28.33, 30.13]

An Algebraic Method

6. Both the graphical zooming method and the "numerical zooming" method using a table can usually provide only approximate solutions for intersection points. An algebraic method can provide an exact solution. Because the ski jumper lands at a point where the height of the jump equals the height of the hill, this point has an x-coordinate for which y_1 and y_2 are equal. Thus, we can write $y_1 = y_2$.

   Replace y_1 and y_2 in this equation by their respective expressions in x. We can think of this as substituting the expression for y_1 into the second equation for y_2.

6. $86 - 0.0079x^2 = 80 - 0.60x.$

7. You now have an equation in one variable, x, that can be used to locate the landing point (the intersection of the graphs). Solve it to find the x-coordinate of the landing point.

7. $-0.0079x^2 + 0.60x + 6 = 0$.

   Using the quadratic formula, $x = -8.946210108\ldots$ and $x = 84.8955772\ldots$ The x-coordinate of the landing point is about 84.90 m.

8. As with most quadratic equations, you found two solutions. Only one can be the x-coordinate of the landing point. What is the significance, if any, of the other solution?

8. **A negative result has no meaning in this context because it implies a landing before the takeoff point. $x = -8.94\ldots$ is the x-coordinate of another intersection point of the graphs of y_1 and y_2 but one that has no physical reality in this context.**

9. You now know the x-coordinate of the landing point. Because the two functions have equal value at this point, you can find the y-coordinate of the landing point by evaluating either function for your value of x. Write the coordinates of the landing point.

9. **Using the linear function for the hill surface, $y = 80 - 0.60(84.8955772) = 29.0626\ldots \approx 29.06$ m. Therefore, the coordinates of the landing point are $(x, y) = (84.90, 29.06)$.**

The solution to a system of two equations is an ordered pair or pairs (x, y) that satisfy both equations in the system. Each equation given in Discovery 6.5 can be satisfied by an unlimited set of such ordered pairs, but only two pairs satisfy both. And only the pair (84.90, 29.06) represents a possible landing point for the ski jump.

The algebraic method used in items 6 and 7 of Discovery 6.5 is called a **substitution method** because it involves substituting an expression in one variable for another variable in an equation. The substitution results in a single equation that combines the information contained in *both* original equations. Any ordered pair that is a solution of this combined equation will satisfy both of the original equations. Each such pair of values of x and y is a solution of the system of equations.

> The assumption of a parabolic trajectory for ski jumping is an idealization and ignores air effects. In reality, a human body meets with considerable air resistance, distorting the path from that of a true parabola. Jumpers try to streamline their bodies in order to improve their aerodynamics and minimize air resistance. In recent years, they have switched from keeping their skis parallel during a jump to spreading them in the shape of a V, using the air in a positive way to increase buoyancy and prolong their jumps.

This method can usually be used to solve a system of two equations, provided that at least one of the equations can be solved for one of the variables by algebraic methods.

Example 13

Solve the system $y + 1 = 2x^2 + 7x$ and $x + y = 9$ by substitution.

Solution:

Because the second equation is linear, solve it for y. Then substitute the resulting expression for y in the first equation and solve the new equation for x.

$y + 1 = 2x^2 + 7x$ and $x + y = 9$	Original system
$y = 9 - x$	Solve the second equation for y.
$(9 - x) + 1 = 2x^2 + 7x$	Substitute $(9 - x)$ for y in the first equation.
$2x^2 + 8x - 10 = 0$	Standard quadratic form
$2(x + 5)(x - 1) = 0$	Factor the left side.
$x = -5$ or 1	Solve for x.

The corresponding y-values can be found by substituting each of these x-values into either of the original equations. You can use the equation $x + y = 9$. Using $x = -5$ results in $y = 14$, whereas using $x = 1$ results in $y = 8$. The solutions of the given system of equations are the two (x, y) pairs $(-5, 14)$ and $(2, 8)$.

Example 14

Solve the system $x + \dfrac{6}{y} = 5$ and $4x - 3y = 2$.

Solution:

$x + \dfrac{6}{y} = 5$ and $4x - 3y = 2$	Original system
$x = 5 - \dfrac{6}{y}$	Solve the first equation for x.
$4\left(5 - \dfrac{6}{y}\right) - 3y = 2$	Substitute $\left(5 - \dfrac{6}{y}\right)$ for y in the second equation.
$20 - \dfrac{24}{y} - 3y = 2$	Distributive property
$20y - 24 - 3y^2 = 2y$	Multiply both sides by y.
$3y^2 - 18y + 24 = 0$	Standard quadratic form
$3(y - 4)(y - 2) = 0$	Factor the left side.
$y = 4$ or 2	Solve for y.

When $y = 4$, $x = \left(5 - \dfrac{6}{4}\right) = \dfrac{7}{2}$, so one solution is $(x, y) = \left(\dfrac{7}{2}, 4\right)$.

When $y = 2$, $x = \left(5 - \dfrac{6}{2}\right) = 2$, so another solution is $(x, y) = (2, 2)$.

Example 15

Use a graph to solve the system $2x - 5y = 6$ and $\dfrac{1}{x-3} = y - 1$ to the nearest hundredth.

Solution:

The graphical method from items 4 and 5 of Discovery 6.5 can be used. In order to enter the equations into a graphing calculator, each equation must first be solved for y. The functions $y = \dfrac{2}{5}x - \dfrac{6}{5}$ and $y = \dfrac{1}{x-3} + 1$ are entered (see **Figure 6.32**).

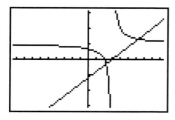

$[-9.4, 9.4] \times [-3.1, 3.1]$

Figure 6.32

The graphs intersect at two points, so there are two solutions to the system. By zooming in on each intersection point, the solutions can be estimated (see **Figures 6.33a** and **6.33b**).

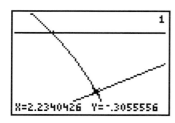

X=2.2340426   Y=-.3055556

$[1.8, 2.6] \times [-0.4, 0,1]$

Figure 6.33a

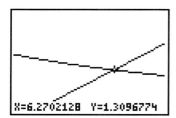

X=6.2702128   Y=1.3096774

$[5.6, 6.6] \times [1.1, 1.6]$

Figure 6.33b

One solution is $(x, y) \approx (2.23, -0.31)$. The other solution is $(x, y) \approx (6.27, 1.31)$.

Systems of Linear Equations

There are many applications that involve systems of two linear equations. Substitution works especially well in these cases because each equation can always be solved for either variable.

Example 16

Solve the system $y = 3x - 2$ and $x + 4y = 20$.

Solution:

The solution to the system can be found by substituting the expression $3x - 2$ for the y in the second equation:

$y = 3x - 2$ and $x + 4y = 20$	Original system
$x + 4(3x - 2) = 20$	Substitute $(3x - 2)$ for y in the second equation.
$x + 12x - 8 = 20$	Distributive property
$13x = 28$	Simplify.
$x = \dfrac{28}{13}$	Divide both sides by 13.

When $x = \dfrac{28}{13}$, $y = 3\left(\dfrac{28}{13}\right) - 2 = \dfrac{58}{13}$, so the solution is $(x, y) = \left(\dfrac{28}{13}, \dfrac{58}{13}\right)$.

The solution can be verified by substitution into the other original equation:

$$x + 4y = 20$$

$$\frac{28}{13} + 4\left(\frac{58}{13}\right) \stackrel{?}{=} 20$$

$$\frac{260}{13} \stackrel{?}{=} 20$$

$$20 = 20 \checkmark$$

Of course, the solution can be approximated, for instance, as $(2.15, 4.46)$. But we must remember that when any solution is approximated, it no longer exactly satisfies the system:

$2.15 + 4(4.46) = 19.99$, which to the accuracy of the solution is approximately 20.

Example 17

Solve the system $5x - 2y = 8$ and $3x - 4y = -12$.

Solution:

Solve one of the equations for either variable. For instance, the first equation can be solved for x:

$5x = 2y + 8$ and $3x - 4y = -12$ Original system

$$x = \frac{2}{5}y + \frac{8}{5}$$ Solve the first equation for x.

$$3\left(\frac{2}{5}y + \frac{8}{5}\right) - 4y = -12$$ Substitute $\left(\frac{2}{5}y + \frac{8}{5}\right)$ for x in the second equation.

$$\frac{6}{5}y + \frac{24}{5} - 4y = -12$$ Distributive property

$$-14y + 24 = -60$$ Multiply both sides by 5.

$$-14y = -84$$ Subtract 24 from both sides.

$$y = 6$$ Divide both sides by -14.

$$5x - 2(6) = 8$$ Substitute into an original equation to find x.

$$x = 4$$ Solve for x.

The solution is $(x, y) = (4, 6)$. Be sure to check your solution in the original equation.

Example 18

In previous sections of this chapter, we have examined characteristics of structural beams. The 10-foot-long beam shown in **Figure 6.34** supports a load of 1000 pounds at a location 3 feet from its left end. In such a case, the supports at the ends of the beam must exert different forces F_1 and F_2 (called *reaction forces*) on the beam in order to hold it up.

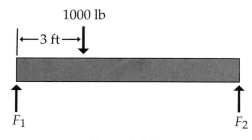

Figure 6.34

The forces F_1 and F_2 must satisfy the following system of equations:

$F_1 + F_2 = 1000$

$3F_1 - 7F_2 = 0$

Find the values of F_1 and F_2.

Solution:

$F_1 + F_2 = 1000$ and $3F_1 - 7F_2 = 0$ Original system.

$F_2 = 1000 - F_1$ Solve the first equation for F_2.

$3F_1 - 7(1000 - F_1) = 0$ Substitute $(1000 - F_1)$ for F_2 in the second equation.

$3F_1 - 7000 + 7F_1 = 0$ Distributive property

$10F_1 - 7000 = 0$ Simplify.

$10F_1 = 7000$ Add 7000 to both sides.

$F_1 = 700$ Divide both sides by 10.

$700 + F_2 = 1000$ Substitute 700 for F_1 in one of the original equations.

$F_2 = 300$ Solve for F_2.

The reaction forces are $F_1 = 700$ pounds at the left end of the beam and $F_2 = 300$ pounds at the right end.

Example 19

Concrete is a mixture of cement and aggregate, which is typically sand or gravel. The construction engineer on a high-rise building project wants to use concrete that is 64% cement and 36% aggregate. Each 600-pound batch must therefore contain $0.64(600\text{ lb}) = 384$ lb of cement and $0.36(600\text{ lb}) = 216$ lb of aggregate.

Supplies of two types of concrete are on hand, one from supplier A, which contains 70% cement and 30% aggregate, and another from supplier B, which contains 45% cement and 55% aggregate.

In order to produce the required 600-pound batch, the following system of equations must be solved (with A representing the weight in pounds of concrete from supplier A, and B representing the weight in pounds of concrete from supplier B):

$0.70A + 0.45B = 384$ Weight of cement in a 600 lb batch

$0.30A + 0.55B = 216$ Weight of aggregate in a 600 lb batch

How much of each type of available concrete should be mixed to produce a single batch of the required composition?

Solution:

Solve one of the equations for either variable. For instance, the first equation can be solved for B, and the resulting expression is then substituted in the second equation.

$0.70A + 0.45B = 384$ and $0.30A + 0.55B = 216$	Original system
$B = 853.3 - 1.556A$	Solve the first equation for B.
$0.30A + 0.55(853.3 - 1.556A) = 216$	Substitute $(853.3 - 1.556A)$ for B in the second equation.
$0.30A + 469.3 - 0.856A = 216$	Distributive property
$-0.556A + 469.3 = 216$	Simplify.
$-0.556A = -253.3$	Subtract 469.3 from both sides.
$A \approx 456$	Divide both sides by -0.556.
$0.70(456) + 0.45B = 384$	Substitute 456 for A in one of the original equations.
$B \approx 144$	Solve for B.

The required blend is 456 pounds of concrete from supplier A and 144 pounds from supplier B. Notice that in this particular problem there are actually three equations with two unknowns, because we could have also written $A + B = 600$.

The systems of equations we have considered have been limited to those containing only two unknowns. Systems containing more than two unknowns can sometimes be solved as long as the number of equations is at least as large as the number of unknowns. A modified form of the substitution method can be used to solve some of these larger systems, but other methods are frequently more efficient.

Exercises 6.5

I. Investigations *(SE pages 183–187)*

1. A small computer components company produces hard drives and rewritable CD drives. A detailed time study has shown that assembly of each hard drive takes 40 minutes, whereas assembly of a CD drive takes 53 minutes. Testing the assembled product takes 16 minutes for a hard drive and 9 minutes for a CD drive. During an average workday, 6640 total minutes of assembly labor and 1680 minutes of testing labor are available. This information is summarized in **Table 6.19.**

	Assembly	Testing
Hard Drive	40	16
CD Drive	53	9
Total Available Minutes	6640	1680

Table 6.19

How many units of each product would the company have to produce daily to keep both the assembly and testing shops operating at full capacity? In order to answer this question, we must begin by formulating a mathematical model of the company's labor resources.

a) The limitation on the time available for assembly is an example of a constraint. Write a sentence describing this constraint that contains the word *equals.*

1. **a) Total time available for assembly equals 6640 minutes.**

b) If we let H be the number of hard drives produced, write an algebraic expression for the total time spent assembling hard drives.

b) $40H$

c) Write a similar expression for the total time spent on CD drive assembly.

c) $53C$

d) Use your answers to (b) and (c) to rewrite your answer to (a) as an algebraic equation.

d) $40H + 53C = 6640$.

e) Write a sentence describing the constraint regarding total available testing time.

e) Total time available for testing equals 1680 minutes.

f) Rewrite this constraint as an algebraic equation in H and C.

f) $16H + 9C = 1680$.

g) Solve the system of equations you have written, and interpret the solution.

g) 40H + 53C = 6640 and 16H + 9C = 1680. Solving the second equation for H results in H = 105 – 0.5625C. Substitute for H in the first equation:

40(105 – 0.5625C) + 53C = 6640.

C = 80.

Substitute C = 80 into H = 105 – 0.5625C, and H = 60. The assembly and testing shops will operate at full capacity if 80 CD drives and 60 hard drives are produced each day.

2. Two circular gears are to be designed so that their centers are 4.2 cm apart horizontally and 1.6 cm apart vertically (see **Figure 6.35**). The lowest points on the rims of both gears are to be at the same level, and the circles representing the gears are tangent to each other. How large should the gears be?

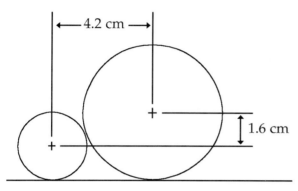

Figure 6.35

a) Start by drawing a right triangle with its hypotenuse as the line joining the centers of the gears. The legs should be vertical and horizontal lines.

2. a)

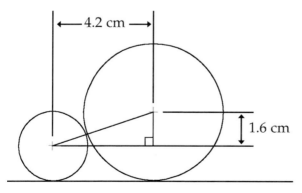

b) Write the length of the hypotenuse of the right triangle in terms of dimensions of the gears. (Let R = the radius of the larger gear and r = the radius of the smaller gear.)

b) The hypotenuse equals the sum of the radii, which can be written (R + r).

c) Use the Pythagorean theorem to write an equation containing your answer to (b).

c) $(1.6)^2 + (4.2)^2 = (R + r)^2$, **or** $(R + r)^2 = 20.2$.

d) By examining the drawing from (a), write another equation containing R and r.

d) $R - r = 1.6$.

e) Solve the system of equations you have written to find the sizes of the gears.

e) Solving the second equation for R yields $R = r + 1.6$. Substitute $(r + 1.6)$ for R in the first equation:

$$(r + 1.6 + r)^2 = 20.2$$

$$(2r + 1.6)^2 = 20.2$$

$$4r^2 + 6.4r + 2.56 = 20.2$$

$$4r^2 + 6.4r - 17.64 = 0.$$

Alternatively, take the square root of both sides of the first equation: $R + r = \sqrt{20} \approx 4.49$. Then solve the resulting linear system.

Solutions are −3.05 and 1.45. The radius cannot be negative, so the smaller gear has a radius of 1.45 cm, whereas the larger has a radius of 1.45 cm + 1.6 cm = 3.05 cm.

3. In Exercise 3 of Section 5.6, a simplified two-dimensional model of the Global Positioning System was introduced. In the model, GPS transmitters were located 500 km apart at A and B (see **Figure 6.36**). A receiver at point P determined its distances from A and B to be 384 km and 267 km, respectively.

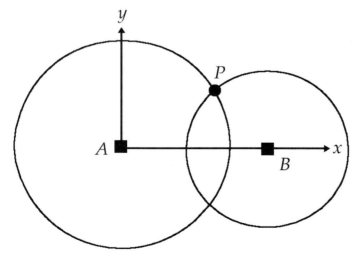

Figure 6.36

a) If A is located at $(0, 0)$, then the equations of the two circles in Figure 6.36 are $x^2 + y^2 = 147{,}456$ and $x^2 - 1000x + 250{,}000 + y^2 = 71{,}289$. Because point P lies on both circles, this is a system of equations for determining the coordinates of P. Find the location of point P correct to the nearest meter, which is the accuracy of GPS. (Notice that the expression $(x^2 + y^2)$ appears in both equations. Use this fact to help you solve the system.)

3. a) From the first equation, $x^2 + y^2 = 147{,}456$. Substitute $147{,}456$ for $(x^2 + y^2)$ in the second equation, which can be written
$$x^2 + y^2 - 1000x + 250{,}000 = 71{,}289.$$

$$147{,}456 - 1000x + 250{,}000 = 71{,}289$$

$$326{,}167 = 1000x$$

$$x = 326.167.$$

Substitute this value into $x^2 + y^2 = 147{,}456$ to find y:

$$(326.167)^2 + y^2 = 147{,}456$$

$$y \approx 202.660.$$

The coordinates of P are $(326.167, 202.660)$.

b) Your answer to (a) is not unique. In this two-dimensional version of GPS, there is one other point that satisfies the requirements. Where is it?

b) The other point where the circles intersect is $(326.167, -202.660)$.

4. We have seen that a system of two linear equations can have a single solution that consists of a pair of values of the two variables. Such a solution gives the coordinates of the point of intersection of the graphs of the two equations. But there are some systems for which a unique solution does not exist.

a) Consider the following system:

$$x - 5y = 20.$$

$$2x + 8 = 10y.$$

Solve the first equation for x and substitute the resulting expression for x in the second equation. What happens when you try to solve the system?

4. a) $\qquad\qquad x = 5y + 20$

$$2(5y + 20) + 8 = 10y$$

$$10y + 32 = 10y$$

$$32 = 0.$$

This equation makes no sense. The system can't be solved.

b) This is an example of a system that is called **inconsistent,** because the information about x and y contained in the first equation is not consistent with the information in the second equation. The system has no solution. To see this result in another way, graph both equations and describe what you see.

b) **The lines are parallel. They have no points in common.**

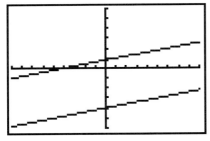

[–9, 9] x [–6, 6]

c) Now consider the following system:

$3x - 4y = 7.$

$12y + 21 = 9x.$

Try solving this system by substitution. What happens?

c) **Solving for x in the first equation,**

$x = \dfrac{4}{3}y + \dfrac{7}{3}$, **then substituting in the second:**

$12y + 21 = 9(\dfrac{4}{3}y + \dfrac{7}{3})$

$12y + 21 = 12y + 21.$

This equation is true for any value of y.

d) Again, there is not a unique solution. But the system has a solution for every possible value of y. Such a system is called **dependent.** Choose a value (any value) for y and substitute it into each equation to find the corresponding x.

d) **If, for instance, $y = 6$, $x = 22$ in each equation. Every possible value for y will give a solution for both equations.**

e) To see this result in another way, graph both equations and describe what you see.

e) Both equations graph as the same line. Every point that is on one line is also on the other.

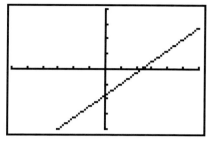

[–6, 6] x [–4, 4]

5. The substitution method is not the only method that can be used to solve linear systems of equations. Another method, called the **elimination method,** can also be used on such systems. Extensions of this method are the basis for computer solutions that can be adapted to much larger systems of equations.

Consider the following system:

$2x + 5y = 20$

$4x - 3y = 1$

Notice that the coefficient of x in the second equation is exactly twice the coefficient of x in the first equation.

a) Multiply both sides of the first equation by –2.

5. **a) (–2)(2x + 5y) = (–2)(20)**

$$-4x - 10y = -40.$$

b) Write a new system consisting of your answer to (a) and the second equation in the original system.

b) –4x – 10y = –40

4x – 3y = 1

Notice that the coefficient of x in the first equation is now the opposite of the coefficient of x in the second equation.

c) Add the left-hand side of the second equation in this new system to the first equation, and do the same for the right-hand side.

c) –13y = –39.

Because you have added equal expressions to both sides of the first equation, the resulting equation must be true. Notice that this result has combined the information in the original equations into a single equation in such a way as to eliminate the variable x.

d) Solve the equation from (c).

d) $y = 3$.

e) As with the substitution method, you can now substitute the value you found for y into either of the original equations to find the corresponding value of x. Do so, and find the solution for the system.

e) **Using the first original equation, $2x + 5(3) = 20$,**

$$2x + 15 = 20$$
$$2x = 5$$
$$x = 5/2.$$

The solution of the system is (3, 5/2).

f) Now consider the following system:

$$5x - y = 4$$
$$3x + 4y = 30$$

Multiply both sides of the first equation by a number that will result in a new equation with a coefficient of y that is equal and opposite to the coefficient of y in the second equation.

f) $4(5x - y) = 4(4)$

$$20x - 4y = 16$$

g) Add this new equation to the second equation in the original system, and find the solution for the system.

g) **Adding $20x - 4y = 16$ to $3x + 4y = 30$ results in $23x = 46$, so $x = 2$. Then using the first of the original equations, $5(2) - y = 4$ and $y = 6$. The solution of the system is $(x, y) = (2, 6)$.**

h) The value of y could also have been found using elimination. The coefficients of x in the original equations are 5 and 3. Multiply each term of the first equation by 3 and each term of the second equation by –5. Then add the two resulting equations.

h) $(3)(5x - y) = (3)(4)$

$$15x - 3y = 12 \qquad \text{\textbf{"New" version of first equation}}$$

$(-5)(3x + 4y) = (-5)(30)$

$$-15x - 20y = -150 \qquad \text{\textbf{"New" version of second equation}}$$

Adding the "new" equations yields $-23y = -138$, from which $y = 6$, as before.

i) Use the elimination method to find the values of both x and y in the following system:

$6x + 2y = -5$

$2x - y = 5$

i) **To eliminate x, multiply the second equation by –3, then add the resulting equation to the first equation.**

$-6x + 3y = -15$

$6x + 2y = -5$

When these equations are added, the result is $5y = -20$, so $y = -4$.

To eliminate y, multiply the second equation by 2, then add the resulting equation to the first equation.

$4x - 2y = 10$

$6x + 2y = -5$

When these equations are added, the result is $10x = 5$, so $x = 1/2$.

The solution is $(x, y) = (1/2, -4)$.

j) Use the elimination method to find the values of both x and y in the following system:

$-4x + 2y = 6$

$5x + 7y = 2$

j) **To eliminate x, multiply the first equation by 5 and the second by 4, then add.**

$-20x + 10y = 30$

$20x + 28y = 8$

When these equations are added, the result is $38y = 38$, so $y = 1$.

To eliminate y, multiply the first equation by –7 and the second by 2, then add.

$28x - 14y = -42$

$10x + 14y = 4$

When these equations are added, the result is $38x = -38$, so $x = -1$.

The solution is $(x, y) = (-1, 1)$.

6. In this section, we have seen a variety of systems consisting of two equations. The graph of a system of two linear equations has exactly one point of intersection, except for inconsistent and dependent systems as discussed in Exercise 4. But Discovery 6.5 and Examples 13, 14, and 15 showed that systems containing nonlinear equations can have more than one point of intersection.

a) Consider a two-equation system consisting of the rational function $y = \dfrac{1}{x}$ (with the graph shown in **Figure 6.37**) and a linear function. It is possible for such a system to have different numbers of intersections, depending on the form of the linear function. In this case, the system can have either no solution, one solution, or two solutions. Draw sketches illustrating possible graphs for each of these possibilities.

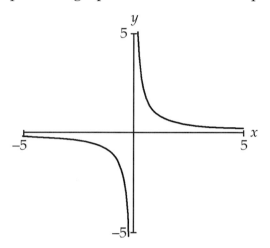

Figure 6.37

6. **a) There can be zero, one, or two intersections. Sample graphs:**

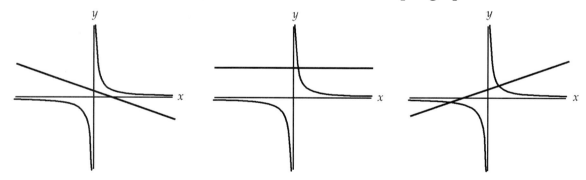

b) Now consider a system consisting of a quadratic function and the equation of a circle. Draw sketches illustrating the different possible numbers of intersections for a parabola and a circle.

b) There can be from zero to four intersections.

II. Projects and Group Activities *(SE page 188)*

7. Materials: small straws, scissors, ruler

   a) Measure the length of a straw, then cut it approximately in half and measure the lengths of the cut parts.

7. a) If the straw is 6 inches long, the parts should be close to 3 inches in length.

   b) Calculate the ratio of the length of the original straw to the length of the larger part (that is, if one is larger than the other). Also calculate the ratio of the length of the larger part to the length of the smaller part.

b) $\dfrac{6\text{ in}}{3\text{ in}} = 2, \dfrac{3\text{ in}}{3\text{ in}} = 1$. There will be some variation, but students' ratios should be similar to these.

   c) Do you think you can cut a straw into two parts so that the ratio of the length of the whole straw to the length of the larger part is equal to the ratio of the lengths of the two parts? Try it, and see what the ratios are.

c) Ratios should be close to 1.6 (depending on students' estimation skills).

   d) If your ratios are close to each other, you have determined an approximation of what is known as the **golden ratio** (symbolized by ϕ, and sometimes called the *golden mean* or *golden section*). Its exact value can be found by solving a system of equations. Consider the sketch of a straw in **Figure 6.38,** with the lengths of the larger and smaller cut parts denoted by y and x, respectively. Write a linear equation expressing the relationship between the lengths x and y based on this figure.

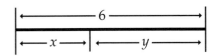

Figure 6.38

d) $x + y = 6$.

e) The golden ratio results when the ratio of the length of the whole straw to the length of the larger part is equal to the ratio of the length of the larger part to the length of the smaller part. Write an equation that expresses this equality as a proportion.

e) $\dfrac{6}{y} = \dfrac{y}{x}$.

f) Your answers to (d) and (e) form a system of two equations. Solve the system.

f)

$x = 6 - y$ **Solve the equation from (d) for x.**

$\dfrac{6}{y} = \dfrac{y}{6-y}$ **Substitute $(6 - y)$ for x in the equation from (f).**

$y(6-y)\left(\dfrac{6}{y}\right) = y(6-y)\left(\dfrac{y}{6-y}\right)$ **Multiply both sides by $y(6 - y)$.**

$(6 - y)(6) = y^2$ **Simplify.**

$y^2 + 6y - 36 = 0$ **Standard form**

$y = -9.708...$ or $y = 3.708...$ **Quadratic formula**

A negative result has no meaning in this context because lengths must be positive. So $y \approx 3.708$ in, and $x = (6 - 3.708) = 2.292$ in.

g) The golden mean ϕ is the ratio $\dfrac{6}{y}$ or $\dfrac{y}{x}$. Use your answer from (f) to find the golden mean.

g) $\dfrac{6}{3.708} = \dfrac{1.618}{1}$, $\dfrac{3.708}{2.292} = \dfrac{1.618}{1}$. **The exact value is $\phi = \dfrac{1+\sqrt{5}}{2} \approx 1.618033989...$**

h) A rectangle whose length and width are in the same ratio as ϕ is called a **golden rectangle.** It is commonly thought to be an ideal sort of rectangle, in that its proportions seem to be particularly pleasing to the human eye. Examples of the golden rectangle appear widely in architectural design, even in ancient Greek buildings like the Parthenon (although the ancient Greeks were probably not familiar with the mathematical properties of the golden ratio). Find a rectangle whose length L and width W form a ratio $\dfrac{L}{W}$ that equals ϕ. Draw the rectangle.

h) The rectangle should have this shape:

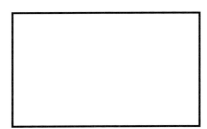

III. Additional Practice *(SE pages 189–195)*

8. **Figure 6.39** shows a schematic diagram of an electrical circuit. The currents I_1 and I_2 in two loops of the circuit must satisfy the following system of equations:

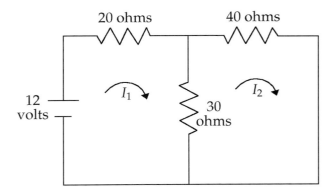

Figure 6.39

$50I_1 - 30I_2 = 12$

$-30I_1 + 70I_2 = 0$

Find the values of the currents.

8. **Solving the second equation for** I_2**,** $I_2 = \dfrac{3}{7}I_1$**. Substitute this expression into the first equation:** $50I_1 - 30(\dfrac{3}{7}I_1) = 12$**.** $I_1 = \dfrac{21}{65} \approx 0.32$ **amperes. Then**

$I_2 = \dfrac{3}{7}\left(\dfrac{21}{65}\right) = \dfrac{9}{65} \approx 0.14$ **amperes.**

9. The 200 kg weight in **Figure 6.40** is held up by wires A and B attached to the ceiling as shown. The tensions T_A and T_B (measured in newtons) in the wires must satisfy the following system of equations:

$$0.643T_A + 0.940T_B = 1960$$
$$0.766T_A = 0.342T_B$$

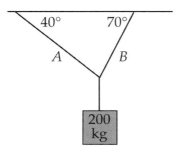

Figure 6.40

9. **Solving the second equation for T_A, $T_A = 0.4465T_B$. Substitute in the first equation: $0.643(0.4465T_B) + 0.940T_B = 1960$, and $T_B = 1597$ newtons. Then $T_A = 0.4465(1597) = 713$ newtons.**

10. In order to produce the correct operating temperature of 220°C in a soldering iron (see **Figure 6.41**), the control circuit must have a current of 0.35 milliamperes. This is accomplished by designing a voltage divider using two resistors R_1 and R_2 (measured in ohms) that must satisfy the following system of equations:

$$1.19(R_1 + R_2) = 5R_1$$
$$0.00035(R_1 + R_2) = 5$$

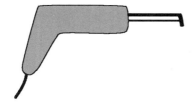

Figure 6.41

Find the values of R_1 and R_2.

10. **Solving the first equation for R_2, $R_2 = 3.20R_1$. Substitute in the second equation: $0.00035(R_1 + 3.20R_1) = 5$, and $R_1 = 3401 \approx 3400$ ohms. Then $R_2 = 10{,}884 \approx 10{,}900$ ohms.**

11. Inventory systems are often costly for large companies. For many items, inventory costs can be minimized when the total expense connected with placing and receiving orders is equal to the total expense for storage of items after purchase.

The manager of a hospital laboratory unit determines that the total yearly cost C for placing and receiving orders of syringes is related to the number N of boxes of syringes ordered at a time by the function $C = \dfrac{104{,}400}{N}$. The total yearly storage cost is $C = 7.25N$. What order size N will result in minimum total expense for the syringe inventory system?

11. $C = \dfrac{104{,}400}{N}$ and $C = 7.25N$. **Substitute 7.25N for C in the first equation.**

$$7.25N = \dfrac{104{,}400}{N}$$

$$7.25N^2 = 104{,}400$$

$$N^2 = 14{,}400$$

$$N = 120 \text{ boxes of syringes.}$$

12. Find two numbers x and y whose sum is 17 and whose product is 53.

12. **$x + y = 17$ and $xy = 53$. Substitute $(17 - x)$ for y in the second equation to get $x(17 - x) = 53$, which in standard form is $-x^2 + 17x - 53 = 0$. $x \approx 4.11$ and 12.89. When $x = 4.11$, $y = 12.89$; when $x = 12.89$, $y = 4.11$. The only pair of numbers that satisfies the requirements is 4.11 and 12.89.**

13. Find two consecutive positive integers m and n whose product is 462.

13. **$mn = 462$. If the numbers are consecutive, then $n = m + 1$. Substitute $(m + 1)$ for n in the equation $mn = 462$, and $m(m + 1) = 462$ or $m^2 + m - 462 = 0$. The factored form is $(m + 22)(m - 21) = 0$, from which $m = -22$ or 21. (Most students would probably use the quadratic formula to get the same result.) The problem requires positive numbers, so $m = 21$ and $n = 22$.**

14. Three identical braces are to be cut from the piece of wood shown in **Figure 6.42**. Find the lengths x and y if all acute angles are 45°.

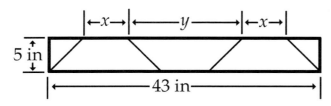

Figure 6.42

14. **The small right triangle on either end of the piece of wood is isosceles, so the horizontal leg of the triangle is also 5 inches. From the top edge of the wood, we have $2x + y + 2(5) = 43$, and from the bottom edge, we have $x + 2y = 43$. The system simplifies to $2x + y = 33$ and $x + 2y = 43$. The solution is $(x, y) = (23/3, 53/3)$, so $x = 23/3 \approx 7.7$ in and $y = 53/3 \approx 17.7$ in.**

15. Computer memory chips have to be designed so that they do not overheat. **Figure 6.43** represents a simplified version of a flat square chip on which temperatures at four nodes must be found. The constant temperatures at the edges of the plate are shown in units of °C. Nodes 1 and 2 are both at temperature T_{12}, while nodes 3 and 4 are at temperature T_{34}. The temperatures must satisfy the following system of equations:

$$T_{34} - 3T_{12} + 100 + 500 = 0$$

$$100 + T_{12} + 100 - 3T_{34} = 0$$

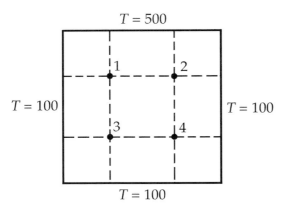

Figure 6.43

15. **The system simplifies to $-3T_{12} + T_{34} = -600$ and $T_{12} - 3T_{34} = -200$. The solution is $T_{12} = 250°C$ and $T_{34} = 150°C$.**

16. The length of a diagonal of a rectangular computer monitor screen is 20 inches. If the perimeter of the screen is 56 inches, what are the dimensions of the screen?

16. **If L and W are the dimensions, then $L^2 + W^2 = 20^2$ and $2L + 2W = 56$. Solving for L in the second equation and substituting the result in the first gives $(28 - W)^2 + W^2 = 20^2$, which simplifies to $2W^2 - 56W + 384 = 0$. $W = 12$ or 16, and the dimensions are 12 in by 16 in.**

17. **Figure 6.44** shows a 15-foot crane arm supporting a 150-pound load. The arm itself weighs 45 pounds. Find the magnitudes of the forces F_1 and F_2 that support the crane arm.

$$0.423F_1 + F_2 - 45 - 150 = 0$$

$$7.5F_2 - 3.17F_1 + 1125 = 0$$

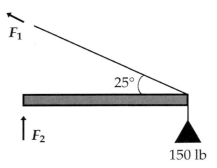

Figure 6.44

17. **The system simplifies to $0.423F_1 + F_2 = 195$ and $-3.17F_1 + 7.5F_2 = -1125$. The solution is $(F_1, F_2) \approx (408, 22)$.**

18. A motorcycle daredevil takes off from one ramp and lands on another (see **Figure 6.45**). The path through the air followed by the rear wheel of the motorcycle is $y = -0.07x^2 + 1.5x - 4.4$. The landing ramp is 3 meters high and has the equation $y = -\frac{1}{3}x + 7$. At what point does the rear wheel touch down on the landing ramp? (All distances are measured in meters.)

Figure 6.45

18. **Substitute $(-\frac{1}{3}x + 7)$ for y in the first equation: $-\frac{1}{3}x + 7 = -0.07x^2 + 1.5x - 4.4$. Standard form for the quadratic is $-0.07x^2 + 1.83x - 11.4 = 0$. Solutions are $x = 10.2$ and $x = 15.9$. The corresponding y-values are 3.6 and 1.7. Because the ramp is only 3 meters high, the only feasible solution is $(x, y) \approx (15.9, 1.7)$.**

19. A certain strain of laboratory mice must be fed a very specific diet. The feed is to be mixed in quantities that include 1.4 kg of protein and 0.6 kg of fat. Two commercial mixes are available: Mix A contains 12% protein and 7% fat, while mix B contains 18% protein and 5% fat. How many kg of each of the commercial mixes should be blended in order to produce the required diet for the mice?

19. Let *A* and *B* represent the number of kg of each mix in the blend. The system of equations is $0.12A + 0.18B = 1.4$ (for protein content in the diet) and $0.07A + 0.05B = 0.6$ (for fat content). The solution is (5.76, 3.94). The blend should contain 5.76 kg of mix A and 3.94 kg of mix B.

20. One form of iron oxide can be treated with hydrogen to change it into a different form and release water according to the following chemical reaction:

$$(A)Fe_2O_3 + H_2 \rightarrow (B)Fe_3O_4 + H_2O$$

In order to balance this chemical equation, numbers *A* and *B* must be found to satisfy the following requirements involving numbers of iron (Fe) and oxygen (O) atoms:

$2A = 3B$ (iron)

$3A = 4B + 1$ (oxygen)

Find the coefficients *A* and *B* in the chemical equation.

20. The solution to the system is *A* = 3 and *B* = 2.

21. A florist has 200 daisies and 150 irises with which to create some floral arrangements. She has decided on two basic combinations: The bargain assortment will contain 8 daisies and 3 irises, while the deluxe assortment will contain 4 daisies and 6 irises. If she is to use all the available flowers, she must solve the following system of equations:

$8B + 4D = 600$

$3B + 6D = 450$

How many of each type of arrangement will use all the available flowers?

21. *B* = *D* = 50; 50 of each type should be created.

22. An old stone archway (see **Figure 6.46**) has the equation $y = 4x - \dfrac{1}{2}x^2$, with the origin of the coordinate system located at the lower left corner of the arch. The wall has developed cracks and will be braced by a wooden beam with equation $y = 1.2x$. Find the coordinates of the upper end of the beam.

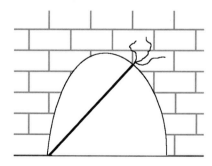

Figure 6.46

22. Replacing *y* in the first equation with 1.2*x* results in $1.2x = 4x - \dfrac{1}{2}x^2$, or $\dfrac{1}{2}x^2 - 2.8x = 0$. Solutions are *x* = 0 and *x* = 5.6, with corresponding *y*-values of 0 and 6.72. (0, 0) is the lower end, so (5.6, 6.72) is the upper end.

23. In an ideal competitive free enterprise system, the price p of a product will approach an equilibrium price based on supply and demand quantities, denoted by the variable q. If the demand equation (determined by what people will pay) for a portable disc player is $5p - 102q = 300$ and the supply equation (determined by what businesses are able and willing to produce) is $4p + 35q = 600$, find the equilibrium price of the disc player. (Here p is measured in dollars and q represents thousands of units.)

23. **The solution of the system is (p, q) = (122.9845..., 3.0874...); 3087 units should be sold at a price of \$122.98 each.**

For 24–42, solve the given systems of equations. If there are no real solutions, write "The system has no real solutions." Round off approximate solutions to the nearest thousandth.

24. $y = x^2 - 3x + 7$

$x + y = 7$

24. **The system has two solutions: (x, y) = (0, 7) and (2, 5).**

25. $3x - 4y = 2$

$6x = y + 11$

25. **The system has one solution: (x, y) = (2, 1).**

26. $y = x^2 + 1$

$2x - 3y = 6$

26. **The system has no real solutions.**

27. $y = -3x^2 + 4x + 6$

$2x - y + 3 = 0$

27. **The system has two solutions: $(x, y) \approx$ (1.387, 5.775) and (−0.721, 1.558).**

28. $\dfrac{3y - 1}{x + 5} = 10$

$3y - x = 24$

28. **The system has one solution: (x, y) = (−3, 7).**

29. $x + 2y = -3$

$2x - y = 4$

29. **The system has one solution: (x, y) = (1, −2).**

30. $y = 0.6x^2 - 1.4x + 2$

   $2x - 5y = 22$

30. The system has no real solutions.

31. $x + \dfrac{25}{y} + 4 = 0$

   $5x - y = 10$

31. The system has two solutions: $(x, y) = (1, -5)$ and $(-3, -25)$.

32. $8x + 3y = 7$

   $2x - 5y = 1$

32. The system has one solution: $(x, y) = (19/23, 3/23)$.

33. $3x + y^2 = 28$

   $x - y + 4 = 0$

33. The system has two solutions: $(x, y) = (1, 5)$ and $(-12, -8)$.

34. $2x - 3y = -7$

   $x + y = 4$

34. The system has one solution: $(x, y) = (1, 3)$.

35. $\dfrac{y}{x} = \dfrac{2x}{y} + 1$

   $3x + y = 1$

35. The system has two solutions: $(x, y) = \left(\dfrac{1}{2}, -\dfrac{1}{2}\right)$ and $\left(\dfrac{1}{5}, \dfrac{2}{5}\right)$.

36. $4x + y = 5$

   $7x - 4y = 26$

36. The system has one solution: $(x, y) = (2, -3)$.

37. $2x - 3y = 12$

   $2x + 7y = 8$

37. The system has one solution: $(x, y) = \left(\dfrac{27}{5}, -\dfrac{2}{5}\right)$.

38. $7x = 2y$

   $5x - 6y = 16$

38. The system has one solution: $(x, y) = \left(-1, -\dfrac{7}{2}\right)$.

39. $2x + y = 10$

$\quad x - 3y = -2$

39. The system has one solution: $(x, y) = (4, 2)$.

40. $\dfrac{1}{x} + \dfrac{3}{y} = 4$

$\quad x + 2y = 1$

40. The system has no real solutions.

41. $0.34x + 1.25y = 21$

$\quad 2.07x - 0.79y = 16$

41. The system has one solution: $(x, y) \approx (12.811, 13.315)$.

42. $5y = 3x + 2$

$\quad 9x = 2y + 1$

42. The system has one solution: $(x, y) \approx (0.231, 0.538)$.

For 43–44, use a graph to solve the system. Round off approximate solutions to the nearest hundredth.

43. $y = \dfrac{2}{3}x^2 + \dfrac{5}{6}x - 1$

$\quad 11x - 6y + 6 = 0$

43. Solve the second equation for y: $y = \dfrac{11}{6}x + 1$. **Then graph both equations.**

The first sample graph shows both solutions:

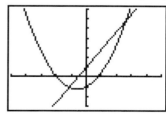

[−5, 5] **x** [−3, 7]

Zoom in on each point of intersection:

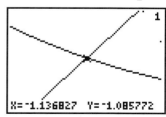

[−1.33, −0.95] **x** [−1.34, −0.85]

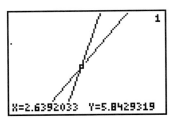

[2.34, 2.98] **x** [5.55, 6.19]

The system has two solutions: $(x, y) \approx (-1.14, -1.09)$ and $(2.64, 5.84)$.

44. $-5x + 3y = 2$

    $4x + 2y = 5$

44. Solve each equation for y: $y = \dfrac{5}{3}x + \dfrac{2}{3}$ and $y = -2x + y = -2x + \dfrac{5}{2}$. Then graph both equations. Sample graph:

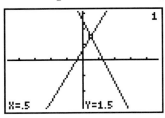

[−4.7, 4.7] **x** [−3.1, 3.1]

This window shows the single exact solution $(x, y) = (0.5, 1.5)$ or $\left(\dfrac{1}{2}, \dfrac{3}{2}\right)$.

Chapter 6 Summary

Modeling with rational functions

Evaluating rational functions

Identifying asymptotes of rational functions

Reducing rational expressions to lowest terms

Adding, subtracting, multiplying, and dividing rational expressions

Solving rational equations

Solving systems of equations

Chapter 6 Review (SE pages 197–199)

1. Write a summary of the important mathematical ideas found in Chapter 6.

1. Answers will vary. Following are some of the important ideas that should be listed:

Rational functions can be used to model real-world situations.

Rational functions can be evaluated by replacing variables in the expression with numbers as long as the denominator of the fraction does not equal zero.

Some rational functions have asymptotes. Asymptotes can be horizontal, vertical, or oblique.

A function exhibits asymptotic behavior when the graph of the function gets closer and closer to a line.

Rational expressions can often be simplified or reduced to lowest terms. This is done by factoring the numerator and denominator and then dividing the numerator and denominator by the common factors.

Rational expressions can be added, subtracted, multiplied, and divided.

To add or subtract rational expressions, the expressions must have a common denominator.

To solve rational equations, first eliminate the fractions by multiplying both sides of the equation by a common denominator.

Solutions to rational equations should be checked carefully as extraneous solutions may have been introduced when the equations were cleared of fractions.

Systems of equations can be used to model real-world situations.

Systems of equations may be solved simultaneously using algebraic methods and graphical methods.

2. Consider the function $y = \dfrac{3x}{x-4}$.

    a) Identify vertical asymptotes on the graph of the function by inspecting the function.

2. a) $x = 4$ **is an excluded value (causes the function to be undefined), so there is a vertical asymptote at** $x = 4$**.**

    b) Identify horizontal asymptotes by examining the graph with a graphing calculator.

    b) There is a horizontal asymptote at $y = 3$**.**

---

3. When a person tests positive for a disease, it is possible that the positive result is a false positive, that is, the person really doesn't have the disease. Consider an institution with a population of 1000 people, all of whom are being tested for HIV. The expression $\dfrac{hr}{rh + s - sh}$ can be used to calculate the fraction of those testing positive that actually have HIV. The variable h represents the fraction of the population that are HIV-infected, r represents the fraction of the HIV-infected people whose test gives a positive result, and s represents the fraction of the non-HIV-infected people who test positive.

Assume that for the given population, $h = 0.01$, $r = 0.98$, and $s = 0.1$.

   a) What fraction of those testing positive are actually HIV-infected?

3. **a)** $\dfrac{(0.01)(0.98)}{(0.98)(0.01) + 0.1 - (0.1)(0.01)} \approx 0.09 \text{ or } \dfrac{9}{100}$.

   b) What fraction of those testing positive have false positive tests?

   b) If 9/100 of the positive tests are actually positive, then 1 − 9/100 or 91/100 of the positive tests are false positives.

4. A baseball player's batting average is defined as the ratio of the number of hits to the number of official times at bat.

   a) Calculate a professional player's batting average if he has been at bat 135 times and has 25 hits.

4. **a) 25/135 = 0.185.**

   b) Suppose the same player hits safely in his next t times at bat. Write an expression for his new batting average.

   b) $\dfrac{25 + t}{135 + t}$

5. a) Write an expression for the average cost of n units if materials cost $2.45 per unit and there is a fixed delivery charge of $25.

5. **a)** $\dfrac{2.45n + 25}{n}$

   b) Find the average cost of 10 units, 100 units, and 1000 units.

   b) $\dfrac{2.45(10) + 25}{10} = \$4.95; \dfrac{2.45(100) + 25}{100} = \$2.70; \dfrac{2.45(1000) + 25}{1000} = \$2.48.$

6. Use the quadratic formula to solve the equation $2t^2 - 4t = 3$ for t. Be sure to simplify your solution.

6. $$2t^2 - 4t = 3$$
    $$2t^2 - 4t - 3 = 0$$
    $$t = \frac{-(-4) \pm \sqrt{(-4)^2 - 4(2)(-3)}}{2 \cdot 2}$$
    $$= \frac{4 \pm \sqrt{40}}{4}$$
    $$= \frac{4 \pm 2\sqrt{10}}{4}$$
    $$= \frac{2 \pm \sqrt{10}}{2}.$$

7. In **Figure 6.47,** find the ratio of the area of the shaded rectangle to the area of the larger, unshaded rectangle.

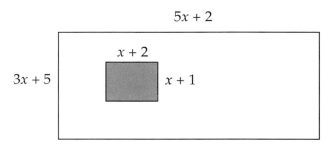

Figure 6.47

7. $$\frac{\textbf{Area of the shaded rectangle}}{\textbf{Area of the unshaded rectangle}} = \frac{(x+2)(x+1)}{(5x+10)(3x+5)} = \frac{(x+2)(x+1)}{5(x+2)(3x+5)} = \frac{x+1}{5(3x+5)}.$$

8. The expression $\dfrac{E^2R^2 - E^2r^2}{R^2 + 2Rr + r^2}$ shows the power output for a voltage source of strength E and internal resistance r acting across an input resistance R. Write the expression in simplest form.

8. $$\frac{E^2R^2 - E^2r^2}{R^2 + 2Rr + r^2} = \frac{E^2(R-r)(R+r)}{(R+r)(R+r)} = \frac{E^2(R-r)}{R+r}.$$

For 9–10, reduce each rational expression to lowest terms.

9. $\dfrac{24r^3s^2}{30r^2s^3}$

9. $\dfrac{4r}{5s}$

10. $\dfrac{z^2 + z - 12}{z^2 - 7z + 12}$

10. $\dfrac{z+4}{z-4}$

11. For what value(s) is the expression $\dfrac{m^2 - 25}{m^2 - 10m + 25}$ undefined?

11. The expression is undefined when $m = 5$.

12. Find two different pairs of rational expressions whose product is $\dfrac{t^2 - 5t + 6}{t^2 + 2t - 24}$.

12. $\dfrac{t-3}{t+6}$ **and** $\dfrac{t-2}{t-4}$ **or** $\dfrac{t-2}{t+6}$ **and** $\dfrac{t-3}{t-4}$

For 13–16, perform the indicated operations. Reduce answers to lowest terms.

13. $\dfrac{x^2 - 16}{x^3} \cdot \dfrac{x^2 - 3x}{x^2 + x - 12}$

13. $\dfrac{x-4}{x^2}$

14. $\dfrac{25x^4y^3}{36z^2} \div \dfrac{35x^3y^3}{20xz^3}$

14. $\dfrac{25x^4y^3}{36z^2} \div \dfrac{35x^3y^3}{20xz^3} = \dfrac{25x^4y^3}{36z^2} \cdot \dfrac{20xz^3}{35x^3y^3} = \dfrac{25x^2z}{63}.$

15. $\dfrac{x^2y}{3z^2} + \dfrac{2z}{6xy^2}$

15. $\dfrac{x^2y}{3z^2} + \dfrac{2z}{6xy^2} = \dfrac{x^2y}{3z^2} \cdot \dfrac{2xy^2}{2xy^2} + \dfrac{2z}{6xy^2} \cdot \dfrac{z^2}{z^2} = \dfrac{2x^3y^3 + 2z^3}{6xy^2z^2} = \dfrac{x^3y^3 + z^3}{3xy^2z^2}.$

16. $\dfrac{1}{m-2} - \dfrac{3}{m+3}$

16. $\dfrac{1}{m-2} - \dfrac{3}{m+3} = \dfrac{(m+3) - 3(m-2)}{(m-2)(m+3)} = \dfrac{-2m+9}{(m-2)(m+3)}.$

17. Solve for t. Be sure to check for extraneous solutions.

$$\dfrac{5}{t-5} = t + \dfrac{t}{t-5}$$

17. $\dfrac{5}{t-5} = t + \dfrac{t}{t-5}$; $\ 5 = t(t-5) + t$; $\ 0 = t^2 - 4t - 5$; $\ 0 = (t-5)(t+1)$.

$t = -1$ **is a solution; $t = 5$ is an extraneous solution.**

18. Solve for a. Be sure to check for extraneous solutions.

$$\frac{5a - 28}{a - 6} = a$$

18. $a = 4$ or $a = 7$

19. When two resistors R_1 and R_2 are connected in parallel, the reciprocal of the total resistance R is equal to the sum of the reciprocals of the individual resistances. That is, $\frac{1}{R} = \frac{1}{R_1} + \frac{1}{R_2}$. Solve this equation for R.

19. $\frac{1}{R} = \frac{1}{R_1} + \frac{1}{R_2}$; $RR_1R_2\left(\frac{1}{R}\right) = RR_1R_2\left(\frac{1}{R_1} + \frac{1}{R_2}\right)$; $R_1R_2 = RR_2 + RR_1$; $R_1R_2 = R(R_2 + R_1)$;

$R = \frac{R_1R_2}{R_2 + R_1}$.

20. Is the ordered pair (1, 3) a solution of the system $x + 3y = 10$ and $3x - 6y = 15$? Explain.

20. No, the ordered pair (1, 3) is not a solution of the system. It does make the equation $x + 3y = 10$ true, but not $3x - 6y = 15$. And to be a solution of the system, it must make both equations true.

21. Use a graph to solve the system $2x + y = 1$ and $6x - y = 15$.

21. Enter the equations $y = -2x + 1$ and $y = 6x - 15$ into the calculator or computer. The graphs intersect at the point (2, –3). Hence, the solution of the system is (2, –3).

22. Solve the system $\frac{3x}{y} - \frac{2}{x} = 1$ and $y - x = 4$.

22. The system has two solutions: (4, 8) and (–1, 3).

Chapter 7—Probability

Goals of the Chapter

- To understand the meaning of probability references in the news and everyday occurrences
- To become familiar with some typical workplace uses of probability
- To understand sampling as the basis of inference
- To be aware of potential uses and abuses of statistics

Preparation Reading

"The probability of rain tomorrow is 80%."

"The odds against your winning the lottery are five million to one."

"I think I have a 50-50 chance of making it to the top of the mountain."

"There is a 30% chance of finishing this project on time."

"A perpetual motion machine can't be built."

"I'm absolutely sure you'll receive it by tomorrow."

"A coin will land either heads or tails."

"This control system has a 99.9% reliability."

"The chance of an accident at that nuclear plant is one in a million."

All these statements have one thing in common: They indicate how likely it is that some event may happen. That likelihood ranges from being impossible, or at least very unlikely, to being absolutely certain to occur. We use the word **probability** to refer to a number that measures the degree of likelihood of an event.

Some probabilities, like the chance of a coin flip resulting in heads or tails, can be determined exactly without any experimentation. Such logic-based probabilities are often called **theoretical probabilities.** In other cases (for example, in the determination of defect rates for quality control), data must be collected before probabilities can be estimated. This is because no rules can be applied to find them. These are called **experimental probabilities.** A third category consists of **subjective probabilities.** These may be based on educated guesses or intuition. Subjective probabilities are sometimes partially based on experimental data. But they can also be rather arbitrary and unreliable ("I think I have a 50-50 chance of making it to the top of the mountain.").

Probability and statistics are closely related. We use the term **statistics** to refer to numbers that are either collected data or directly calculated from data. Statistics can be used to help us draw conclusions about probabilities. On the other hand, a knowledge of probability can be useful in designing experiments that involve statistical data collection.

Probability is a commonly used branch of mathematics. We see and hear probability information with regularity in the news. Weather forecasting, reported risk of disease, investment risks, and the chance of a disaster (flood, nuclear meltdown) or of winning a lottery all involve probabilities. Familiarity with a few basic probability concepts can help us understand what is actually meant by various probability statements.

Many occupations rely on probability for decision making. Businesses take probability into account for strategic planning. Modern quality control is based on probability. The reliability of electronic and mechanical systems is a type of probability. Accurate data transmission is dependent on probability-related error correction. Civil engineering makes extensive use of probability in design of structures and municipal systems. In this chapter, we will examine some typical uses of probability, both in the workplace and in society at large.

Reflect and Discuss *(SE page 201)*

1. What decisions do you sometimes make that depend on estimating a probability?

1. **Whether to carry an umbrella on a cloudy day; which route to take when driving to work or school; how early to arrive at a movie theater to be sure of getting a good seat; and so on.**

2. Arrange the following events in order from most likely to least likely. Explain how you arrived at your order.

   - Being killed by lightning
   - Having type O blood
   - Choking to death
   - Being born left-handed
   - Winning the lottery
   - Being killed in a car accident
   - Winning the table centerpiece at a wedding
   - A large asteroid hitting Earth
   - Being killed by a terrorist in a foreign country
   - An airline flight taking off on time
   - Having two children that are both girls
   - Dying from heart disease

2. Student answers will vary. The approximate chance of each event happening is given.

An airline flight taking off on time	3 out of 4
Having type O blood	1 in 2
Having two children that are both girls	1 in 4
Winning the table centerpiece at a wedding	1 in 8 (variable)
Being born left-handed	1 in 10
Dying from heart disease	1 in 400
Being killed in a car accident	1 in 6000
Choking to death	1 in 70,000
A large asteroid hitting Earth	1 in a million
Being killed by a terrorist in a foreign country	1 in 2 million
Being killed by lightning	1 in 4 million
Winning a weekly lottery	1 in 5 million (varies by state)

Section 7.1 The Meaning of Probability

What You Need To Know

- How to construct a line graph
- How to express fractions as equivalent decimals and percentages

What You Will Learn

- To calculate the probability of a simple event

Materials

- Penny or other coin

Most people, if asked to define the term *probability,* would say that it is the chance of something happening. They would probably also be aware that probabilities can have a numerical basis. But many are not familiar with how the numerical value of a probability is determined or what kinds of numbers can be used to represent probabilities. In this section, we will consider what is meant by probability and how it can be measured.

Suppose that you want to buy a pair of gardening shears for someone that you don't know well. If the store has both right-handed shears and left-handed shears, which would you buy? Not knowing whether the intended recipient of the shears is right- or left-handed, you would probably choose right-handed shears. Because most people are right-handed, this choice would be correct in most cases. We say that there is a greater probability that someone is right-handed rather than left-handed. But in any particular instance, this choice might be incorrect. In the following activity, we will examine how probabilities are related to actual occurrences.

Most large manufacturing companies have a department of quality control. It is the job of a quality control professional to determine whether the company's products work correctly when they are produced. But it is not practical to test every item to make sure it works. Rather, an inspector tests a few items and then interprets the results in terms of the manufacturing process. Quality control specialists must know how the overall quality of a product is related to the probability of finding a certain number of defective items in a sample.

Activity 7.1 Heads or Tails? *(SE pages 203–204)*

We often base simple decisions on the toss of a coin. If you and a friend are going to play a game in which one player must start before the other, you may determine who goes first by having one person call the result of a coin toss as heads or tails. Heads and tails are the only two possible **outcomes** when a coin is tossed.

1. What is meant by the statement "The chance of winning the coin toss is 50-50"?

1. **We expect that in half (or 50%) of all coin tosses, the side of the coin with a person's head on it will land facing upward. The other 50% of tosses will result in the other side (the tail side) facing upward.**

2. Toss a coin 10 times and record the result as H for heads and T for tails. In order to investigate whether a 50-50 split actually occurs, we will focus on only one of the two possible outcomes. If we choose to look at the fraction of tosses resulting in heads, then we are studying the **event** that a coin toss results in heads. What fraction of the tosses resulted in heads?

2. **Answers will vary randomly, with the fraction $\dfrac{\text{Number of heads}}{\text{Total number of tosses}}$ usually being between 3/10 and 7/10.**

3. Again toss the coin 10 times. What is the fraction out of the total of 20 tosses that resulted in heads?

3. **In one sample, the first two trials of 10 tosses both resulted in 3 heads, for a fraction of 6/20 = 3/10 heads for the 20 tosses.**

4. Repeat the process of tossing the coin 10 times and recording the number of heads and the fraction of heads for each group of 10 tosses until you have completed a total of 200 tosses. (Each set of 10 tosses can be considered a separate **trial** of the coin-tossing **experiment.**) Then complete a table, such as **Table 7.1,** showing the total number of heads as the total number of tosses increases from 10 to 200. Include the fraction of heads at each stage, expressed both as a proper fraction and a decimal. Then construct a line graph showing the fraction of heads vs. the total number of tosses (see **Figure 7.1**).

Total Tosses	Number of Heads	Fraction of Heads	Decimal Equivalent
10			
20			
30			
etc.			

Table 7.1

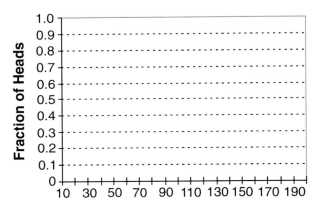

Total Tosses

Figure 7.1

4.

Trial	Number of Heads	Fraction of Heads
1	3	3/10
2	3	3/10
3	6	6/10
4	6	6/10
5	5	5/10
6	6	6/10
7	7	7/10
8	5	5/10
9	1	1/10
10	3	3/10
11	5	5/10
12	6	6/10
13	7	7/10
14	5	5/10
15	6	6/10
16	5	5/10
17	7	7/10
18	8	8/10
19	4	4/10
20	3	3/10

Total Tosses	Number of Heads	Fraction of Heads	Decimal Equivalent
10	3	3/10	0.3
20	6	6/20	0.30
30	12	12/30	0.40
40	18	18/40	0.45
50	23	23/50	0.46
60	29	29/60	0.48
70	36	36/70	0.51
80	41	41/80	0.51
90	42	42/90	0.47
100	45	45/100	0.45
110	50	50/110	0.45
120	56	56/120	0.47
130	63	63/130	0.48
140	68	68/140	0.49
150	74	74/150	0.49
160	79	79/160	0.49
170	86	86/170	0.51
180	94	94/180	0.52
190	98	98/190	0.52
200	101	101/200	0.51

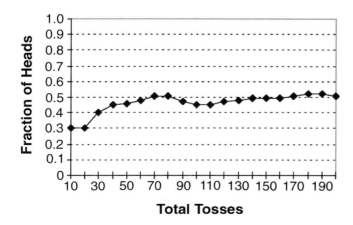

5. What do you observe about the total fraction of heads as compared with the fraction of heads in individual groups of 10 tosses?

5. In groups of 10 tosses, the fraction of heads varies quite a bit and is exactly half only about one-fourth of the time. As the total number of tosses increases, the fraction of heads stays fairly close to 0.50.

6. Pool the data from your entire class and determine the fraction of heads for the class.

6. The fraction of heads should be very close to one-half.

Because tossing a coin has only two possible outcomes, and we assume that the chances that it will fall heads or tails are equal, we predict that one-half of all coin tosses should result in heads. We then say that "the event that a coin toss results in heads has a probability of $\frac{1}{2}$." This is called a **theoretical probability** because we do not actually have to toss a coin in order to determine the probability of heads, but we can predict it from our knowledge of the way a coin should fall. In general, a probability expresses the fraction of times a particular event should occur when an experiment (such as tossing a coin) is repeated. However, as the activity demonstrated, a theoretical probability cannot predict exactly what will happen in any one instance. If the experiment is repeated many times, the fraction of the time that the particular event (such as heads) actually occurs will eventually be close to the prediction.

> The **probability** that a particular event will occur as a result of an experiment is a ratio equal to the fraction of the time that the event will occur if a very large number of repetitions (trials) of the experiment is performed.

Extend the Activity *(SE page 204)*

7. What would you expect to happen to the fraction of tosses that result in heads if you continued to toss the coin until you reached a total of 1000 tosses?

7. The fraction of heads would be very close to one-half, but the total number of heads would probably not be exactly 500.

8. a) When a six-sided die (a cube with the numbers 1 through 6 on its faces) is rolled, what is the probability that a "1" faces upward?

8. a) 1/6

   b) If you were to roll a six-sided die 12,000 times, what can you say about the number of rolls that would result in a "1" facing upward?

b) The fraction of rolls resulting in a 1 should be close to one-sixth. The number of such rolls should be about $\frac{1}{6}(12{,}000) = 400 = 2000$.

9. a) If the vowels *a, e, i, o,* and *u* are printed on slips of paper and you randomly draw one of the slips of paper before looking at it, what is the probability of drawing the letter *u*?

9. a) 1/5

   b) If the random drawing of one of the slips of paper is repeated 2000 times, how many times will the letter *u* be drawn?

b) The fraction of draws resulting in a *u* should be close to 1/5. The number of draws resulting in *u* should be about $\frac{1}{5}(2000) = 400$.

10. Consider again the decision about whether to buy right-handed or left-handed gardening shears from the beginning of this section. About 90% of the human population is right-handed. In other words, the probability is 0.9 that a randomly selected person is right-handed.

   a) If you were to survey 10 strangers, what can you say about the number that are right-handed?

10. a) Most should be right-handed. But there may not be exactly 9 righties. Different groups of 10 might have more or fewer than 9 right-handed people.

   b) The population of Tucson, Arizona, is about 800,000. What can you say about the number of residents of Tucson that are right-handed?

b) There are approximately (0.9)(800,000) = 720,000 right-handed people in Tucson.

Section 7.2 Probabilities for Compound Events

What You Need To Know

- How to find the probability of a simple event

What You Will Learn

- To calculate the probability of a compound event

- To express probabilities in different ways

- To know the difference between a deterministic model and a probabilistic model

- To use the Basic Multiplication Principle in counting

Materials

- Penny or other coin

- Paper clips

- Cardboard

- Handout 7.1

- Handout 7.2

An event can consist of just one outcome or a collection of different outcomes. Any event that is absolutely certain to occur has a probability of 1. For instance, the probability that the next president of the United States will be over 30 years old, the legal requirement of the office, is 1. An impossible event, like a single coin toss resulting in both heads *and* tails, has a probability of 0. Because a particular event cannot happen either more than all of the time or less than never, the value of a probability can never be greater than 1 or less than 0.

The probability that a coin toss will result in heads is given by the fraction $\frac{1}{2}$. This means that if a coin is tossed many times, the fraction of tosses that result in heads will be close to $\frac{1}{2}$. Because decimals and percents are alternative ways of writing fractions, the probability of heads can also be expressed as 0.5 or as 50%.

In order to determine a theoretical probability for any experimental situation, it is essential that all possible outcomes are considered. We will investigate the importance of this requirement in the next discovery.

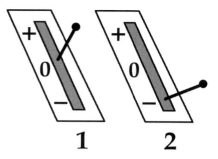

Figure 7.2

Imagine that two people are on a special bumper car ride at an amusement park. They are both in the same car, and each has a control stick with three positions, labeled "+," "0," and "−" (see **Figure 7.2**). Further imagine that neither person can see the other's control stick and that the car will move only when the sticks are set at different positions. If each person randomly chooses a position for the stick, what is the probability that the car will move?

1. Investigate this question by making two simple three-position spinners from Handout 7.1. Cut out the circles and attach them to cardboard. Then stick a pencil through a paper clip at the center of one of the circles and spin the paper clip around the pencil point. If you spin one of the spinners many times, what fraction of the spins do you expect would result in a "+"? What is the probability of a spin resulting in a "+"?

1. **Approximately one out of three, or 1/3, of the spins should be a "+," so the probability of a "+" is 1/3.**

2. Have one person spin one of the spinners and someone else spin the other, and record the position of each. If these were the control sticks of the bumper car, would the car move?

2. **Answers will vary, depending on spinner positions. The car would move only if the spinner readings are different.**

3. Now record 100 spins of both spinners in a table, such as **Table 7.2**. For each pair of spins, indicate whether the car would move.

Trial Number	Spinner #1	Spinner #2	Car Moves? (Y/N)
1			
2			
3			
etc.			

Table 7.2

3. **Answers will vary. Typical data are summarized in item 4.**

4. Use **Table 7.3a** and **Table 7.3b** to summarize the results from item 3.

Spinner #1	Spinner #2	Frequency	Fraction
+	+		
+	0		
+	–		
0	+		
0	0		
0	–		
–	+		
–	0		
–	–		

Table 7.3a

State of Motion	Frequency	Fraction
Car Moves		
Car Doesn't Move		

Table 7.3b

4. **Sample answer:**

Spinner #1	Spinner #2	Frequency	Fraction
+	+	12	12/100
+	0	8	8/100
+	–	12	12/100
0	+	11	11/100
0	0	11	11/100
0	–	13	13/100
–	+	13	13/100
–	0	7	7/100
–	–	13	13/100

State of Motion	Frequency	Fraction
Car Moves	64	64/100 = 16/25
Car Doesn't Move	36	36/100 = 9/25

5. You can use a visual model to help you find a theoretical probability for whether the car will move. Draw a square and let the top edge represent control stick (or spinner) #1. Divide the edge into three equal parts labeled "+," "0," and "−." Then let the left edge represent stick (or spinner) #2 and divide it in the same way. Draw in horizontal and vertical lines based on these divisions so that the square is subdivided into smaller squares. How many smaller squares are there?

5. There are nine smaller squares.

Spinner #1

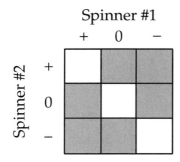

Each of the smaller squares can represent one of the possible combinations of positions of the control sticks. The three positions on each of the sticks are **equally likely,** which means that all three have equal probabilities of being selected. Therefore, all nine combinations of positions of the two sticks (represented by the nine small congruent squares) are also equally likely.

6. Shade in the small squares on your diagram that represent combinations of stick positions that will result in movement of the car. How many of these squares are there?

6.

Spinner #1

There are six such squares.

7. Use your geometrical model to make a statement on theoretical probabilities that explains the experimental results you summarized in Table 7.3b.

7. There are nine possible spinner (or control stick) combinations. Only six of these are combinations that result in movement of the bumper car, so the theoretical probability that the car will move is 6/9, or 2/3.

8. How would you design a single spinner so that it would produce only two possible results—"car moves" or "car doesn't move"—with the correct probabilities of occurrence?

8. Make the regions for "car moves" and "car doesn't move" unequal in size, and proportional to the probabilities of these two results.

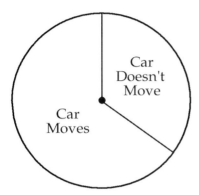

Or another possibility might be to divide the circle into nine equal parts corresponding to the nine spinner combinations and assign different colors to the groups that contain the "moves" and "doesn't move" outcomes.

As we saw in Discovery 7.1, determining a theoretical probability in an experiment requires identifying all of the possible outcomes, or results, of the experiment. But these outcomes must all have the same chance of occurring. That is, they must be equally likely.

Computing Theoretical Probabilities

The probability that a particular event will occur when an experiment is performed is found by first identifying and counting all the different possible outcomes of the experiment:

$$\text{Probability of an event} = \frac{\text{Total number of ways the event can occur}}{\text{Total number of equally likely outcomes}}$$

An event can consist of a single outcome. In an experiment in which a coin is tossed, the event that the toss results in heads consists of a single outcome. But sometimes the description of an event obscures the fact that it actually contains several equally likely outcomes, as in Discovery 7.1. If we compare the events "car moves" and "car doesn't move," we find unequal probabilities. But if we examine the underlying outcomes that produce the events "car moves" and "car doesn't move," we find that the probabilities of the nine individual combinations of control stick positions are all equally likely. Unless we identify those equally likely outcomes, we cannot correctly determine the probabilities of the events that describe states of motion of the car.

> Events may be referred to as simple events or compound events. A **simple event,** like a coin toss resulting in "heads," is one that consists of only one outcome. A **compound event** consists of two or more simple events. In the bumper cars example of Discovery 7.1, "car doesn't move" is a compound event that includes three outcomes: both control sticks at the "+" position, both control sticks at the "0" position, and both control sticks at the "−" position.

Example 1

Use a visual model to find the probabilities that if two coins are tossed, (a) they will both turn up heads; (b) they will both turn up tails; (c) they will turn up one heads and one tails.

Solution:

Draw a square with the top edge evenly divided into heads and tails, representing the first coin. Then divide the left edge into heads and tails to represent the second coin (see **Figure 7.3**). There are four equally likely two-coin outcomes. Because only one of these consists of both coins turning up heads, the probability of two heads is 1/4.

First Coin

	Heads	Tails
Second Coin — Tails	1st Heads 2nd Tails	Both Tails
Second Coin — Heads	Both Heads	1st Tails 2nd Heads

Figure 7.3

Similarly, there is only one two-coin outcome consisting of two tails, so the probability of two tails is also 1/4. However, there are two outcomes that involve one coin turning up heads and the other coin turning up tails, so the probability of such an event occurring is 2/4 = 1/2.

The Basic Multiplication Principle

As we saw in Discovery 7.1 and Example 1, accurate determination of the probability of an event requires an exact counting of all the possible equally likely outcomes that may result when a probability experiment is performed. For the bumper car situation in Discovery 7.1, each control stick could be in any of three possible positions. The total number of possible two-stick combinations is 3 x 3 = 9. Similarly, for the twin coin toss in Example 1, each coin could land in one of two orientations. The total number of possible two-coin combinations is 2 x 2 = 4. Each experiment consists of two simpler parts. In each case, the total number of possible experimental outcomes is equal to the product of the numbers of outcomes for each part. Such "counting by multiplication" can help in finding the number of equally likely outcomes, especially when it is difficult to list all the possibilities.

According to the **Basic Multiplication Principle,** if two events A and B can occur in a and b different ways, respectively, then a compound event containing both event A and event B can occur in (a x b) different ways.

Example 2

Consider a game in which the player rolls a six-sided die and also flips a coin. Determine the number of outcomes in the game.

Solution:

A die can fall in six ways, and a coin can fall in two ways. So there are 6 x 2 = 12 ways the two objects can fall. Because the number is small, it is reasonable to verify this result by listing all 12 combinations (see **Figure 7.4**).

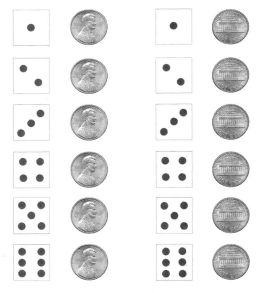

Figure 7.4

The Basic Multiplication Principle can be used for counts involving any number of events. It is only necessary to know how many ways each event can occur.

Example 3

There are many circumstances that require the use of a personal identification number (PIN). Some involve monetary transactions, such as accessing bank accounts at ATMs or transferring money between bank accounts over the phone. Others are related to computer security and privacy issues. How many different four-digit PINs are there if any number from 0 through 9 is allowed in each position?

Solution:

Each digit can be thought of as an event that can occur in 10 possible ways. There are four such events that make up the four-digit PIN. Therefore, there are $10 \times 10 \times 10 \times 10 = 10,000$ different PINs.

Deterministic and Probabilistic Models

The mathematical models we examined prior to this chapter are primarily **deterministic models.** That is, they are assumed to be exact predictors of the behavior of some system. When we write that the trajectory of a ski jumper is $y = 86 - 0.0079x^2$, it is assumed that the equation exactly predicts the location of the jumper at any point in the jump, provided the initial velocity is known precisely. Air resistance can be included to refine such a model if necessary. Writing the cost of buying CDs as $y = 15x + 10$ or expressing the amount of pollutant in a lake as $\dfrac{50 + 0.1t}{200,000 + 150t}$ are other examples of the use of deterministic models.

However, many situations for which predictions are desirable contain built-in uncertainty about the result, which cannot be eliminated. Examples include calculating the expected load on the floor of an office building, determining how long a copying machine will run without needing repairs, or estimating the chance of success of a business investment strategy. These are all cases in which **probabilistic models** are the most appropriate ones. Probabilistic models never allow us to be sure about the outcome of a particular individual case. They can only indicate, as you found in Activity 7.1, the fraction of the number of trials that a particular event will occur if an experiment is repeated many times.

As with other models we have considered, probability models can be either theory-driven or data-driven. The probability models for coin tossing and bumper car control explored in Activity 7.1 and in Discovery 7.1 are examples of theory-driven models, which can be constructed without experimentation. In Section 7.3, we will examine some data-driven probability models.

Even though some models may appear to be deterministic, they deal with quantities that cannot be known exactly. For example, structural engineers use formulas based on size and strength of beams to predict the deflection of a beam that supports a given load in a building. These formulas appear to be deterministic. But in reality, beam strength is always somewhat uncertain due to ever-present variability in the beam fabrication process. And expected loads cannot be predicted with absolute certainty either. Probabilistic models would be more realistic in such cases. Structural codes have built-in safety factors to allow for these variations.

Exercises 7.2

I. Investigations *(SE pages 212–215)*

1. The visual models used in Discovery 7.1 and Example 1 helped us list the equally likely outcomes in an experiment that consists of compound events. An alternative pictorial method for doing this is called a **tree diagram.** A tree diagram uses sets of tree branches to identify the different outcomes of the parts of an experiment. For example, **Figure 7.5** shows a tree diagram representing the outcomes of the tossing of two coins discussed in Example 1. The two branches in the first stage correspond to the two possible outcomes (heads or tails) for one coin. At the end of each branch, another set of branches is drawn. Each of these sets in the second stage of the tree shows the two possibilities for the other coin. Each path through the tree, from the starting point on the left, represents one of the four possible combinations for the two-coin experiment.

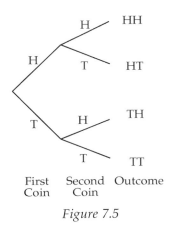

Figure 7.5

   a) Draw a tree diagram for the coin-and-die-tossing experiment discussed in Example 2. Draw the die outcomes in the first stage. How many possible outcomes are there?

1. a)

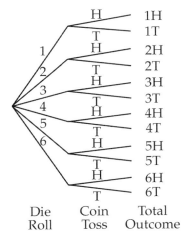

There are 12 possible outcomes.

---

b) Tree diagrams can also be used to visualize possible outcomes for experiments that contain more than two stages. Extend the tree diagram of Figure 7.5 to show the possible outcomes when three coins are tossed. (Add a third stage including sets of branches to represent the outcomes when the third coin is tossed.) How many equally likely outcomes are there? What is the probability of tossing three heads? What is the probability of tossing two heads and one tail?

b)

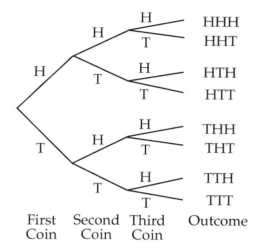

First Second Third   Outcome
Coin Coin Coin

There are eight equally likely outcomes. Because only one of these contains three heads, the probability of tossing three heads is 1/8. Because there are three outcomes that contain two heads and one tail (HHT, HTH, THH), the probability of tossing two heads and one tail is 3/8.

c) A general contractor is planning a project and must pick subcontractors for parts of the job. She can choose any one of three electrical contractors, one of four masonry contractors, and one of two plumbing contractors. Use a tree diagram to determine how many different groupings of subcontractors are possible for this project. If subcontractors are chosen randomly (so that each electrical contractor has the same chance of being chosen, and so on), what is the probability that a particular group of one electrical, one masonry, and one plumbing contractor is chosen?

c) **Draw a tree diagram in which the first stage has three branches, representing the three electrical contractors; the second stage has four branches at the ends of each first-stage branch, representing the four masonry contractors; and the third stage has two branches at the end of each second-stage branch, representing the two plumbing contractors.**

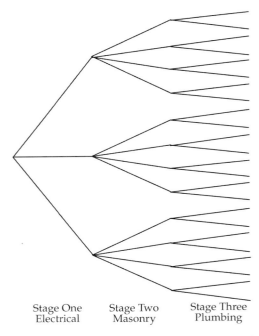

Stage One
Electrical

Stage Two
Masonry

Stage Three
Plumbing

The third stage of the tree has 24 branches, so there are 24 different possible combinations of subcontractors. If each group is equally likely, the probability of any particular group being chosen is 1/24.

2. Many states have daily or weekly lotteries that provide sources of additional revenue. Such lotteries are always managed so that they result in net income for the state, even though some individuals win large cash prizes. On the average, people who play lotteries lose more than they win. Daily lotteries are often mathematically fairly simple. But calculating the probability of winning many weekly types of lotteries is an interesting exercise in counting.

The state of Delaware holds a lottery in which a player picks any six of the numbers from 1 to 30. Then the state lottery commission holds a drawing in which six numbers from 1 to 30 are randomly chosen. If those numbers exactly match any player's set of six, that player wins a large cash prize. In order to compute the probability of winning the Delaware lottery, we must count the number of possible groups of six numbers that can be selected.

a) Imagine that you are filling out a lottery ticket for the Delaware lottery. How many choices do you have for selecting the first number in your group of six?

2. a) 30 choices

b) In lotteries of this type, no number can be repeated. After your first number is chosen, how many numbers remain from which to make your second selection?

b) 29

c) After making your first two selections, in how many ways can you choose a third number?

c) 28

d) Continuing in this way, determine how many ways there are to make your fourth, fifth, and sixth selections?

d) 27, 26, 25

e) Use the Basic Multiplication Principle to compute the total number of ways that you can fill out your lottery ticket.

e) $30 \cdot 29 \cdot 28 \cdot 27 \cdot 26 \cdot 25 = 427{,}518{,}000$ ways.

f) When your ticket is submitted, it doesn't matter in what order you entered your six numbers. For example, the group {1, 2, 3, 4, 5, 6} is the same group, for lottery purposes, as {6, 5, 4, 3, 2, 1} or any other arrangement of these six numbers. Reasoning in a manner similar to parts (a)–(e), determine how many ways a lottery ticket could be filled out using the same six numbers. Assume you already know which six numbers you will choose. Write down the number of ways of selecting one of these numbers to fill in on the form first, then the number of ways of filling in the second of your six numbers, and so on. How many ways are there of filling out the ticket for your six chosen numbers?

f) $6 \cdot 5 \cdot 4 \cdot 3 \cdot 2 \cdot 1 = 720$.

g) If you divide your answer to (e) by your answer to (f), you will have the total number of different *groups* of numbers that can be played in the Delaware lottery. How many such groups are there?

g) $\dfrac{427{,}518{,}000}{720} = 593{,}775$.

h) When the lottery drawing is held, only one of those groups will be chosen (again, order doesn't matter). What is the probability that it might be the group you picked? In other words, what is the chance of winning the Delaware lottery?

h) $\dfrac{1}{593{,}775} \approx 1.684 x 10^{-6} = 0.000001684$.

i) In the Oregon lottery, players must choose any six numbers from 1 to 44. Otherwise, this lottery is similar to the Delaware lottery. What is an individual's chance of winning the Oregon lottery?

i) $\dfrac{6 \cdot 5 \cdot 4 \cdot 3 \cdot 2 \cdot 1}{44 \cdot 43 \cdot 42 \cdot 41 \cdot 40 \cdot 39} = \dfrac{720}{5{,}082{,}517{,}440} = \dfrac{1}{7{,}059{,}052} \approx 1.417 x 10^{-7} = 0.000000417$.

j) Find out if your state has a similar kind of weekly lottery, as well as for what purpose the state uses the lottery revenue. What is the probability of an individual player's lottery pick winning your state's lottery? (Consider only lotteries structured similarly to those discussed in this exercise, which don't involve variations like special "powerballs.")

j) Answers will vary depending on the state in which the student lives.

3. We often hear the term **odds** used in connection with the chance of winning a game or competition. For example, prior to the 2001 Masters golf tournament, it was stated that the odds were 6 to 5 against Tiger Woods winning the tournament. (He did in fact win, and in doing so became the youngest golfer in history to have won the four major tournaments of golf.)

Odds are always stated as a ratio, which equals the ratio of two probability numbers. Odds of 6 to 5 against Tiger Woods means that the ratio of the probability that any of the other golfers wins to the probability that Woods wins is 6/5. This can only be true if the probability of Woods winning the tournament is 5/11 and the probability of anyone else winning is 6/11. The probabilities must add to 1, because someone must win, and $\dfrac{6/11}{5/11} = \dfrac{6}{5}$.

Odds can be stated *for* or *against* a person or team. If the Yankees are 5 to 3 favorites to beat the Braves in baseball's World Series, then the probabilities that the Yankees or Braves will win are judged to be 5/8 and 3/8, respectively. Therefore, the odds are 5 to 3 *against* the Braves.

a) If the odds against a runner winning a gold medal at the Olympics are 3 to 1, what is the probability that the runner will win a gold medal?

3. **a) 1/4**

b) Suppose that you and four friends have pooled your money to buy a TV set, and you decide to draw straws to see whose room the TV will be kept in. Four short straws and one long straw are mixed, and each person draws one of the straws, with the long straw determining the winner. What are the odds against your drawing the long straw, and what is the probability of your drawing the long straw?

b) 4 to 1; 1/5 or 0.2

c) A construction company has bid on a contract for a new office building. The company estimates that it has a 0.4 probability of winning the contract, based on past experience. What are the odds against the company winning the contract?

c) The probability of winning the contract is 2/5, and the probability of not winning is 3/5. The odds against winning are $\dfrac{3/5}{2/5} = \dfrac{3}{2}$, or 3 to 2.

4. Does 1 + 1 = 2? The answer, in many cases, is "Maybe." If the "1s" are measured numbers, then neither number is exactly 1, and their sum is not exactly 2. For example, what if you pour some water into a bucket and estimate (without measuring) that the volume of the water is about a gallon. Assuming that you are accurate only to the nearest gallon, the actual amount of water in the bucket might be anywhere between 0.5 and 1.5 gallons. If you do the same with another bucket and combine the two amounts, the total amount of water could be as little as 1 gallon and as large as 3 gallons. So 1 + 1 is not necessarily always 2 when measured quantities are combined.

a) We could ask, "What is the probability that 1 + 1 = 2 if the 1s represent quantities that are rounded off to 1?" To answer this question, an area model can be helpful. Draw a graph like that in **Figure 7.6,** where the axes represent two measurement scales. A measured quantity with a value estimated to be 1 might have an actual value somewhere between 0.5 and 1.5 as indicated on the x-axis. A second quantity with an estimated value of 1 can be represented by a segment from 0.5 to 1.5 on the y-axis. Every point in the square on the graph represents a possible pair of actual values for the two numbers that are both estimated to be 1. Point A has coordinates (0.5, 1.5), which add to 2. What are the coordinates and sums represented by points B, C, and D?

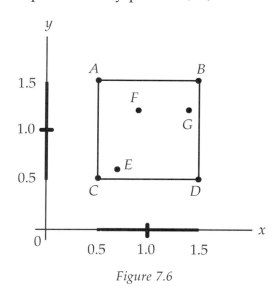

Figure 7.6

4. **a)** *B: (1.5, 1.5), 3; C: (0.5, 0.5), 1; D: (1.5, 0.5), 2*

b) Point E is (0.71, 0.58). What is the sum of the coordinates, and what integer does it round off to?

b) 1.29 ≈ 1.

c) Point F is (0.931, 1.217) and G is (1.4, 1.2). What integers do their sums round off to?

c) *F: 2, G: 3*

d) Shade in the region of the square that contains all the points whose coordinate sums round off to 2.

d)

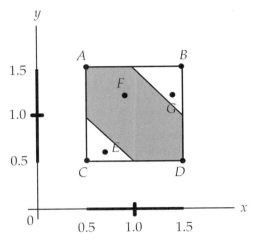

e) Divide the area of the shaded portion by the area of the entire square to find the probability that $1 + 1 = 2$.

e) 0.75/1 = 0.75, or 3/4.

II. Projects and Group Activities *(SE pages 216–218)*

5. Materials: cardboard squares or circles

   You probably know that blood type is one of the many characteristics you inherit from your parents. There is actually a fairly large number of blood types, some of which involve minor genetic differences. However, the main groupings can be classified as Type O, Type A, Type B, and Type AB. Every person's chromosomes (**Figure 7.7**) contain two genes for blood type, one from each parent. Each gene can be one of three varieties (called "alleles" by geneticists): A, B, or O. A person with two O genes will have Type O blood, while someone with two A genes will have Type A blood, and likewise for Type B. And someone with one each of A and B genes will have a hybrid Type AB blood. However, a person with an A gene and an O gene will still have Type A blood; we say that the A gene is "dominant" over the O gene, which is called "recessive": The presence of the A gene masks the presence of the O gene, in the sense that an AA individual and an AO individual both have the same blood type, Type A.

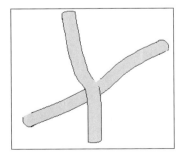

Figure 7.7

In order to explore some basic probability concepts, you can conduct a simulated genetics experiment involving the chances of inheriting certain blood types.

Write the letter O on four small circles or squares of cardboard, which will represent O genes. Do the same for A genes and B genes.

a) Place two of the O genes in one cup and another two O genes in a different cup. Each cup represents a parent having two genes for blood type. In this case, both parents have a "genotype" (or gene set) that can be symbolized OO. What blood type would each parent have?

5. **a) Each parent would have Type O blood.**

b) Draw one "gene" from each "parent" cup without looking. The pair of genes you have drawn represents the genotype for a child of those parents. What is the genotype, and what is the blood type for this child?

b) genotype: OO; blood type: O

c) If you return the genes to the "parent" cups and draw again, and repeat the process many times, what will you find?

c) Genotype will always be OO. Blood type will always be O.

d) This is an experiment for which we can be certain of the outcome. This certainty can be expressed numerically as a percent. What percent of the time will parents who both have Type O blood have a child with Type O blood? (Remember that people with Type O blood always have the OO genotype.)

d) 100% of the time

e) Now replace one of the OO parents with an AA parent. Repeat the drawing process, and for this situation express the percentage of children you expect to have Type A blood and Type O blood. (You may find a visual model helpful in answering this and later parts.)

e) Because an A gene will always be drawn from one parent and an O gene from the other, the child's genotype will always be AO, resulting in 100% of children having Type A blood. None, or 0%, will have type O blood.

f) Replace the AA parent with an AO parent (which, you remember, will also have Type A blood due to the dominance of the A gene). Without drawing any genes from the cups, predict the probability (as a percentage) that a child of these parents would have Type A blood.

f) Half of the offspring would have genotype AO, resulting in Type A blood. The other half will have OO genotype and Type O blood. Probability of Type A blood is 50%.

g) In (f), you determined a theoretical probability for which it was not necessary to perform an experiment. To test your theory, draw one gene from each parent and record the genotype and blood type of the "child." Return the genes to the "parents," have someone shake the cups to randomize the next selection, and draw again. Repeat the same process many times, and record the result each time. Then construct a line graph of the percentage of Type A children versus the total number of children. What do you observe?

g)

Trial	Number of AOs	Percentage of AOs
1	1	100%
2	2	100%
3	3	100%
4	3	75%
5	3	60%
6	4	67%
7	5	71%
8	6	75%
9	6	67%
10	7	70%
11	7	64%
12	8	67%
13	9	69%
14	10	71%
15	11	73%
16	11	69%
17	11	65%
18	11	61%
19	11	58%
20	12	60%

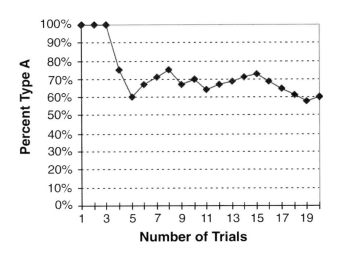

Results will vary, but this table and graph represent one possibility. Theory predicts 50% AO genotype (Type A blood), but actual counts for a sample of 20 may range from about 6 to 14. If larger samples are taken, the percentage should approach 50%.

h) Create other parent genotypes by placing different combinations of the A, B, and O genes in the two cups, and repeat (f) and (g) for each situation. (It may not always be easy to predict the theoretical probabilities. In such a case, collect the data first and then see if the graph helps you find the probability.) Summarize your findings.

h) There are six possible genotypes for each parent: OO, AO, BO, AA, BB, AB. There are 21 possible "parent" pairings, each of which may yield up to four different genotypes and as many as three different blood types. Results will be variations on the findings from (f) and (g).

6. Materials: Handout 7.2, penny, ruler

   Examine Handout 7.2, which is a grid of squares. If you toss a penny onto the grid, what do you think might be the chance that the penny will land so that it is not touching any of the lines in the grid? You will answer this question by first gathering some data and then by looking for a theoretical probability.

   a) Toss the penny onto the grid and note whether it touches any lines. Then repeat the process many times (ignoring any tosses that don't land within the perimeter of the grid), and develop a graph as you go along that shows the percentage of tosses not touching a line versus total number of tosses. When the graph begins to level off and approach a steady value, make an estimate of the probability that a tossed penny will not touch a line of the grid.

6. **a) A graph of sample data is shown.**

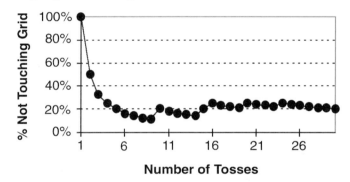

The graph appears to level off at about 20%, which is the estimated probability that a tossed penny will not touch one of the lines.

   b) To determine a theoretical probability, first consider another hypothetical case. What if the grid were just large enough so that a penny would barely touch all four sides of a square when placed in its center? What would be the probability of tossing a penny so that it would not touch a line of the grid?

   b) 0

c) Now, what if each side of the hypothetical square were doubled in length? What would be the probability that a tossed penny would not touch a line of this grid? (Hint: Think about where the center of the penny would have to land so that the penny would not touch the sides of a square.)

c)

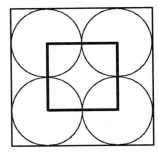

The center of the penny would have to land inside a square with each side half the length of the side of a square on the grid. This square has one-fourth the area of a square on the grid. The probability that a penny lands in such a square is 1/4.

d) Finally, measure the diameter of the penny and the length of a side of a square on the grid. Reasoning the same way you did for (c), determine the probability that a tossed penny would not touch a line of the grid. Is your answer consistent with your experimental observation in (a)?

d) The diameter of a penny is 1.8 cm; grid squares are about 3.4 cm on each side. The area of a grid square is $(3.4 \text{ cm})^2 = 11.56 \text{ cm}^2$. A penny would not touch a line if its center were within a square that is 3.4 cm − 1.8 cm = 1.6 cm on a side in the center of a grid square. The allowed area is $(1.6 \text{ cm})^2 = 2.56 \text{ cm}^2$. The probability that the penny would not touch a line is equal to $2.56 \text{ cm}^2 / 11.56 \text{ cm}^2 = 0.22145... \approx 0.22$.

e) Use a similar analysis to solve the following problem: An empty work room has the dimensions shown in **Figure 7.8**, with six electrical outlets. Two of the outlets are located near the floor at the centers of the shorter walls, and the others are equally spaced along the longer walls. Each outlet is $4\frac{1}{2}$ inches across. An employee is asked to pile some heavy boxes containing computer equipment against a wall in a single stack. Each box is a 3-foot cube. If the employee doesn't pay attention to the electrical outlets, what is the probability that the boxes will be placed so that they don't cover up any part of the outlets?

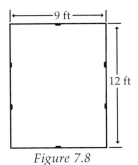

Figure 7.8

---

e) The center of the bottom box will be $1\frac{1}{2}$ ft from the wall. The possible positions of the center of the box therefore lie on a 9 ft-by-6 ft rectangle, which has a perimeter of 30 ft = 360 in. The box will block an outlet if its center is along a line of length $\left(3 \text{ ft} + 4\frac{1}{2} \text{ in}\right) = 40\frac{1}{2}$ in on this rectangle and in front of the outlet. Because there are six outlets, a total of $6\left(40\frac{1}{2} \text{ in}\right) = 243$ in of the rectangle's perimeter is not allowed if the outlets are to remain unobstructed. This leaves 360 in − 243 in = 117 in for placing the box's center out of the total of 360 in. The probability of placing the box so that no outlet is obstructed is (117 in)/(360 in) = 0.325, or 3/8.

III. Additional Practice (*SE pages 218–221*)

7. A company holds a contest to win a new car. Six keys are shown to the group of six finalists. Only one of the keys will start the car. One of the finalists selects a key and keeps the car if the chosen key starts it. Otherwise, the key is discarded and the next finalist selects a key, and so on, until a winner is found. Is there an advantage to being the first finalist to select a key?

7. **No, all keys are initially equally likely to start the car, so everyone has a 1/6 probability of winning. The order of selection doesn't matter. (Problem is taken from the column "Ask Marilyn" in *Parade* magazine.)**

For 8–11, indicate whether the groups of events described are equally likely.

8. The number that results when a single six-sided die is rolled

8. **Yes**

9. The sum of the two numbers that result when a pair of six-sided dice is rolled

9. **No, the number of outcomes resulting in different sums is variable. For example, a sum of 2 can occur in only one way (1 on each die), whereas a sum of 4 can occur in three different ways (1 on the first die and 2 on the second, 2 on the first die and 1 on the second, or 2 on both dice).**

10. The possible months during which a randomly selected person was born

10. **No, the number of days in a month is variable.**

11. The possible days of the week on which a randomly selected person was born

11. **Yes**

For 12–18, indicate whether the given number could be the probability of occurrence for some event. Explain.

12. 0.002

12. **Yes, 0.002 is between 0 and 1.**

13. 3/2

13. **No, 3/2 is greater than 1.**

14. 2/3

14. Yes, 2/3 is between 0 and 1.

15. 43%

15. Yes, 43% is between 0% and 100%.

16. –1/7

16. No, –1/7 is less than 0.

17. 200%

17. No, 200% is greater than 100%.

18. 0

18. Yes, 0 is the probability of an impossible event.

For 19–23, express the indicated probability as a fraction in lowest terms, as a decimal, and as a percent.

19. A parts cabinet in an electronics laboratory has a drawer that contains resistors of various sizes, all mixed together. The drawer's contents include 40 one hundred ohm resistors, 20 two hundred ohm resistors, and 30 five hundred ohm resistors. If a student reaches into the drawer and randomly picks one resistor, what is the probability that it will be a one hundred ohm resistor?

19. 40/90 = 4/9 ≈ 0.444 = 44.4%.

20. The median of a numerical data set is the middle number of the set when it is arranged in numerical order. In a large data set, what is the probability of randomly choosing a value that is below the median?

20. 1/2 = 0.5 = 50%, because half the numbers are below the median.

21. Veterans Day is one of the few nonreligious holidays that is celebrated on its original date (November 11) rather than always on a Monday. In any arbitrary year, what is the probability that Veterans Day will occur on a weekend?

21. 2/7 ≈ 0.222 = 22.2%.

22. If you guess randomly at the answer to a multiple-choice test question for which there are five choices, what is the probability that you will guess correctly?

22. 1/5 = 0.2 = 20%.

23. Of 366 possible birthdates, what is the probability that a December date will be randomly chosen in a lottery, such as the military draft lotteries that the United States has sometimes used?

23. 31/366 ≈ 0.085 = 8.5%.

For 24–26, use a visual model to help you determine the indicated probabilities.

24. Four married couples get together to play bridge. Teams are determined randomly by having each man pick one of four cards, which are labeled "Ace," "King," "Queen," and "Jack." Each woman picks from a second set of similar cards. The man and woman who each draw an Ace will be one team, the two drawing Kings will be another team, and so forth. What is the probability that a particular husband and wife pair will be on the same team?

24.

Husband

	A	K	Q	J
A	AA	AK	AQ	AJ
K	KA	KK	KQ	KJ
Q	QA	QK	QQ	QJ
J	JA	JK	JQ	JJ

Wife

There are 16 possible combinations of cards, only 4 of which are identical pairs. The probability that husband and wife are partners is 4/16 = 1/4 = 0.25 = 25%.

25. Two spinners as shown in **Figure 7.9** are spun simultaneously. What is the probability that the sum of the numbers on the spinners will be 3?

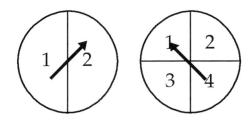

Figure 7.9

25.

2nd Spinner

	1	2	3	4
1	2	3	4	5
2	3	4	5	6

1st Spinner

(sums are inside squares)

2/8 = 1/4 = 0.25 = 25%.

26. A cubic (six-sided) die numbered 1 through 6 and a tetrahedral (four-sided) die numbered 1 through 4 are tossed (see **Figure 7.9**). What is the probability that the sum of the numbers facing down on the two dice will be 5?

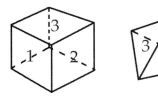

Figure 7.9

26.

Cubic Die

		1	2	3	4	5	6
	1	2	3	4	5	6	7
	2	3	4	5	6	7	8
	3	4	5	6	7	8	9
	4	5	6	7	8	9	10

Tetrahedral Die

(sums are inside squares)

4/24 = 1/6 ≈ 0.167 = 16.7%.

For 27–31, use the Basic Multiplication Principle to do the counting.

27. How many nine-digit student identification numbers are possible if each digit can be any number from 0 through 9?

27. 10^9 = 1,000,000,000.

28. The number of telephone area codes in the United States has been recently increased, primarily to meet the demand for new phone lines for cellular phones and computer connections. Currently, the first digit of an area code can be anything but 0 or 1. The second digit can be any number from 0 through 9. The third digit cannot be 0.

a) How many different area codes are currently allowed?

28. a) (8)(10)(9) = 720.

b) If all numbers from 0 through 9 are allowed for the remaining seven digits, how many possible phone numbers are there in the United States (including the area code)?

b) $(720)(10^7) = 7.2 \times 10^9 = 7.2$ billion.

29. Suppose you plan to buy components for a stereo system. If you are considering three types of receivers, four types of CD changer, and three types of speakers, but only one tape deck, how many different systems are possible?

29. $3 \cdot 4 \cdot 3 \cdot 1 = 36.$

30. How many six-character automobile license plates are possible in a state where the first three characters are numbers and the last three are letters?

30. $10 \cdot 10 \cdot 10 \cdot 26 \cdot 26 \cdot 26 = 17{,}576{,}000.$

31. A strand of DNA in a gene is a sequence of chemically joined nucleotides. There are four building blocks of the sequence, denoted by A (for adenine), T (for thymine), C (for cytosine), and G (for guanine). A gene may contain hundreds of these building blocks. How many possible 100-nucleotide sequences are there?

31. $4^{100} \approx 1.6 \times 10^{60}.$

For 32–36, indicate whether the model described is deterministic or probabilistic.

32. The number of sixteen-penny nails needed to build walls containing different numbers of studs if each stud requires four nails

32. deterministic

33. The number of hours needed to build one of the stud walls in Exercise 32

33. probabilistic

34. The amount of mercury contained in tuna of various weights caught in the North Atlantic Ocean

34. probabilistic

35. The temperature at which water in an open container boils at various altitudes

35. deterministic

36. The amount of rain in St. Louis in various months

36. probabilistic

Section 7.3 Finding Probabilities from Data

What You Need To Know

- How to calculate the probability of an event

- How to express probabilities in different forms

What You Will Learn

- To obtain probability information from relative frequencies in data

- To find the probability of the complement of an event

- To compare the relative likelihood of common and rare events

Materials

- Deck of playing cards

- Stopwatch

- Penny

- Thumbtacks

- Handout 7.3

The probabilities associated with coin flipping, lotteries, genetic traits, and card games can all be calculated without actually performing any experiments to collect data. They are sometimes called rule-based probabilities because we can use logic and mathematical rules like the Basic Multiplication Principle to compute them. But in many, and perhaps most, workplace situations where knowledge of probability is important, no rule or theory is available. In such cases, only the collection of data can provide the information necessary to determine probabilities.

Measuring the time it takes to perform a task provides useful information for many industries. The time required to cut a machine part, to cook a hamburger, to assemble an electronic switch, or to prepare an intravenous drug solution is a variable quantity. Sometimes variations of even a few seconds in the performance of such routine tasks can have significant effects on a business's productivity and profit. Operations analysts conduct work measurement studies (often called time and motion studies) in order to accumulate data on these tasks. Decisions based on the data can help companies run more efficiently.

Scientific principles were first applied to management decision making in the late 19th century by Frederick W. Taylor. His "shovel study" examined the productivity of workers shoveling ore. Although it had been thought that these men should be required to use the largest loaded shovel they could carry, Taylor showed that workers were more productive in the long run using smaller loads. This is because they could work faster without getting tired. Taylor, who essentially founded the profession of industrial engineering, is often called the "father of scientific management."

Discovery 7.2 Work Measurement and Probability *(SE page 223)*

How long does it take to deal a deck of cards? What is the probability that it might take longer than 30 seconds? These are questions that can't be answered without collecting data. We could guess the answers to these questions based on a knowledge of how many cards are in a deck and imagining a "thought experiment" in order to estimate the time required. But there is no basis for a theoretical probability determination here.

In this discovery, you will use the act of dealing a deck of cards in a specified manner to investigate probabilities associated with a typical time study. Probability values that are determined by experimentation (that is, through data collection) are called experimental probabilities or **empirical probabilities.**

1. Take a standard deck of 52 playing cards and have someone use a stopwatch to measure how many seconds it takes for you to deal the cards into 12 piles equally spaced around a circle that is about a foot and a half in diameter. Record the time.

1. **Typical times are from 20 to 40 seconds.**

2. Repeat the experiment with a new dealer, and continue recording data until everyone in the class has dealt the cards in the prescribed manner.

2. **Sample data:**

Time (s)	19	21	22	23	24	25	27	28	31	34	36
Frequency	1	2	1	2	4	2	3	2	2	1	1

3. How many total time measurements were recorded?

3. **21, based on the sample data in item 2**

4. How many of these times were longer than 30 seconds?

4. **4, based on the sample data in item 2**

5. What fraction of the times were longer than 30 seconds?

5. **4/21, based on the sample data in item 2**

6. Your answer to item 5 is a statement of fact about the time it takes members of your class to deal cards. But what if someone were to visit your classroom and ask a randomly chosen student to deal the cards? Your answer to item 5 can now be interpreted as a probability. It is the probability that the randomly selected student would take longer than 30 seconds to deal the cards. Use your answer to item 5 to make a probability statement.

6. **There is a probability of 0.19 that a randomly selected student will take longer than 30 seconds to deal the cards, based on the sample data in item 2.**

7. Express the probability as a proper fraction, as a decimal, and as a percent.

7. **4/21 ≈ 0.19, or 19%, based on the sample data in item 2**

8. In a similar manner, determine the probabilities that a randomly selected student would deal the cards in (a) less than 20 seconds, (b) no more than 30 seconds, (c) between 25 and 30 seconds, and (d) more than 40 seconds.

8. **a) 1/21 ≈ 0.05, or 5%, based on the sample data in item 2**

   b) 17/21 ≈ 0.81, or 81%, based on the sample data in item 2

   c) 7/21 ≈ 0.33, or 33% (including the 25-second time), based on the sample data in item 2

   d) 0, based on the sample data in item 2

9. If your class is large, your answers to item 8 could possibly have more general applicability. In the workplace, a time study of a large sample of workers might be used to establish characteristics of a task for *any* worker, not only those who were studied. But before generalizing such results, it is important to determine whether the workers in the study were typical. For instance, the distribution times expected for experienced machinists to grind a part would be different from those expected for inexperienced machinists. Can you think of circumstances under which the probabilities for card dealing found from an experiment like the one you performed in this discovery might not be valid for college students in general?

9. **If most of the students in the class were regular card players, they might be accustomed to dealing cards and have dealing times that are unusually short.**

As was pointed out in Discovery 7.2, if we use data to predict probabilities, the data must be typical for the situation being studied. In order to have confidence in your probability conclusions, you might need to sample more people than you did in that Discovery. More advanced statistical methods can be used to determine how large a sample is necessary to support such conclusions. But provided enough data is collected, experimental fractions or percentages are good approximations for probabilities. And because they are arrived at by a somewhat different process than theoretical probabilities, we will summarize that process as follows.

Statisticians usually use the word **population** when referring to a particular group being studied. In Discovery 7.2, the population consisted of the students in your class. But in a study of lake pollution, the population would be a group of lakes, and in a traffic study (see Discovery 7.4), the population might be a number of time intervals during which cars were counted. The word **proportion** also has a special meaning in statistical contexts. It is used to refer to a fractional part of a population or data set. This is a different usage of "proportion" than when it refers to an equation consisting of two equal ratios.

> **Experimental Probability**
>
> The experimental probability that a particular event will occur is found by collecting data. The probability that a randomly selected member of a group has a certain characteristic is equal to the fraction of the group that has that characteristic.
>
> $$\text{Experimental probability} = \frac{\text{Number of members with the characteristic}}{\text{Total number of members in the group}}$$

The key to the validity of any probability found from experiment is random selection. If the hypothetical visitor in item 6 of Discovery 7.2 were to select a student that was known to be an avid card player (and therefore probably an experienced dealer), the probabilities you determined in that discovery would not be expected to apply.

Example 4

Of every 205 babies born in the world, 100 are girls. What is the probability that a randomly selected baby is a girl?

Solution:

The probability is $100/205 = 20/41 \approx 0.488$, or 48.8%.

Example 5

Data on long-duration military missions in close quarters, such as in submarines, show that in a year there is a 6% chance that a person will require emergency medical care. Of 532 people on such missions, how many are likely to require emergency care in a year?

Solution:

Six percent of 532 is $(0.06)(532) = 31.92$. About 32 out of 532 people on such missions will require emergency care in a year.

The Sum of Probabilities in an Experiment

There is a simple relationship between the probabilities of all the possible outcomes of any experiment. One of the outcomes *must* happen. And we know that the probability of a *certain* event is 1. Therefore, the sum of the probabilities of all possible outcomes of an experiment is 1.

We know that when we toss a coin, there are only two possible outcomes—heads or tails. The total probability that either heads or tails turns up must equal one because there are no other possible outcomes when a coin is tossed. Therefore, the probability of tails must be equal to 1 minus the probability of heads. We already know this to be true because each of these events has a probability of $\frac{1}{2}$.

The Complement of an Event

If we identify a particular event as one outcome of an experiment, then the **complement** of the event consists of all the other outcomes taken together. When a coin is tossed, the event consisting of tossing tails is called the complement of the event of tossing heads. Similarly, when a six-sided die is rolled, the complement of rolling a 1 is rolling any number from 2 through 6.

We can generalize this result to include the probability of the complement of any event.

> The probability of the **complement** of an event is the same as the probability that the event does not happen.
>
> Probability of the complement of an event = 1 − (probability of the event)

In Example 5, it was stated that there is a 6% chance that a person on a long-duration military mission in close quarters will require emergency medical care. The complement of requiring care is *not* requiring care. Therefore, the probability that a person on a long-duration military mission in close quarters will not require emergency medical care is (1 − 0.06) = 0.94, or 94%.

Example 6

According to the National Highway Traffic Safety Administration, statistics on seat belt use have shown that approximately 70% of passenger vehicle occupants wear seat belts. What is the probability that a randomly selected passenger in such a vehicle does *not* wear a seat belt?

Solution:

The probability that a randomly selected passenger does wear a seat belt is 0.70. Not wearing a seat belt is the complement of wearing a seat belt. So the probability that a passenger does not wear a seat belt is (1 − 0.70) = 0.30, or 30%.

Example 7

A 1992 inspection of 500 steel rods reinforcing the ceiling of a tunnel under Boston Harbor found that 18 of the rods were cracked and needed replacing. What is the probability that a randomly selected rod would *not* need to be replaced because of cracking?

Solution:

The probability that a rod must be replaced is 18/500 = 0.036, so the probability that a rod would not need to be replaced is (1 − 0.036) = 0.964, or 96.4%.

Exercises 7.3

I. Investigations *(SE pages 227–229)*

1. For events that happen very rarely, probability estimates can usually be made only if many years of data are available. Bridges are often designed to withstand a "100-year flood," meaning a flood of such a magnitude that it is expected to happen only once every 100 years, on the average. Alternatively, a 100-year flood can be defined as a flood with a yearly probability of occurrence of 1/100.

   a) Most buildings are designed to withstand a "50-year wind" without suffering any damage. What would be the probability that a 50-year wind might occur in any given year?

1. a) 1/50 = 0.02

   b) The probability of being killed by lightning is about 1 chance in 4 million, or $0.00000025 = 2.5 \times 10^{-7}$. If the population of the world is 6 billion, about how many of those people would be expected to be killed by lightning?

b) $(2.5 \times 10^{-7})(6 \times 10^{9})$ = 1500 people

   c) The graph in **Figure 7.10** shows how often asteroids of various sizes strike Earth. (Notice that the scales on the axes are not uniform.) The dashed part of the graph represents estimates of collisions that are too rare to have been observed. Indicated on the dashed line is a point corresponding to a hypothetical asteroid collision with a strength equivalent to over 100,000 megatons (10^{11} tons) of TNT. Such a collision would result in global catastrophe. Use the graph to estimate the yearly probability of such an asteroid hitting Earth.

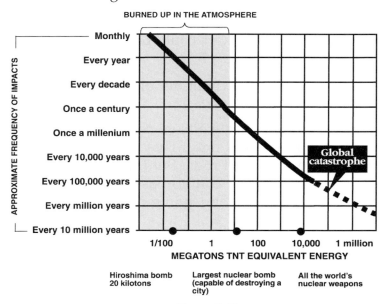

Figure 7.10

---

c) Such an asteroid should hit Earth approximately once every 300,000 years. The one-year probability of such a collision is 1/300,000 $\approx$ 3 x 10^{-6}. Students may estimate the interval as up to 500,000 years, with a probability of 2 x 10^{-6}.

2. The **reliability** of a manufactured product is the probability that it will perform its intended function satisfactorily for a specified time (or some other measure of use). For example, if an electronic control unit for a spacecraft has a 30-day reliability of 0.98, there is a 0.98 probability that it would be working at the end of a 30-day mission. This also means that there is a 0.02 probability of failure, and an average of 2% of all such units could be expected to fail during the first 30 days of use.

a) A floppy disk drive for a computer has a 200-hour reliability (for continuous use) of 0.996. What is the probability of failure of one of the drives before the end of 200 hours of continuous use?

2. a) **Failure probability is 1 – 0.996 = 0.004.**

b) Of 5000 such drives, how many could be expected to fail during the first 200 hours of continuous use?

b) **Number of failures is (0.004)(500) = 2.**

c) A front wheel roller bearing for a particular make of automobile has a 10,000-mile reliability (with respect to distance traveled) of 0.999. If 60,000 of these cars are sold, each having two front wheel roller bearings, how many bearings could be expected to fail in the first 10,000 miles of travel?

c) **Failure probability is 1 – 0.999 = 0.001. Number of bearing failures is (0.001)(60,000)(2) = 120.**

d) If a family watches television for an average of 6 hours a day for a year, they will watch about 2200 hours of television in a year. Suppose a brand of television has a 2200-hour reliability (without need for servicing) of 0.992. If 7500 such sets are sold, how many could be expected to require service within the first 2200 hours of use?

d) **Failure probability is 1 – 0.992 = 0.008. Number of failures is (0.008)(7500) = 60.**

3. Up to this point, most of the probabilities we have considered involved things that occur in whole number values only. They are variables that can be counted, like coin tosses, people, and steel rods. But there are many kinds of variables whose values can range along a continuous scale. The units for such variables can be subdivided indefinitely, depending on the measuring instruments used.

Time, which was the variable examined in Discovery 7.2, is measured in standard units of seconds. But common laboratory instruments can measure in milliseconds, and high precision timers measure to even smaller subunits. Time has no natural smallest counting unit and is an example of a **continuous variable.** Variables that do have natural countable units, on the other hand, are called **discrete variables.** In more advanced work with probability and statistics, the distinction between discrete variables and continuous variables is important for deciding on how to analyze the data.

Classify each of the following as discrete or continuous:

a) The strength of a steel beam

3. **a) Continuous**

b) The number of errors in a transmitted message

b) Discrete

c) The operating temperature of a computer CPU

c) Continuous

d) The diameter of a propeller shaft

d) Continuous

e) A person's white blood cell count

e) Discrete

4. In Activity 7.1, you compared experimental data on tossing a penny with the theoretical probability of obtaining 50% heads and 50% tails. You found that as the number of coin tosses increased, the observed fraction of heads (or tails) approached the theoretical value. Will the same results hold true if the penny is spun instead of tossed? Is there any reason to think that the fraction of heads might not be 50% in this case? You can experiment to find out.

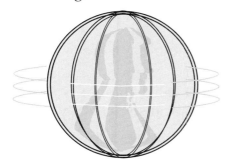

Figure 7.11

a) Use one finger to hold a penny vertically so its edge touches a table (or other horizontal surface). "Flick" the penny with a finger of your other hand so that it spins (see **Figure 7.11**). Notice which side lands face up. Repeat for a total of 10 spins and record your results. What is the fraction of heads?

4. **a) There are many possibilities. The fraction of heads is usually between 2/10 and 7/10.**

b) Just as you did in Activity 7.1, collect data for many sets of 10 spins each. Then construct a graph of "fraction heads" vs. "number of spins" as the number of spins increases from 10 to 20 to 30, and so on. Continue to collect data until your graph appears to level off. Then identify the experimental probability that a spinning penny will land heads up.

b) Answers will vary; one experiment with 84 trials resulted in 33 heads, for a fraction heads of 33/84 ≈ 39%.

c) Share your results with other members of your class to see if you agree on the experimental probability that a spinning penny will land heads up.

c) Heads should result significantly less than 50% of the time.

5. Design and conduct an experiment to determine the experimental probability that an ordinary thumbtack will land with its point facing up when dropped on a horizontal surface (see **Figure 7.12**).

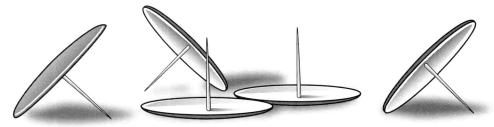

Figure 7.12

5. A sample experiment consisted of dropping 40 tacks at a time and counting the number of tacks landing point up. In 16 trials, a total of 398 of 640 tacks landed point up, for an experimental probability of about 0.62, or 62%, that a tack lands point up. The number that landed point up out of 40 in a trial varied from 21 to 30.

II. Projects and Group Activities *(SE pages 230–231)*

6. Materials: Handout 7.3

In this activity, you will collect data that will allow calculation of experimental probabilities for characteristics of a typical student in your class.

a) First, record the number of students participating in this activity.

6. a) Answer depends on class size.

b) Have each student answer all the survey questions in Handout 7.3. Then collect the surveys, and summarize the Yes responses in **Table 7.4.** Note that responses are separated into columns based on the answer to the first question.

Characteristics	Younger Students	Older Students
Full-time student		
Live on campus		
Drive or ride in a car to classes		
Own a car		
Own a home		
Own a cellular phone		
Work at least part time		
Work full time		
English as a first language		
More than 7 hours of sleep at night		
Play organized college sports		
Voted in the last presidential election		
Watch MTV		
Eat breakfast		
More than 4 live concerts		
Motor vehicle accident in past year		
Drink coffee in the morning		
Prefer beach to mountain vacation		
Listen more to music than to words		

Table 7.4

b) **Answers vary depending on students in class.**

c) Now, for each question in (b), determine the experimental probability that a randomly selected student in your class would answer either Yes or No. Add the values in both columns of the table to get the total number of Yes answers, and subtract from the answer to (a) to get the No answers. Record these probability values as percentages (to the nearest percent) in another table.

c) **Answers will vary, but probability percentages should reflect the relative frequencies from (b).**

d) Your class might be considered a sample of the general population of college students. Consider each of the questions in your table. Do you think it's reasonable to expect that the probabilities developed for a randomly selected student in your class might be representative of college students in general? Why or why not?

d) Answers will vary. Some reasons for nonrepresentative results could include whether college is a four-year or two-year institution, whether class is a day or evening class, whether or not college is in a city, where in the country it is located, and so on.

e) Record the total numbers of "younger" and "older" students. These values, as well as the survey summary, will be used in Exercise 5(f) in Section 7.4.

e) The determination of "younger" vs. "older" students should be made so that the numbers are roughly equal.

III. Additional Practice *(SE pages 231–233)*

For 7–27, express each answer as a fraction, as a decimal, and as a percent.

7. During 1997–98, there were about 1,660,000 students enrolled in distance education courses. Of these, 710,000 were in two-year colleges. What is the probability that a randomly selected student enrolled in a distance education course during 1997–98 was taking that course at a two-year college?

7. (710,000)/(1,660,000) = 71/166 ≈ 0.43 = 43%.

8. At the start of the 1999–2000 season, 46 out of 319 players in the National Basketball Association were at least 7 feet tall. What was the probability that a randomly selected NBA player was at least 7 feet tall?

8. 46/319 ≈ 0.144 = 14.4%.

9. According to recent statistics, 59,700,000 Americans have one or more types of cardiovascular disease. In a total population of 275,000,000, what is the probability that a randomly selected American has cardiovascular disease?

9. (59,700,000)/(275,000,000) = 597/2750 ≈ 0.217 = 21.7%.

10. Between 1993 and 1997, there were 278,078 drivers involved in fatal car crashes on U.S. roads. Of these, 38,374 did not produce a valid driver's license. What was the probability that in a randomly selected fatal car crash from 1993–97, the driver did not produce a valid driver's license?

10. (38,374)/(278,078) ≈ 0.14 = 14%.

11. A typical defect rate in U.S. electronics industries in the early 1990s was 26,000 parts per million (ppm) defective. For a company with a 26,000 ppm defect rate, what was the probability that a randomly selected item would be defective?

11. (26,000)/(1,000,000) = 13/500 = 0.026 = 2.6%.

12. One out of every 700 children gets cancer before age 15. What is the probability that a randomly selected child will get cancer before age 15?

12. 1/700 ≈ 0.0014 = 0.14%.

13. Of 544,000 new workers entering the New England labor market in a six-year period, 396,000 were women. What is the probability that a randomly selected new worker during that period was a woman?

13. (396,000)/(544,000) = 99/136 ≈ 0.73 = 73%.

14. Of 5300 miles of rivers and streams in Vermont in 1994, 1200 miles were either at risk or unable to support swimming, fishing, or aquatic life. What is the probability that a randomly selected mile of a river or stream in Vermont was at risk or unable to support swimming, fishing, or aquatic life in 1994?

14. 1200/5300 = 12/53 ≈ 0.23 = 23%.

15. Of 3,880,894 babies born in the United States in 1997, 104,137 were born as a member of a pair of twins, according to the National Center for Health Statistics. What is the probability that a randomly selected baby born in the United States in 1997 was a twin?

15. (104,137)/(3,880,894) ≈ 0.0268 = 2.68%.

16. From 1993 to 1996, 23,831 Florida pedestrians were involved in single-vehicle accidents, and 1550 of them died as a result, according to the National Highway Traffic Safety Administration. What is the probability that a pedestrian involved in a single-vehicle accident in Florida during those years died?

16. (1550)/23,831) ≈ 0.0650 = 6.50%.

17. The probability of drawing a heart from a well-shuffled deck of cards is 1/4. What is the probability of drawing a card from any other suit but hearts?

17. 1 − 1/4 = 3/4 = 0.75 = 75%.

18. When a car approaches a certain traffic signal, there is a probability of 0.53 that the signal will be green and a probability of 0.39 that it will be red. What is the probability that the light will be yellow?

18. 1 − (0.53 + 0.39) = 0.08 = 8/100 = 2/25 = 8%.

19. If an American household is selected at random, there is a probability of 0.57 that it contains no more than two people. What is the probability that a randomly selected American household contains more than two people?

19. 1 − 0.57 = 0.43 = 43/100 = 43%.

20. Eighteen percent of all American farms cover at least 500 acres in area. What is the probability that a randomly selected American farm covers less than 500 acres?

20. 100% − 18% = 82% = 0.82 = 82/100 = 41/50.

21. Only 0.19% of young American women are taller than 5 feet 11 inches, according to the U.S. Center for Health Statistics. What is the probability that a randomly selected young American woman is not taller than 5 feet 11 inches?

21. 100% − 0.19% = 99.81% = 0.9981 = 9981/10,000.

22. About two billion prescriptions are written each year in the United States for all illnesses, but only about 80% of them are ever filled. What is the probability that a randomly selected prescription issued in the United States will go unfilled?

22. 100% − 80% = 20% = 0.20 = 1/5.

23. According to the U.S. Department of Transportation, in 1998 San Francisco's airport had the most delays of all major U.S. airports, with 32% of flight arrivals delayed. What is the probability that a randomly selected 1998 arrival at San Francisco's airport was on time?

23. 100% − 32% = 68% = 0.68 = 68/100 = 17/25.

24. Tourette's syndrome, a neurological disorder, affects 1 out of every 1000 children. What is the probability that a randomly selected child does not have Tourette's syndrome?

24. 1 − 1/1000 = 999/1000 = 0.999 = 99.9%.

25. In 1996, 77% of American infants 19–35 months old had all their childhood immunizations. What is the probability that a randomly selected American infant aged from 19 to 35 months in 1996 did not have all of his or her childhood immunizations?

25. 100% − 77% = 23% = 0.23 = 23/100.

26. In 1999, 22% of births in the United States were by Cesarean section, according to the National Center for Health Statistics. What is the probability that a randomly selected birth in 1999 was not by Cesarean section?

26. 100% − 22.0% = 78.0% = 0.78 = 78/100 = 39/50.

27. For apples grown by conventional methods in the United States during 1996–97, there was a 10.8% chance that a 44-pound child would consume an unacceptably high dose of pesticides from a 3.5-ounce serving, according to the Consumers Union of the United States. What is the probability that such a child would not have consumed an unacceptable dose of pesticides from a 3.5-ounce serving of apples during 1996–97?

27. 100% − 10.8% = 89.2% = 0.892 = 892/1000 = 223/250.

Section 7.4 Combining Probabilities

What You Need To Know

- How to compute probabilities of individual events

What You Will Learn

- To find joint probabilities for independent events
- To find joint probabilities for mutually exclusive events

Materials

- Container of plastic chips in two colors
- Coin

There are many types of situations in which a system contains multiple components, all of which have some degree of uncertainty associated with their operation. Anyone who has ever made extended use of a photocopying machine is familiar with the possibility of machine breakdown due to various causes such as misfeeding, overheating, and running out of ink. Electronic systems usually contain a large number of components, each of which has a certain probability of working correctly. The design of a system must take these probabilities into account. In Discovery 7.3, we will consider one such system.

Discovery 7.3 System Reliability *(SE pages 234–236)*

Consider a pumping system that supplies water to a five-story building from a nearby reservoir (see **Figure 7.13**). If the building is 50 feet high, then a pressure of about 22 pounds per square inch (psi) is necessary to lift water to the top of the building. A single pump that has a reliability of 70% for one year can be used. This means that there is a probability of 70%, or 0.7, that the pump will still function correctly to pump water one year after installation.

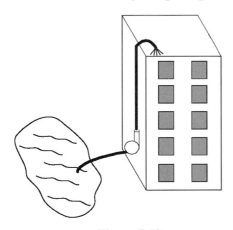

Figure 7.13

1. What is the probability that the pump will malfunction at some time during the first year?

1. 30%, or 0.3

2. Use an area model to represent the reliability of this pump. Draw a small square whose area will represent a probability of 1. Draw a vertical line that divides the square into rectangles that contain 70% and 30% of the total area of the square. The areas of these rectangles represent, respectively, the probability of correct functioning and the probability of failure for the pump.

2.

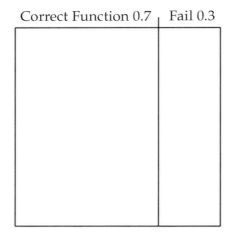

Correct Function 0.7 | Fail 0.3

3. One alternative is to install two less powerful pumps that are each more reliable than the single pump. Each pump is capable of producing water pressures of 11 psi. These pumps would be connected in series (the outlet hose of one is the inlet hose of the other) so that together they produce the required 22 psi of pressure. Assume that one of these pumps, the more expensive of the two, has a reliability of 90% and that the other has a reliability of 80%. Use an area diagram to determine the overall reliability of this system, as follows.

Draw a square whose area will represent a probability of 1. Divide the top edge of the square proportionally to represent the probabilities of correct functioning and failure for the more reliable pump, which are 0.9 and 0.1, respectively. Then draw a vertical line at this location to split the square into rectangles representing these probabilities, just as you did for the single pump system in item 2. Divide the left edge of the square in a similar manner to represent the corresponding probabilities for the less reliable pump, which are 0.8 and 0.2. Finally, split the square at this location with a horizontal line.

3.

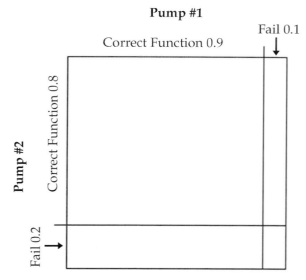

4. The square is now subdivided into four rectangles. The largest rectangle represents the event that both pumps function correctly for one year. Label each of the four rectangles according to the combination of events that it represents.

4.

Pump #1
Fail 0.1
Correct Function 0.9

Both Pumps Function Correctly

Pump #2 Works, #1 Fails

Pump #2 Correct Function 0.8

Fail 0.2

Pump #1 Works, #2 Fails

Both Fail

5. Calculate the areas of the rectangles and complete **Table 7.5.**

State of the System	Probability
Both pumps function correctly	
Pump #1 works, pump #2 fails	
Pump #1 fails, pump #2 works	
Both pumps fail	

Table 7.5

5.

State of the System	Probability
Both pumps function correctly	0.72
Pump #1 works, pump #2 fails	0.18
Pump #1 fails, pump #2 works	0.08
Both pumps fail	0.02

6. Each of the probabilities in the table is called a **joint probability** because it is a combination of the probabilities of two events. In each case, the state of the system depends on two separate events both occurring. Examine the way you calculated each joint probability, and write a rule for determining the probability that two separate events both occur.

6. To find the probability that two separate events both occur, multiply the probabilities of the individual events.

7. In a series system like the one discussed here, both pumps would have to function correctly in order for water to be delivered to the top floor of the building. What is the one-year reliability of this pumping system? Is it more reliable than the single pump system considered in item 1?

7. The system reliability is 0.72, or 72%, slightly better than the 70% for the single-pump system.

8. Consider one additional alternative. An extra backup pump, identical to the single 22 psi pump from item 1 (and with a 70% reliability), could be installed and configured to automatically take over if the first pump fails. This type of system is called a parallel system: both pumps would have a common input and a common output so that either one operating alone could deliver water to the fifth floor of the building. Using your rule from item 6 and an area model, complete **Table 7.6.**

State of the System	Probability
Both pumps function correctly	
Pump #1 works, pump #2 fails	
Pump #1 fails, pump #2 works	
Both pumps fail	

Table 7.6

8.

State of the System	Probability
Both pumps function correctly	(0.7)(0.7) = 0.49
Pump #1 works, pump #2 fails	(0.7)(0.3) = 0.21
Pump #1 fails, pump #2 works	(0.3)(0.7) = 0.21
Both pumps fail	(0.3)(0.3) = 0.09

Pump #1

Correct Function 0.7 | Fail 0.3

	Correct Function 0.7	Fail 0.3
Pump #2 Correct Function 0.7	0.49	0.21
Fail 0.3	0.21	0.09

9. This system would deliver water to the top floor even if only one pump were operational at any given time. Therefore, the system only fails if both pumps fail. What is the overall system reliability in this case? How does it compare to the other systems considered in this discovery?

9. **The system reliability is (0.49 + 0.21 + 0.21) = 0.91, or 91%, which is significantly better than either the single pump or the series system.**

The area diagrams you drew in Discovery 7.3 allowed you to determine joint probabilities of two **independent events** both occurring. Just as areas of rectangles are found by multiplying their dimensions, the probability that two independent events both occur is found by multiplying their individual probabilities. This result can be generalized to situations involving the occurrence of any number of independent events.

The Multiplication Rule for Probabilities

The probability that two or more independent events occur is equal to the product of the probabilities of the individual events.

The notion of independence has a very specific and important meaning in many situations. The choice of one item must in no way be influenced by, or linked to, the choice of another item. For example, in Discovery 7.3, it was assumed that pump failures for the two-pump systems were independent events. This means that the failure of one pump is not dependent upon, or influenced by, the condition of the other pump. In a real pumping system, it might be the case that the failure of one pump would result in additional stress to the second pump. That would make it more likely to fail than under normal circumstances. If so, the two pump failures would *not* be independent events. And the probability that both fail would *not* be equal to the product of the separate failure probabilities. In fact, the joint failure probability for both pumps would in such a case be greater than the product of their individual probabilities. The assumption of independence is often valid. But it is important to verify that such an assumption is justified before using the Multiplication Rule.

Example 8

In Example 4 of Section 7.3, we found that the probability of a baby being born a girl is 0.488. What is the probability that four randomly selected newborn babies are all girls?

Solution:

Assuming that the births are independent of each other, the probability that all four babies are girls is found by multiplying the individual probabilities, which in this case are all equal to 0.488. The probability that all four babies are girls is $(0.488)(0.488)(0.488)(0.488) = (0.488)^4 = 0.0567...$, or about 5.7%.

Addition of Probabilities

The Multiplication Rule for Probabilities applies to many situations in which we want to know the probability of one event *and* a second event both occurring. Another type of joint probability involves whether one event *or* another event occurs.

An analysis of accidents involving pedestrians and single vehicles during the years 1993–97 in Florida showed the following:

- The probability of a pedestrian fatality was 0.065.

- The probability of non-fatal but incapacitating injuries was 0.270.

- The probability of non-incapacitating injuries was 0.388.

- The probability of no pedestrian injury was 0.277.

What is the probability that an accident resulted in either death or an incapacitating injury?

Because it is not possible for an injury to be both fatal and nonfatal, these events cannot both occur. But the probability that one or the other occurs is just the sum of their individual probabilities. Therefore, the probability of either death or an incapacitating injury is $0.065 + 0.270 = 0.335$. Similarly, the probability that an accident resulted in either an incapacitating or a non-incapacitating injury is $0.270 + 0.388 = 0.658$.

This example can be generalized to provide a rule that applies to situations involving the occurrence of any one of several possible events. It requires that only one of the events can actually happen within the context of the experiment.

> When only one of a group of events can occur, the events are called **mutually exclusive** events. Such events can be thought of as non-overlapping. For example, the categories of fatal and incapacitating injuries for the pedestrian accident data are non-overlapping categories. When events overlap, the Addition Rule as presented here does not apply (but see Exercise 3 for a treatment of such a situation). It is important to verify that events are mutually exclusive before applying the Addition Rule.
>
> Complementary events are by definition mutually exclusive. But the converse is not necessarily true. That is, the fact that two events are mutually exclusive does not mean that they must be complementary.

The Addition Rule for Probabilities

The probability that any one of two or more possible events occurs is equal to the sum of the probabilities of the individual events, provided that it is not possible for more than one of the events to occur on any single trial of an experiment.

Example 9

Table 7.7 shows the probability of randomly picking various colored candies from a package of plain M&Ms.

Color	Brown	Red	Yellow	Green	Orange	Blue
Probability	0.3	0.2	0.2	0.1	0.1	0.1

Table 7.7

What is the probability that a randomly chosen M&M is either red or yellow?

Solution:

An M&M can be either red or yellow, but it can't be both. "Being red" and "being yellow" are mutually exclusive, so the Addition Rule for Probabilities applies. The probability that an M&M is either red or yellow is equal to the sum of the individual probabilities that the candy is either red or yellow: 0.2 + 0.2 = 0.4.

Example 10

A project manager planning the work schedule for developing a new product is uncertain about how long the test-marketing phase of the project will take. Based on past experience, she estimates that the probability of finishing this phase in three weeks is about 10%; in four weeks, 30%; in five weeks, 40%; and in six weeks, 20%. What is the probability that the test-marketing phase will be finished in less that five weeks?

Solution:

The only possibilities that are less than five weeks are three weeks or four weeks. Therefore, the question can be rephrased as, "What is the probability that the test-marketing phase will be finished in either three weeks or four weeks?" Because it must be one or the other, the events are mutually exclusive. The Addition Rule applies, and the probabilities for three weeks and four weeks are added: 10% + 30% = 40%.

Example 11

When two six-sided dice are rolled, the sum of the values on the dice can be any number from 2 to 12. What is the probability that the sum will be in the range from 2 to 11?

Solution:

There must be a unique result when the dice are rolled, so the Addition Rule applies. We can find the answer by adding the probabilities of all the individual sums from 2 through 11. However, the various results are not equally likely (see Exercises 8–11 of Section 7.2), so this would require many separate calculations. But the question "What is the probability that the sum will be in the range from 2 to 11?" is equivalent to the question "What is the probability that the sum will not be 12?" Therefore, we can find the probability that the complement of the required event occurs and subtract the result from 1. The probability of rolling 12 is the same as the probability of rolling two 6s, or $\left(\dfrac{1}{6}\right)\left(\dfrac{1}{6}\right) = \dfrac{1}{36}$. So the probability of rolling a sum from 2 through 11 is the same as the probability of not rolling 12, or $\left(1 - \dfrac{1}{36}\right) = \dfrac{35}{36}$.

Exercises 7.4

I. Investigations *(SE pages 240–244)*

1. On January 28, 1986, the space shuttle *Challenger* exploded over Florida. Many authorities believe this tragedy could have been prevented if closer attention had been paid to the laws of probability. The rocket that carried the shuttle aloft was separated into sections that were sealed by large rubber O-rings. Experts believe that at least one of the O-rings leaked burning gases that caused the rocket to explode. Studies after the tragedy found that the probability that a single O-ring would work was 0.977. (This can be thought of as the *mission reliability* for an O-ring.)

   a) If the six O-rings were truly independent of one another, what is the probability that all six would function properly during the mission?

1. a) $(0.977)^6 \approx 0.870$.

   b) The game of Russian roulette is played by spinning the cylinder of a revolver containing a single cartridge and then pulling the trigger. What is the probability the gun will fire if it has six cylinders?

b) $1/6 \approx 0.167$.

   c) Compare the probability that all six O-rings will function properly with the probability that the gun will not go off in Russian roulette.

c) The probability that the gun will not go off is $(1 - 0.167) = 0.833$. The probability that all six O-rings function properly is about 87%, compared with 83% that the gun will not go off in Russian roulette.

2. People are often surprised by coincidences. They sometimes think that unexpected coincidences are somehow magical or mystical phenomena. But when viewed in terms of probabilities, many coincidences are in fact fairly likely to occur. A famous example is the so-called "Birthday Problem."

   a) Suppose that you are in a lab group with four people. What is the chance that any two (or more) of those people share a birthday? Because there are so many possible combinations of people sharing birthdays, it will be easier to determine the chance that none of the birthdays are the same. Consider one of the people in the group. What is the probability that that person has a birthday sometime during a 365-day year? (For simplicity, we will ignore the possibility of a February 29 birthday because it does not have the same likelihood as the other 365 dates.)

2. a) $\dfrac{365}{365} = 1$.

   b) What is the probability that the second person has a birthday on a different day than the first person? (To find the numerator of the probability fraction, think about how many days are left.)

b) $\dfrac{364}{365}$

c) What is the probability that the third person has a different birthday from the first two?

c) $\dfrac{363}{365}$

d) What is the probability that the fourth person has a different birthday from the first three?

d) $\dfrac{362}{365}$

e) Using the Multiplication Rule for Probabilities, find the probability that none of the four people has the same birthday.

e) $\dfrac{365}{365} \cdot \dfrac{364}{365} \cdot \dfrac{363}{365} \cdot \dfrac{362}{365} \approx 0.9836.$

f) What is the probability that at least two people do share a birthday?

f) The event that at least two share a birthday is the complement of the event that none share a birthday, and its probability is equal to $-0.9836 = 0.01634$.

g) By trying different group sizes, determine how large a group is necessary for there to be at least a 50% chance that two or more people in the group share a birthday.

g) With 23 in the group, probability of not sharing a birthday = $\dfrac{365}{365} \cdot \dfrac{364}{365} \cdot \dfrac{363}{365} \cdots \dfrac{344}{365} \cdot \dfrac{343}{365} \approx 0.4927$. The probability that two or more share a birthday = $1 - 0.4927 = 0.5073$.

h) The previous results hold only if particular birthdays are not specified. That is, the original question implied whether people in a group shared *any* birthday. It is much less probable that there is even one person in a group who was born on a *particular* date. For example, what if we ask whether anyone was born on April 24? Using the group size you found in (g), first find the probability that no one was born on April 24. Then subtract from 1 to get your answer.

h) The probability that none of 23 people were born on April 24 = $(364/365)^{23} \approx 0.9388$. The probability that someone was born on April 24 = $1 - 0.9388 = 0.0612$.

3. The Addition Rule for Probabilities requires that the events being considered are mutually exclusive; that is, only one of the events can actually occur. For example, in item 7 of Discovery 7.3, we considered a two-pump backup (or parallel) system in which each pump had a one-year reliability of 0.7. The system reliability was equal to the probability that at least one pump would be operating after a year. If the Addition Rule were used in this case, we would find that the probability of Pump #1 or Pump #2 working after a year is 0.7 + 0.7 = 1.4. This is an impossible result because a probability value can never be greater than 1. The problem occurs because the events that each pump works after a year are not mutually exclusive; that is, it is possible for both pumps to be operational after a year.

An area model (see **Figure 7.14**) makes this clear. The region labeled 0.49 represents the probability that both pumps work after one year. The correct answer of 0.91 for the probability of at least one pump working can be found by adding 0.49 + 0.21 + 0.21. When two events are not mutually exclusive, the probability that at least one occurs can be found by using an area model like Figure 7.14. An alternative method is to subtract from 1 the probability that neither event occurs.

Pump #1

Figure 7.14

a) The probability of a lake in a certain part of the country being polluted by phosphates is 0.18, and the probability of chromium pollution is 0.12. What is the probability of a lake being polluted by at least one of these two pollutants? (Assume that pollution by phosphate and by chromium are independent of each other.)

3. a)

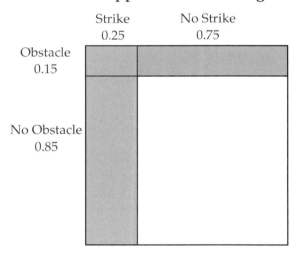

Phosphate 0.18   No Phosphate 0.82

Chromium 0.12   0.0216   0.0984

No Chromium 0.88   0.1584   0.7216

Probability of at least one pollutant being present = (probability of phosphate only) + (probability of chromium only) + (probability of both phosphate and chromium) = 0.1584 + 0.0984 + 0.0216 = 0.2784.

b) A high-rise construction project is scheduled to be completed after 10 months. Management estimates that there is a one in four chance that the project will be delayed by a labor strike. Past experience also suggests a 15% probability that digging for the foundation will be delayed due to unexpected obstacles. No other possible delays are expected. What is the probability that the project will be delayed?

b) A strike and encountering obstacles are not mutually exclusive events; both could happen. The area diagram illustrates the probabilities.

Strike 0.25   No Strike 0.75

Obstacle 0.15

No Obstacle 0.85

The shaded region represents the probability that the project will be delayed, (0.15)(0.25) + (0.15)(0.75) + (0.85)(0.25) = 0.3625. Alternatively, the probability of a delay = 1 − (probability of no delay) = 1 − (0.85)(0.75) = 1 − 0.6375 = 0.3625.

c) A high school senior has applied to the state university and also to her local community college. Her guidance counselor has estimated that she has an 80% chance of being accepted by the community college but only a 30% chance of getting into the university. What is the probability that she will be accepted by the college?

c)

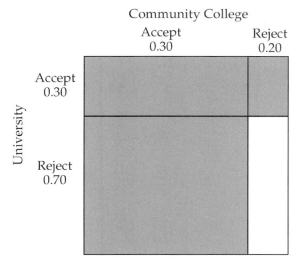

Probability of acceptance by a college = area of shaded region = (0.30)(0.80) + (0.30)(0.20) + (0.70)(0.80) = 0.86. Alternatively, probability of acceptance by a college = 1 − (probability of rejection by both colleges) = 1 − (0.70)(0.20) = 1 − 0.14 = 0.86.

4. When a basketball player is fouled, he or she is allowed to shoot free throws. Over the years, there have been several different formats for awarding free throws. Use your knowledge of probability rules to determine the probability that an 80% free-throw shooter will make two free throws under each of the following circumstances. (Assume that free throws are independent events.)

a) A player fouled in the act of shooting is awarded two free throws.

4. a) Probability of making both free throws = (0.80)(0.80) = 0.64, or 64%.

b) A player is allowed one free throw, and then can try another only if the first one is made. (This is called a one-and-one situation, and is currently not used above the college level.) It will be helpful if you first list all the possible outcomes for this scenario.

b) There are three possible outcomes, which are not equally likely:

Miss the first shot, no second shot; probability 0.20

Make the first shot *and* miss the second shot; probability (0.80)(0.20) = 0.16

Make the first shot *and* make the second shot; probability (0.80)(0.80) = 0.64, or 64%

c) A player is given three chances to make two shots. If the first two are made, there is no third shot. (This format was used in the NBA for a few years and then discarded.) A tree diagram (see Exercise 1 of Section 7.2) may be helpful, but the branches in any stage will not have equal probabilities.

c)

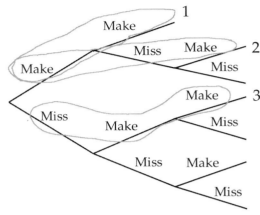

First Shot   Second Shot   Third Shot

There are three paths through the tree diagram that result in two made free throws. Because the paths are mutually exclusive (only one of them can happen), the probability of making two free throws equals the sum of the probabilities along the three paths, or (0.80)(0.80) + (0.80)(0.20)(0.80) + (0.20)(0.80)(0.80) = 0.896, or about 90%.

5. The Multiplication Rule for Probabilities requires that events be independent. There are many situations where that is not true. In such a case, the occurrence of one event may alter the probability of a second event, making it either more or less likely.

    a) A study of 19,112 men at Brigham and Women's Hospital in Boston looked for links between baldness and heart ailments. The results are summarized in **Table 7.8.** Answer each question based on the table.

	Heart Problems	No Heart Problems
Severe Baldness	127	1224
Mild/Moderate Baldness	771	8831
No Baldness	548	7611

Table 7.8 Source: Archives of Internal Medicine.

Of all the men in the study, what is the probability that a randomly selected man had a heart problem?

5. a) $\dfrac{\textbf{Number with heart problems}}{\textbf{Total number in the study}} = \dfrac{\textbf{1446}}{\textbf{19,112}} \approx \textbf{0.076.}$

b) What is the probability that a randomly selected man was severely bald?

b) $\dfrac{\textbf{Number severely bald}}{\textbf{Total number in the study}} = \dfrac{1351}{19{,}112} \approx 0.071.$

c) What is the probability that a man with severe baldness had a heart problem?

c) $\dfrac{\textbf{Number of severely bald with heart problems}}{\textbf{Total number of severely bald}} = \dfrac{127}{1351} \approx 0.094.$

Notice that there is a difference in the probabilities in (a) and (c). If a man is bald, there is a greater probability that he will have heart problems than if he were not bald. We say that the **conditional probability** that a man will have heart problems, given that he is bald, is 0.094. The unconditional probability that a man will have heart problems is 0.076. Baldness is considered a **risk factor** for heart problems, meaning that there is statistical evidence that the risk of heart problems is increased for bald men.

Baldness and heart problems are not independent events. If we thought baldness and heart problems were independent, we might use the Multiplication Rule to calculate the answer to "What is the probability that a man is severely bald and has a heart problem?" as (0.071)(0.076) = 0.0054. The correct answer is found by using the conditional probability of a bald man having heart problems: (0.071)(0.094) = 0.0067.

d) Studies have shown that approximately 70% of passenger vehicle occupants in the United States wear seat belts. In 1999, 15,000 of the 41,000 people killed in crashes were wearing seat belts. What is the conditional probability that a passenger vehicle occupant who was killed in a car crash wore a seat belt?

d) $\dfrac{\textbf{Number killed who wore seat belts}}{\textbf{Total number killed}} = \dfrac{15{,}000}{41{,}000} \approx 0.37.$

e) About 10% of all people are left-handed. When two parents are left-handed, however, about half of their biological children are left-handed. What is the probability that in a family with one child, all three people will be left-handed?

e) **The answer equals (probability that mother is a lefty)(probability that father is a lefty)(probability that child is a lefty). If we assume that left-handedness in mother and father are independent events (that is, lefties are no more likely to marry lefties or righties), then the first two of these are both 0.1. But the conditional probability that a child will be a lefty if both parents are lefties is 0.5. The answer is (0.1)(0.1)(0.5) = 0.005, or 0.5%.**

f) One way of defining independence is to say that two events are independent if their conditional probabilities equal their unconditional probabilities. If your class did Exercise 6 of Section 7.3, look at the data in your table from that exercise. Consider the first characteristic in the table, "Full-time student." The probability that someone in your class is a full-time student was computed when you completed that exercise. That probability is an unconditional probability because it made no distinction among students on any other basis. What fraction of "older" students are full-time students? This can be considered a conditional probability. It is the conditional probability that a student is full time, given that he or she is an "older" student. If the two probabilities are (roughly) equal, then we can say that "being an older student" and "being a full-time student" are independent events. What did you find?

f) Answers will vary, but it would not be surprising if older students are more likely to be part-time students, which would mean that the events are not independent.

g) In a similar manner, analyze probabilities for "older" and "younger" students for the other characteristics in the survey. Identify characteristics that are independent of age and characteristics that depend on age.

g) Answers will vary but should reflect the relative frequencies in the data.

6. If you are tested for a disease and the test results are positive, should you be worried? The answer depends on the test and also on how many people actually have the disease. Tests for such things as medical conditions and drug usage are usually not 100% accurate. And even if a test has a high accuracy, the results can often be misleading.

A PSA (prostate-specific antigen) test is often used to test for prostate cancer in men over the age of 50. It is approximately 80% accurate in correctly identifying the presence of cancer. In other words, if a man has prostate cancer, there is a 0.80 probability that he will test positive. This is called a true positive test. On the other hand, if he doesn't have cancer, the PSA will be negative (and correct) about 97% of the time.

a) If a test on a healthy person turns out positive for a disease, the result is called a false positive. What is the probability that a healthy man will receive a false positive test result for prostate cancer from a PSA test?

6. a) Because 97% of tests on healthy men are negative, 3%, or 0.03, is the probability of a false positive test.

b) Of all the men who take the PSA test, 4% actually have prostate cancer. What is the probability that a man who takes the test has cancer and has a true positive test?

b) (Probability of having cancer)(probability of a true positive test) = (0.04)(0.80) = 0.032.

c) What is the probability that a man who takes the test doesn't have cancer and tests positive (a false positive)?

c) (Probability of not having cancer)(probability of a false positive test) = (0.96)(0.03) = 0.029.

d) Out of every 1000 men who take the PSA test, how many will have a true positive?

d) (0.032)(1000) = 32 men.

e) Out of every 1000 men who take the PSA test, how many will have a false positive?

e) (0.029)(1000) = 29 men.

f) What is the probability that a man who tests positive for prostate cancer on the PSA test is actually healthy?

f) Out of a total of (32 + 29) = 61 men, 29 men with positive tests are actually healthy, for a probability of 29/61 ≈ 48%.

II. Projects and Group Activities *(SE pages 245–246)*

7. Materials: container of chips, coin

Testing for steroid use, especially among athletes, has become common. Steroid tests are expensive, however. In order to save money, samples from several people can be pooled and tested together. If the test result is negative, it is not necessary to perform individual tests. In this activity, you will simulate testing two pooled samples to investigate whether pooling can reduce the number of tests required.

a) Consider a case in which samples from two people are pooled. If the test on the pooled sample is positive, how many more tests might be necessary in order to correctly identify whether either individual, or perhaps both, have steroids in their samples?

7. a) Two more tests might be necessary if the first individual tests positive. Then the other individual must also be tested. If the first individual tests negative, then the second individual must be positive and a third test is not needed.

b) Work with two other people. The group will need a container of chips. If half of the chips are one color and half are another, you can simulate a population in which 50% of people are using steroids. Drawing a chip of the first color can represent a positive steroid test, while the other color represents a negative test. (Alternatively, you can flip a coin for this part of the activity.)

Two people will play the roles of people being tested for steroids; the third plays the role of the tester.

Each person draws a chip or flips a coin to determine whether they test positive or negative. Always return a chip to the container and mix the chips well between draws to ensure random selection.

After the two being tested have determined whether they are positive or negative, they share their results and report a positive or negative pooled sample to the tester. If the pooled result is negative, the tester records that only one test was necessary.

If a positive pooled sample is reported, the tester must flip a coin to decide which person to test first. If that person reports a negative test, the tester records that two tests were necessary. However, if the first person reports a positive test, the tester records that three tests were necessary.

Repeat until you have 100 trials. (Or if other groups are doing the activity, combine the results.) Construct a table to record the number of times one, two, or three tests occurred. Find the average number of tests needed per trial. If it is less than two, sample pooling has reduced the number of tests needed. Did it?

b) Students should find that the average number of tests needed is slightly more than two. For a 50% positive population, two or three tests will usually be needed for each trial.

c) What is the probability that only one test is needed? That is, what is the probability that two people will both test negative if the probability is 0.5 that they are each negative?

c) (0.5)(0.5) = 0.25.

d) What is the probability that a second test must be done?

d) 1 – 0.25 = 0.75.

e) A third test will not be necessary if the second test (on one of the individuals) is negative. What is the probability that a second test was necessary and that it turns out negative? (This would mean that only two tests are needed.)

e) (0.75)(0.5) = 0.375.

f) What is the probability that a third test is necessary?

f) 1 – (0.25 + 0.375) = 0.375.

g) Use your answers to (c), (e), and (f) to explain your findings from (b).

g) The theoretical probabilities suggest that more than two tests should be needed in the majority of cases.

h) Investigate how the percentage of steroid users in the population affects the number of tests needed when samples are pooled. Mix chips in a different ratio to reduce the percentage of steroid use. Repeat the experiment. Again calculate theoretical probabilities of one, two, and three tests. Write a report on the merits of paired sample pooling for different levels of steroid use.

h) As the percentage of the population using steroids decreases, the benefits of sample pooling increase. For example, for 10% usage, the average number of tests for a pair of individuals is only about 1.2.

III. Additional Practice *(SE pages 246–248)*

8. a) Explain why complementary events are mutually exclusive.

8. a) The complement of an event occurs when the event doesn't occur. Because an event cannot both happen and not happen, an event and its complement are mutually exclusive.

b) Give an example of two events that are mutually exclusive but not complementary.

b) Rolling a 1 or 2 on a die; hitting a home run or striking out in baseball; having an April or December birthday

9. A company's function planner is scheduling two company picnics for the summer. He has heard that there is a 15% chance of rain on any given summer day in his part of the country. What is the probability that it will rain on the days of both picnics?

9. (0.15)(0.15) = 0.0225.

10. About 1 out of every 100 bottles of wine becomes what is called "corked," meaning that bacteria in the cork have ruined the wine. What is the probability of opening two bottles of wine and finding them both corked?

10. $\left(\dfrac{1}{100}\right)\left(\dfrac{1}{100}\right) = \dfrac{1}{10,000}.$

11. A 50-year flood is a flood of such a magnitude that it has a probability of 1/50 of happening in any given year. What is the probability that 50-year floods will happen two particular years in a row?

11. $\left(\dfrac{1}{50}\right)\left(\dfrac{1}{50}\right) = \dfrac{1}{2500} = 0.0004.$

12. What is the probability that in a given year, there will be both a 50-year flood and a 100-year earthquake?

12. Assuming that floods and earthquakes are independent,
$$\left(\dfrac{1}{50}\right)\left(\dfrac{1}{100}\right) = \dfrac{1}{5000} = 0.0002.$$

13. A particular structural beam in a building has a 0.02 probability of failure when subjected to a dynamic load of a certain critical size. If the probability of such a load occurring during the design life of the building is one chance in 100, what is the probability of beam failure during the building's design life?

13. Probability of critical load and beam failure = (0.02)(0.01) = 0.0002.

14. A car door is held to the car's chassis by six welds. Each of the welds has a 0.99 probability that it will hold in a major collision. What is the probability that all six welds will hold in a major collision?

14. (0.99)^6 ≈ 0.941.

15. What is the probability that of three children born to the same family, all three are girls (see Example 8)?

15. $(0.488)^3 \approx 0.116$.

16. What is the probability that of three children born to the same family, all three are boys?

16. $(1 - 0.488)^3 = (0.512)^3 \approx 0.134$.

17. The probability of losing a space shuttle has been estimated at 1/438 by NASA officials. What is the probability that five shuttle missions in a row will return safely?

17. $\left(1 - \dfrac{1}{438}\right)^5 \approx 0.989$.

18. A student takes a 10-question multiple-choice quiz. Each question has five possible choices.

    a) If the student guesses randomly at every answer, what is the probability that the student will get all the questions wrong?

18. a) Probability of none correct = (probability of one incorrect)^{10} = $(4/5)^{10} \approx 0.107$.

    b) What is the probability that the student will get all the questions correct?

**   b) Probability of all correct = (probability of one correct)^{10} = $(1/5)^{10}$ = 1.024×10^{-7}.**

19. A certain chemical mutagen (a substance that causes genetic mutations) has a 0.0016 probability of producing a mutation in a particular base pair of a gene that is exposed to the mutagen. If a gene contains 150 base pairs, all of which are exposed to the mutagen, what is the probability that the gene will not undergo any mutations?

19. Probability of no mutation in a single base pair = 1 − 0.0016 = 0.9984; probability of no mutation in any of 150 base pairs = $(0.9984)^{150} \approx 0.786$.

20. The probability that the soil in a house lot in a particular county contains lead contamination is 0.03. The probabilities of dioxin and phenol contamination are 0.02 and 0.005, respectively. What is the probability that a randomly chosen lot contains all three contaminants? (Assume that contamination by any of the three contaminants occurs independently.)

20. $(0.03)(0.02)(0.005) = 0.000003$ or 3×10^{-6}.

21. What is the probability that a randomly chosen lot in the county of Exercise 20 contains none of the contaminants?

21. Probability of no contaminants = (probability of no lead)(probability of no dioxin)(probability of no phenol) = $(0.97)(0.98)(0.995) \approx 0.946$.

22. Four electronic components are wired in series. Their 100-hour reliabilities are 0.98, 0.995, 0.96, and 0.975. What is the 100-hour reliability of the system; that is, what is the probability that all components will work for 100 hours?

22. $(0.98)(0.995)(0.96)(0.975) \approx 0.913$.

23. A pumping system consists of three pumps, each with a one-year reliability of 0.95. If the system is a parallel system (that is, two of the pumps are backups), what is the one-year reliability of the system? (This is the same as the probability that the three pumps don't all fail during the year.)

23. Probability that all don't fail = 1 − (probability that all fail) = 1 − (failure probability for one pump)^3 = 1 − (1 − 0.95)^3 = 0.999875.

24. A college physical plant manager is concerned about possible delays in a tightly scheduled preparation for an outdoor graduation ceremony. There are three activities—landscaping, tent raising, and chair setup—that must be performed sequentially. If any one of these activities is delayed, the ceremony cannot start on time. The manager estimates that the probabilities of delay for the three activities are 10%, 20%, and 10%, respectively. What is the probability that the ceremony can start on time?

24. Probability of starting on time = probability that none of the activities is delayed = (0.90)(0.80)(0.90) = 0.648, or about 65%.

25. A company that manufactures refrigerator ice-making units has a 3% defect rate on one of its primary production lines. What is the probability that a quality control inspector might examine a sample of five randomly selected units and find no defectives in the sample?

25. Probability of no defects in five units = (probability of a nondefective unit)^5 = (1 − probability of a defective unit)^5 = (1 − 0.03)^5 = (0.97)$^5 \approx 0.859$.

26. Students in a particular math class are studying in a variety of majors. If a student from the class is selected randomly, the probabilities that he or she is a science, business, engineering, or psychology major are 0.25, 0.18. 0.34, and 0.23, respectively. What is the probability that a randomly selected student is either a science or engineering major? (Assume that no student has a major in more than one field.)

26. Probability of being a science or engineering major = (probability of being a science major) + (probability of being an engineering major) = 0.25 + 0.34 = 0.59.

27. The probabilities of picking various colors from a bag of M&M candies are 0.3 for brown, 0.2 for red, 0.2 for yellow, 0.1 for green, 0.1 for orange, and 0.1 for blue. What is the probability of picking either a red, green, or orange M&M?

27. Colors are mutually exclusive (that is, a single candy can't be both red and green, etc.); probability of red, green, or orange = (probability of red) + (probability of green) + (probability of orange) = 0.2 + 0.1 + 0.1 = 0.4.

28. The probabilities of a randomly selected blood sample in the New York Blood Center being type O, type A, type B, or type AB are 0.45, 0.40, 0.10, and 0.05, respectively. What is the probability that a sample is either type A or type B?

28. Blood types are mutually exclusive; probability of type A or type B = (probability of type A) + (probability of type B) = 0.40 + 0.10 = 0.50.

29. A circular spinner for a board game is divided equally into eight sectors containing the numbers from 1 through 8.

   a) What is the probability of spinning a number greater than 5?

29. a) Each number has a probability of 1/8; probability of a number greater than 5 = (probability of 6) + (probability of 7) + (probability of 8) = (1/8) + (1/8) + (1/8) = 3/8.

   b) What is the probability of spinning a number less than 3?

b) Probability of a number less than 3 = (probability of 1) + (probability of 2) = (1/8) + (1/8) = 2/8, or 1/4.

   c) What is the probability of spinning a number that is no larger than 4?

c) Probability of a number no greater than 4 = (probability of 1) + (probability of 2) + (probability of 3) + (probability of 4) = (1/8) + (1/8) + (1/8) + (1/8) = 4/8, or 1/2.

Section 7.5 Probability Distributions

What You Need To Know

- How to calculate probabilities from data

- How to construct and interpret a histogram

What You Will Learn

- To construct a probability distribution

- To create and interpret a graph of a probability distribution

- To find the expected value of a probability distribution

Materials

- Handout 7.4 (optional)

- Stopwatch

- Salt packets

- Laboratory balance

- Graph paper

So far in this chapter we have considered ways of determining probabilities associated with individual events. Sometimes, as in Section 7.4, the events were compound and consisted of more than one component. But in each case, a single probability value was sought. In this section, we will examine situations involving uncertainty from a broader viewpoint.

Probabilities often depend on the value of a variable. For example, the probabilities related to card dealing in Discovery 7.2 depend on time, and the probabilities associated with rolling dice vary with different sums. A variable like the time or the sum in such a situation is called a **random variable** because its value has some degree of randomness, or uncertainty, associated with it. Knowing the probabilities associated with all possible values of the random variable provides a bigger picture that can be helpful in understanding a given problem.

Discovery 7.4 Probabilities in a Traffic Study *(SE pages 250–251)*

Materials: stopwatch, Handout 7.4 (optional)

Suppose that the city or town where your college is located is considering making improvements to the street near the main entrance to the college. As a first step, it would probably be necessary to commission a study of the traffic patterns on the street in order to establish whether a need exists for things like wider or more numerous lanes or a special left-turn signal. Such a traffic study is a common task for a transportation engineer. Questions like "What is the probability that more than eight cars might pass the college entrance in any one-minute time period?" require collection of data, specifically counting cars. Therefore, they involve experimental probabilities.

1. Begin by deciding on a traffic statistic that is appropriate to the situation at your college. It might be, for example, the number of cars per minute that appear near the entrance, the number that pass in a certain direction each minute, or the number that turn left into the entrance in a five-minute period. Regardless of what you decide to measure, it is important to be very precise about exactly what will be measured and how and by whom the measurements will be taken. The more data you collect, the better your probability predictions will be.

1. **Answers will vary. One possibility: Count the numbers of cars approaching the entrance in an eastbound direction during 80 different one-minute intervals. One person times the intervals, another counts, another records.**

2. Prepare a data table, such as **Table 7.9,** that will be sufficient to hold whatever data is anticipated. Then go out and collect the data, recording it in the table.

Count of Cars	Frequency of Count
0	
1	
2	
3	
4	
5	
6	
7	
8	
9	
10	

Table 7.9

2.

Count of Cars	Frequency of Count
0	1
1	4
2	7
3	11
4	13
5	13
6	11
7	8
8	6
9	3
10	2
11	1
12	0
13	0
14	0

3. Calculate the mean, median, mode, and range of your data. How do the mean, median, and mode compare?

3. Mean = 6.15 cars, median = 6 cars; these are very close. There are two modes: 4 and 5 cars, each counted 13 times. Range = 11 − 0 = 11 cars.

4. To represent the data you have collected in a different way, organize it into a **frequency histogram.** The horizontal axis should contain the counts of cars in whatever time interval you used, and the height of each bar in the histogram should reflect the number of time intervals during which a certain count was recorded. (Note: If the range of your car count data is much larger than 10, you may find that a grouped frequency histogram will produce a more useful display of the data. See Handout 7.4 for guidance.)

4.

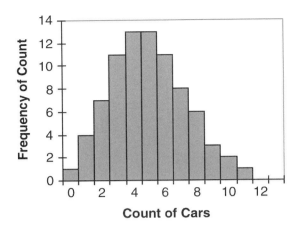

Count of Cars

5. Recall that the probability of an event is defined as follows:

$$\text{Probability of an event} = \frac{\text{Total number of ways the event can occur}}{\text{Total number of possible experimental outcomes}}$$

The number of possible outcomes for your experiment is equal to the number of time intervals examined, because it is not possible to examine anything but a limited number of intervals. For experimental data, the "equally likely" requirement is equivalent to the assumption that each time interval is similar to every other one. In other words, there is no reason to suspect that there will be more cars at certain times than at others. Why might this assumption not be satisfied in some cases for an experiment like yours?

5. If early morning time intervals are compared with midday times, it is possible that the earlier data might be inflated by rush hour traffic of people driving to work. Or there might be natural "bulges" in traffic counts just before class starting times.

6. Assuming that the time intervals you examined are typical of all time intervals, at least as far as the behavior of traffic is concerned, the relative frequencies of the various traffic counts can be interpreted in terms of probabilities. Based on your data, what is the probability that 0 cars will be observed in a given time interval.

6. In the sample data, 0 cars were observed during only one of the 80 intervals. Therefore, the probability that no cars approach the entrance during a one-minute interval is 1/80 = 0.0125. (Obviously, the accuracy of the decimal result is limited. "About 0.01" or "Slightly over one percent" would suffice here.)

7. Add a third column to your table from item 2. Label it "Probability of Count," and compute probabilities for each of the car counts in your table. Then create a second histogram, but make this one a *relative frequency* histogram. It should show probabilities on the vertical axis rather than frequencies.

7.

Count of Cars	Frequency of Count	Probability of Count
0	1	0.0125
1	4	0.05
2	7	0.0875
3	11	0.1375
4	13	0.1625
5	13	0.1625
6	11	0.1375
7	8	0.10
8	6	0.075
9	3	0.0375
10	2	0.025
11	1	0.0125
12	0	0
13	0	0

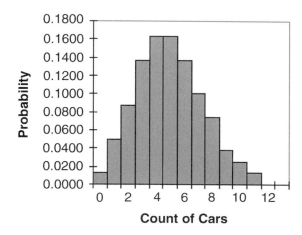

8. From your table, find the probability of observing each of the following under the conditions of your experiment: (a) exactly 8 cars; (b) fewer than 8 cars; (c) more than 8 cars; (d) at least 8 cars.

8. a) 0.075

b) 0.0125 + 0.05 + 0.0875 + 0.1375 + 0.1625 + 0.1625 + 0.1375 + 0.10 = 0.85

c) 0.0375 + 0.025 + 0.0125 = 0.075

d) 0.075 + 0.0375 + 0.025 + 0.0125 = 0.15

9. Locate the largest traffic count in your survey. According to your results, what would you say is the probability of observing a traffic count larger than this value?

9. Largest count is 11 cars; probability of a count greater than 11 is 0.

10. Find the sum of the probabilities in the third column of your table.

10. The sum should be 1, or very close to it (depending on whether individual values are rounded off).

The relative frequency histogram you created from your traffic data in Discovery 7.4 is equivalent to a **probability distribution** for traffic near your college. It provides the probabilities of observing different numbers of cars. The number of cars is the random variable of the distribution.

A probability distribution is based on all possible values of the random variable. Therefore, it includes all possible events for the experiment, and the sum of the probabilities in the distribution must be equal to 1.

The probability distribution in Discovery 7.4 was constructed from experimental data. It is also possible to determine a probability distribution for theoretical probabilities.

Example 12

Find the probability distribution for the sum of the two numbers that result when the spinners in **Figure 7.15** are spun. Construct its graph.

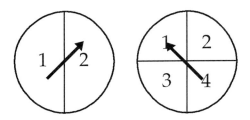

Figure 7.15

Solution:

The possible sums range from 2 to 6. A visual model (see **Figure 7.16**) can be used to identify the probabilities for the different sums.

2nd Spinner

	1	2	3	4
1st Spinner 1	2	3	4	5
2	3	4	5	6

(sums are inside squares)

Figure 7.16

There are eight equally likely outcomes for the two-spinner sums. Sums of 2 or 6 consist of one outcome each, so these sums will occur with probability 1/8. The other possible sums of 3, 4, or 5 each consist of two outcomes. Their probabilities are therefore each equal to 2/8, or 1/4. The probability distribution is listed in **Table 7.10,** and its graph is shown in **Figure 7.17.**

Sum	Probability
2	1/8
3	2/8 = 1/4
4	2/8 = 1/4
5	2/8 = 1/4
6	1/8

Table 7.10

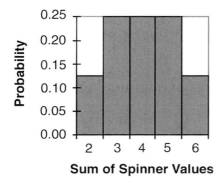

Figure 7.17

The probabilities add to 1, as they must if the probability distribution is complete.

Expected Value of a Probability Distribution

The probabilities listed in your table in item 7 of Discovery 7.4 were based on actual data that you collected. Thus, it was possible to compute an average car count by referring to the data, as you did in item 3. The two-spinner probability distribution in Example 12 is a different type of situation. That distribution was determined theoretically, without spinning the spinners. What would be the average sum on the spinners if they were spun many times? We can predict such an average directly from the probabilities.

A probability is, by definition, equal to the fraction of the time that an outcome will occur over many repetitions of an experiment. Therefore, if the spinners are spun repeatedly, a sum of 2 should occur 1/8 of the time. Out of every eight spins, a sum of 2 should result once, on the average. Similarly, a sum of 6 should result an average of once in every eight spins, while each of the remaining numbers should result an average of twice in eight spins. We can find the mean value of these "average" spins in the usual way, by adding the spin values (with each value multiplied by its average frequency) and dividing the total by 8:

$$\frac{(2 \cdot 1) + (3 \cdot 2) + (4 \cdot 2) + (5 \cdot 2) + 6}{8} = 4.$$

Therefore, we expect that the mean value of many repetitions of the two-spinner experiment will be 4. This number is called the mean, or **expected value,** of the probability distribution.

By separating the fraction in the expected value expression into four simpler fractions, we can write the expected value in terms of probabilities:

$$\frac{2 \cdot 1}{8} + \frac{3 \cdot 2}{8} + \frac{4 \cdot 2}{8} + \frac{5 \cdot 2}{8} + \frac{6 \cdot 1}{8} = 2\left(\frac{1}{8}\right) + 3\left(\frac{2}{8}\right) + 4\left(\frac{2}{8}\right) + 5\left(\frac{2}{8}\right) + 6\left(\frac{1}{8}\right) = 4.$$

The mean of a data set is found by dividing the sum of the data values by the total number of values in the set. A similar division is involved in finding the expected value of a probability distribution, but the denominator is "hidden" in the probability values.

The **expected value** of a probability distribution is found by adding the products formed by multiplying each value of the random variable by its corresponding probability.

Example 13

A company's Internet server handles a large volume of employee connections to the internet. **Table 7.11** shows data that an information technology manager has collected on the distribution of the number of times each employee connects to the server during a workday.

Number of Server Connections	Probability
0	0.13
1	0.31
2	0.26
3	0.18
4	0.09
5	0.03

Table 7.11

Find the expected value of the number of times employees connect to the server.

Solution:

The random variable in this problem is the number of server connections made by an employee. Multiply each value of the variable by the corresponding probability of its occurrence. Then add the products.

$0(0.13) + 1(0.31) + 2(0.26) + 3(0.18) + 4(0.09) + 5(0.03) = 1.88$

The expected value is 1.88 connections per day.

It is important to remember that probabilities given as percentages must be changed to fraction or decimal form before they can be used to compute expected values.

Exercises 7.5

I. Investigations (SE pages 256–259)

1. a) Construct a probability distribution for the result of rolling one six-sided die.

1. a) Each of the numbers 1 through 6 is equally likely.

Number on Die	Probability
1	1/6
2	1/6
3	1/6
4	1/6
5	1/6
6	1/6

b) This kind of probability distribution is called a **uniform distribution.** Explain why.

b) Probabilities for all possible outcomes are the same.

c) Besides rolling a die, there are many situations that involve uniform distributions. A rather simple one is tossing a coin, for which there are two possible outcomes, each with probability 1/2. Name some others.

c) Sample answers: Result of spinning a game spinner if areas for all outcomes are equal; possible wind direction at a particular location on any given day (may not be true everywhere); winning lottery numbers

2. A useful concept that is based on expected value is **expected monetary value, or EMV.** It is calculated by replacing each of the random variable values in a probability distribution with a monetary value and then calculating the expected value as usual. Consider a game in which a person asks you to pick a card randomly from a standard 52-card deck and offers to pay you $10 if the card is an ace, but you will have to pay $1 if any other card is picked. The probability of picking an ace is 1/13, while the probability of picking anything else is 12/13. Your profit from this game can be visualized in **Table 7.12.**

Card Picked	Profit	Probability
Ace	$10	1/13
Other Card	–$1	12/13

Table 7.12

The expected monetary value (or in this case, expected profit) for this game is ($10)(1/13) + (–$1)(12/13) ≈ –$0.15. You would actually have an expected loss of 15 cents each time you play this game. Of course, any one time you play you would either win $10 or lose $1. But if you play the game many times, you can expect an average loss of 15 cents. This is not a good game to play unless you don't mind losing money.

a) Big Al's hot dog stand makes a typical profit of $600 a day except on rainy days, on which the average profit is only $200. If the daily probability of rain is 30%, what is Big Al's expected profit?

2. **a) (0.30)($200) + (0.70)($600) = $480.**

b) Stock market analysts predict that a pharmaceutical stock selling for $12 a share will rise to $20 a share in one year. However, this depends on the success of a new medication for treating hypertension. If the medication fails in a drug study, the stock is predicted to fall to $9 a share. If the probability of failing in the study is 20%, is the stock a good investment?

b) **Expected value of the stock after one year = (0.80)($20) + (0.20)($9) = $17.80. Because this is greater than the purchase price, the investment is likely to be a good one. The amount of profit would be $5.80 per share, minus broker fees.**

c) A contractor is considering bidding for a government project. Based on the size of the bid she thinks she will have to make and past experience with similar projects, she has judged that if she wins the bid, there is a 60% chance of making a $250,000 profit on the project. On the other hand, there is a 10% probability of breaking even and a 30% probability of a $50,000 loss. What is the expected profit if the bid is successful?

c) **(0.60)($250,000) + (0.10)(0) + (0.30)(–$50,000) = $135,000.**

d) The Massachusetts Megabucks lottery pays an average of $3 million to the winner of each jackpot. The chance of winning the jackpot is one in five million. But there is also a one in 25,000 chance of winning $1500, as well as a one in five hundred chance of winning $75 and a 0.027 probability of getting the $1 entry fee back. What is the expected monetary value for playing this lottery?

d) **The sum of all the given probabilities is 0.0290402. But this means that the probability of not winning anything is (1 – 0.0290402) = 0.9709598. The expected winnings are**

$$\left(\frac{1}{5,000,000}\right)(\$3,000,000) + \left(\frac{1}{25,000}\right)(\$1500) + \left(\frac{1}{500}\right)(\$75) +$$

$$(0.027)(\$1) + (0.9709598)(\$0) = \$0.837, \text{ or about 84 cents.}$$

But it costs $1 to play the game, so on the average, players lose 16 cents per play. Of course, this average includes the jackpot winners. The vast majority of lottery players lose $1 almost every time and never win anything more than $75.

3. A typical sample size for many quality control inspections is five. If a company manufactures electrical fuses, a quality control inspection might start with randomly picking five fuses from a production run. Each one would then be tested by passing an overload electrical current through it to see how long it takes for the fuse to blow and stop the current from flowing. If the time is too long, the fuse is defective. However, testing a fuse makes it unusable. (Such a test is called a destructive test.) So even if it were economically feasible to test every fuse made, only a small number can be tested. It is important for the inspector to know the relationship between the number of defective fuses in the sample of five and the actual overall defect rate for the production run, which may contain thousands of fuses. The purpose of this investigation is to find the probability distribution for the number of defectives in a sample of five fuses.

a) For the sake of simplicity, assume that the actual defect rate is 20%, or $\frac{1}{5}$.

This is an unusually high number, indicating that these would not be very good fuses. What is the probability that a sample of five such fuses would contain five defective fuses?

3. a) $\left(\frac{1}{5}\right)^5 = \left(\frac{1}{3125}\right) = 0.00032.$

b) What is the probability that the sample would contain no defectives? In other words, what is the probability that all five are nondefective?

b) $\left(\frac{4}{5}\right)^5 = \left(\frac{1024}{3125}\right) = 0.32768.$

c) Finding the probability of one defective requires a little more thought. The sample consists of five fuses, so one defective means that the other four are non-defective. If the five fuses in the sample are tested one at a time, what is the probability that the first one tested is defective and the others are not?

c) $\left(\frac{1}{5}\right)\left(\frac{4}{5}\right)\left(\frac{4}{5}\right)\left(\frac{4}{5}\right)\left(\frac{4}{5}\right) = \frac{256}{3125} = 0.08192.$

d) There are five different ways for the sample to contain exactly one defective fuse because the defective one could be any of the five. The Addition Rule for Probabilities can be used to find the probability that either the first or the second or the third or the fourth or the fifth fuse is the defective one. Use it to find the probability of exactly one defective in the sample.

d) $0.08192 + 0.08192 + 0.08192 + 0.08192 + 0.08192 = 0.4096.$

e) To find the probability that exactly two fuses are defective, first find the probability that the first two fuses are defective and the last three are non-defective. Then you must count the ways in which the sample could contain two defectives and three non-defectives. A tree diagram (see Exercise 1 of Section 7.2) may be helpful.

e) The probability of the first two defective and the last three fuses being non-defective is $\left(\dfrac{1}{5}\right)\left(\dfrac{1}{5}\right)\left(\dfrac{4}{5}\right)\left(\dfrac{4}{5}\right)\left(\dfrac{4}{5}\right) = \dfrac{64}{3125} = 0.0248.$

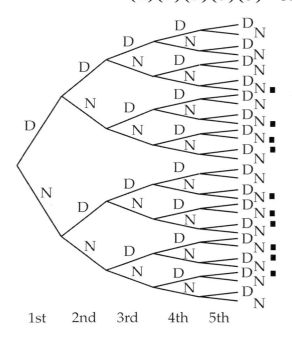

1st   2nd   3rd   4th   5th

There are 32 possible paths through the tree (and therefore 32 possible combinations of defective and non-defective fuses). Of these, 10 combinations (indicated by black squares) have exactly two defectives and three non-defectives. Therefore, the probability of any two of the five fuses being defective while three are non-defective is (10)(0.02048) = 0.2048.

f) Find the probability of exactly three defectives. Then find the probability of exactly four defectives. Make a table and graph of the complete probability distribution.

f) There are also 10 paths through the tree that have three defectives and two non-defectives. The probability of any three being defective is $(10)\left(\dfrac{1}{5}\right)\left(\dfrac{1}{5}\right)\left(\dfrac{1}{5}\right)\left(\dfrac{4}{5}\right)\left(\dfrac{4}{5}\right) = (10)\left(\dfrac{16}{3125}\right) = 0.0512.$ **Similarly, the probability of any four being defective is** $(5)\left(\dfrac{1}{5}\right)\left(\dfrac{1}{5}\right)\left(\dfrac{1}{5}\right)\left(\dfrac{1}{5}\right)\left(\dfrac{4}{5}\right) = (5)\left(\dfrac{4}{3125}\right) = 0.0064.$

g) Write the probability distribution for the number of defectives in a sample of five fuses.

g)

Number of Defectives	Probability
0	0.32768
1	0.4096
2	0.2048
3	0.0512
4	0.0064
5	0.00032

h) What is the expected value of the number of defectives in a sample of five fuses? Does this seem reasonable?

h) (0)(0.32768) + (1)(0.4096) + (2)(0.2048) + (3)(0.0512) + (4)(0.0064) + (5)(0.00032) = 1. This makes sense because if 20% of all fuses are defective, we would expect on the average one out of five defectives.

i) Using your distribution from (g), determine the probability of finding the following: at least two defectives; more than two defectives; less than two defectives; no more than two defectives.

i) Probability of at least 2 defectives = probability of 2, 3, 4, or 5 defectives = 0.2048 + 0.0512 + 0.0064 + 0.00032 = 0.26272.

Probability of more than 2 defectives = probability of 3, 4, or 5 defectives = 0.0512 + 0.0064 + 0.00032 = 0.05792.

Probability of less than 2 defectives = probability of 0 or 1 defective = 0.32768 + 0.4096 = 0.73728.

Probability of no more than 2 defectives = probability of 0, 1, or 2 defectives = 0.32768 + 0.4096 + 0.2048 = 0.94208.

4. It is sometimes more economical to replace items before they need to be replaced. Consider an office building that contains 5000 fluorescent light bulbs. If the lights are mounted on high ceilings, replacement may not be a simple task. Assume that the cost of replacing a single failed bulb, including the bulb itself and the cost of a service person to replace the bulb, is $65. But if all bulbs are replaced at once on a fixed schedule, the service cost is less, so the per bulb cost is only $20, for a total of $100,000 each time all bulbs are replaced. Replacing bulbs all at once, if not done too often, may result in substantial overall savings. Of course, a few bulbs may still fail before a scheduled replacement and would still cost $65 to replace.

a) Consider the probability distribution for bulb failure shown in **Table 7.13**.

Time Until Failure	Probability
First Year of Use	0.1
Second Year of Use	0.3
Third Year of Use	0.4
Fourth Year of Use	0.2

Table 7.13

What is the expected value for bulb failure? Assume (for simplicity) uniform failure rates in a year so that the 0.1 probability in year one can be applied halfway through the year, and so on.

4. **a) (0.5)(0.1) + (1.5)(0.3) + (2.5)(0.4) + (3.5)(0.2) = 2.2.**

b) Because it costs $65 to replace a bulb when it fails, use your answer to (a) to find the average cost per year for replacing a bulb when it fails.

b) $65/(2.2 yr) ≈ $29.55.

c) What is the average yearly cost for replacing all 5000 bulbs only when they fail?

c) 5000($29.55) ≈ $147,750.

d) If all bulbs are replaced every year, (0.1)(5000) = 500 will have failed before replacement, costing (500)($65) = $32,500 in addition to the scheduled replacement cost of $100,000, for a total of $132,500. This is a savings over replacing each bulb as it fails. But this means replacing a lot of good bulbs. What if the scheduled replacement is done every two years? During year one, 500 bulbs will still fail at a total cost of $32,500. How many of these already replaced bulbs will fail during the second year, and at what total cost?

d) (0.1)(500) = 50 bulbs, for a cost of (50)($65) = $3250.

e) How many of the original bulbs will fail during the second year, and at what total cost?

e) (0.3)(5000) = 1500 bulbs, for a cost of (1500)($65) = $97,500.

f) What is the total cost of a two-year replacement schedule?

f) $100,000 + $3250 + $97,500 = $200,750.

g) What is the yearly cost of a two-year replacement schedule?

g) $200,750/2 = $100,375.

h) In a similar manner, determine the yearly cost of a three-year replacement schedule.

h) Replace all bulbs after three years – $100,000

Replace (0.1)(5000) = 500 first-year failures – (500)($65) = $32,500

Replace 10% of first-year replacements during second year – (0.1)(500)($65) = $3250

Replace 30% of first-year replacements during third year – (0.3)(500)($65) = $9750

Replace (0.3)(5000) = 1500 second-year failures – (1500)($65) = $97,500

Replace (0.1)(1500) = 150 second-year replacements during third year – (150)($65) = $9750

Replace third-year failures of original bulbs – (0.4)(5000)($65) = $130,000

Total cost over three years = $100,000 + $32,500 + $3250 + $9750 + $97,500 + $9750 + $130,000 = $382,750

Yearly cost of three-year replacement schedule is $382,750/3 = $127,583.

i) What is the least costly replacement schedule? Why is a four-year schedule not considered?

i) Two-year schedule is the least costly; four-year schedule would result in replacement of all bulbs as they fail, then replacing all again at the end of four years, and would be more expensive than in (b).

II. Projects and Group Activities (SE pages 259–261)

5. Materials: dice

a) The probability of rolling any of the sums from 2 through 12 with a pair of six-sided dice varies with the sum. This is because the various sums are not equally likely. For example, a 2 can be rolled in only one way (1 on both dice), but a 3 can be rolled in two different ways (1 on one die, 2 on the other, or vice versa). Use an area model or a tree diagram to determine the probability distribution associated with rolling two dice. Write the probabilities as exact fractions instead of as rounded decimals.

5. a) There are 36 equally likely outcomes for the two dice.

Sum	Probability
2	1/36
3	2/36 = 1/18
4	3/36 = 1/12
5	4/36 = 1/9
6	5/36
7	6/36 = 1/6
8	5/36
9	4/36 = 1/9
10	3/36 = 1/12
11	2/36 = 1/18
12	1/36

First Die

Second Die	1	2	3	4	5	6
6	7	8	9	10	11	12
5	6	7	8	9	10	11
4	5	6	7	8	9	10
3	4	5	6	7	8	9
2	3	4	5	6	7	8
1	2	3	4	5	6	7

(sums are in squares)

b) Using your probability distribution from (a), find the expected value for rolling a pair of dice.

b) 2(1/36) + 3(1/18) + 4(1/12) + 5(1/9) + 6(5/36) + 7(1/6) + 8(5/36) + 9(1/9) + 10(1/12) + 11(1/18) + 12(1/36) = 7.

c) Test your model by rolling a pair of dice many times to see if the experiment confirms your predictions in (a) and (b).

c) The mean sum should approach 7 as the number of rolls increases.

6. Materials: salt packets, accurate balance, graph paper, Handout 7.4 (optional)

Have you ever noticed that occasionally a bottle of soda has a little more empty space above the liquid than usual? Variability is a natural part of any industrial process. It is impossible to ensure that every bottle of soda contains the same volume of liquid or that every box of cereal contains the same weight of cereal or that steel beams always have exactly the same strength. It is the function of a company's quality control department to measure the variability in a product and take steps to prevent it from becoming unusually large.

a) Examine the balance that you will be using. Does it measure mass or weight? (If it is a metric balance, it probably measures mass in grams, but some balances measure weight in ounces.) What is the precision of the balance (that is, the smallest measurement that can be made with it)?

6. a) If the laboratory balance measures mass in grams, the smallest division should be no larger than 0.01 g, although the smallest labeled values may be 0.1 g.

b) Measure the mass (or weight, depending on the kind of balance available) of one salt packet, to the maximum precision of the balance. Considering the mass of the packet and the precision of the balance, what is the inherent percent error in this measurement? (That is, express the precision of the balance as a percent of the mass of a packet.)

b) **If the mass of one packet is, for example, 0.73 g, the percent error in the measurement is $\left(\dfrac{0.005 \text{ g}}{0.73 \text{ g}}\right) \cdot 100 \approx 0.7\%$.**

c) Now collect data on the masses of 100 packets of salt. Find the mean and the range of the data.

c) Sample data are shown in a tally sheet form:

Mass of Packet (g)	Tally (# of packets)
0.70	/
0.71	
0.72	//
0.73	//
0.74	///// /////
0.75	///// ////
0.76	///// ///// /////
0.77	///// ///// ///// ////
0.78	///// ///// ///
0.79	///// ///// /
0.80	///// ///
0.81	/////
0.82	//
0.83	/
0.84	/
0.85	
0.86	/

mean = 0.77 g, range = (0.86 g − 0.70 g) = 0.16 g

d) Mass (or weight) is a continuous variable. It makes no sense to talk about a packet's mass being "exactly one gram." Instead, it is customary to group continuous data in equal-width groups and count the number of packets with mass in a certain group interval. Record the data in a grouped frequency table. In order to determine appropriate group sizes, first find the range of the data. Then divide the range by 10 to find an approximate group width, and construct groups using this width to include the entire range of data. The first column in your table should list the groups, measured in grams (or some other unit). The second should indicate the number of packets that had masses in each group. Then create a third column showing the percentage of packets that are in each group.

d) **Range/10 = 0.16 g ≈ 0.2 g, so this is a good group width to use. If the lowest group is from 0.695 g to 0.715 g, the group width is 0.2 g and the group accurately reflects the interval of actual packet masses that are recorded as either 0.70 g or 0.71 g.**

Mass (g)	Frequency	%
0.695–0.715	1	1%
0.715–0.735	4	4%
0.735–0.755	19	19%
0.755–0.775	34	34%
0.775–0.795	24	24%
0.795–0.815	13	13%
0.815–0.835	7	7%
0.835–0.855	2	2%
0.855–0.875	1	1%

e) Now construct a relative frequency histogram, with salt packet mass on the horizontal axis and the percentage of packets falling within each group on the vertical axis.

e)

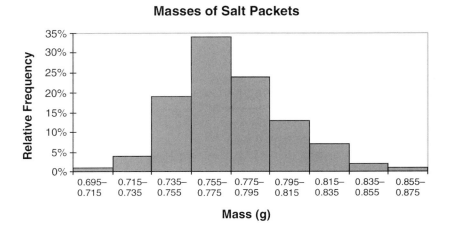

f) If there are other groups of students doing the same experiment, you may want to pool your data at this point in order to work with a larger and more representative data set. If you do this, how does the histogram based on the whole class's data compare with the one based only on your data?

f) For a very large data set, the histogram should be more bell-shaped.

g) It would be very impractical for the company that packaged the salt to accurately measure every packet to make sure it contains the correct amount of salt. But by measuring a sample of the packets the way you have done, it is possible to estimate the probability that the mass of a packet differs significantly from the average. For example, find the number that is equal to one-third of the range of your data, and call it n. Then, denoting the mean of your data as $\bar{x}$, find the mass values equal to $\bar{x} - n$ and $\bar{x} + n$. Finally, find (from either your raw data or your histogram) the probability that the mass of a salt packet is in the interval from $\bar{x} - n$ to $\bar{x} + n$.

g) (range/3) $\approx$ 0.5 g; $(\bar{x} - n)$ = (0.77 g – 0.5 g) = 0.72 g; $(\bar{x} + n)$ = (0.77 g + 0.5 g) = 0.82 g. The probability of a salt packet being in the interval (0.72 g to 0.82 g) is 96% because 96 out of 100 packets are in that range for this sample.

h) What is the probability that the mass of a salt packet is *not* between $\bar{x} - n$ and $\bar{x} + n$?

h) 100% – 96% = 4%

i) Repeat (g) and (h), using the entire class's data. How do the results compare with those based on your data alone?

i) Class results should be very similar to any individual group's values because a sample size of 100 should give fairly reproducible results.

III. Additional Practice *(SE pages 261–267)*

7. **Table 7.14** contains sample data on births in the United States in a year.

Month	Number of Births
January	278,000
February	272,000
March	302,000
April	284,000
May	295,000
June	293,000
July	313,000
August	330,000
September	311,000
October	325,000
November	286,000
December	308,000

Table 7.14

Construct a table of probabilities for the birth month of a randomly selected American baby born in the same year. (This will not be a probability distribution, as it is not based on values of a numerical random variable. It is, however, a type of probability model.)

7. The total number of births is 3,598,000. The probability of a birth in each month equals the fraction of babies born in that month. For example, for January the probability is $\dfrac{278,000}{3,598,000} = 0.077$.

Month	Probability
January	0.077
February	0.076
March	0.084
April	0.079
May	0.082
June	0.081
July	0.087
August	0.092
September	0.086
October	0.090
November	0.079
December	0.086

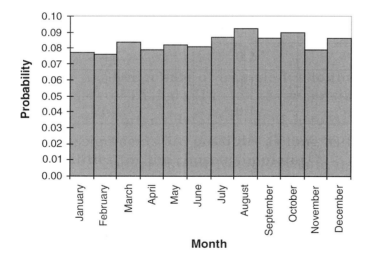

8. There were 3,880,894 babies born in the United States in 1997; 104,137 were born as part of a set of twins, 20,211 were triplets, 2040 were quadruplets, and 400 were members of sets of five or more siblings. Construct a probability distribution for whether a randomly selected 1997 baby was a single birth, a twin, a triplet, a quadruplet, or part of a group of quintuplet size or larger.

8.

Size of Birth Group	Probability
1	0.9673
2	0.0268
3	0.0052
4	0.0005
5 or higher	0.0001

9. **Table 7.15** shows data on the number of full four-year terms served by the first 41 presidents of the United States. Construct a probability distribution for the number of terms served by U.S. presidents.

Number of Terms	Frequency
0	10
1	19
2	11
3	1

Table 7.15

9.

Number of Terms	Probability
0	10/41 = 0.244
1	19/41 = 0.464
2	11/41 = 0.268
3	1/41 = 0.024

10. **Table 7.16** shows a summary of the frequencies of occurrence of major earthquakes (Richter magnitude 7.0 or greater) during the 20th century. Construct a probability model for the occurrence of major earthquakes.

Number of Earthquakes per Year	Frequency
0–4	0
5–9	6
10–14	16
15–19	27
20–24	24
25–29	13
30–34	9
35–39	4
40–44	1

Table 7.16 (Source: United States Geological Survey)

10.

Number of Earthquakes per Year	Probability
0–4	0
5–9	0.06
10–14	0.16
15–19	0.27
20–24	0.24
25–29	0.13
30–34	0.09
35–39	0.04
40–44	0.01

11. The number of trucks arriving per hour at a company's loading dock were monitored over an extended period as part of the research phase for a planned expansion of the loading area. **Table 7.17** shows the data.

Number of Truck Arrivals per Hour	Frequency
0	8
1	24
2	40
3	32
4	27
5	19
6	6
7	2

Table 7.17

Construct a probability distribution for the number of hourly truck arrivals, and graph the distribution.

11.

Number of Truck Arrivals per Hour	Probability
0	8/158 = 0.051
1	24/158 = 0.152
2	40/158 = 0.253
3	32/158 = 0.203
4	27/158 = 0.171
5	19/158 = 0.120
6	6/158 = 0.038
7	2/158 = 0.013

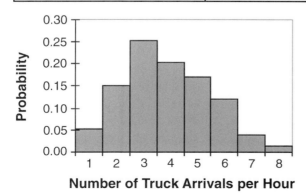

12. Construct a graph of the probability distribution for the number of dots showing on a roll of a single six-sided die (see Exercise 1).

12. Distribution from Exercise 1 is:

Number on Die	Probability
1	1/6
2	1/6
3	1/6
4	1/6
5	1/6
6	1/6

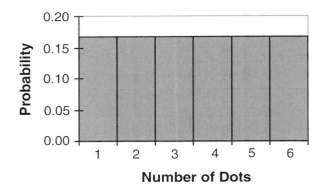

13. Construct a probability distribution for the number of heads that result when three coins are tossed, and then graph the distribution. A tree diagram may be helpful (see Exercise 1 of Section 7.2).

13. From the tree diagram in Section 7.2, Exercise 1(b), there are eight equally likely outcomes for tossing three coins, one of which results in 0 heads, three each resulting in 1 or 2 heads, and one resulting in 3 heads. The probability distribution is:

Number of Heads	Probability
0	1/8 = 0.125
1	3/8 = 0.375
2	3/8 = 0.375
3	1/8 = 0.125

For 14–16, express whether the given table shows a possible probability distribution. If not, explain why.

14. Use **Table 7.18.**

Value of Random Variable	Probability
1	0.34
2	0.16
3	0.29
4	0.21

Table 7.18

14. Yes, it is a possible probability distribution.

15. Use **Table 7.19.**

Value of Random Variable	Probability
0	0.1
1	0.3
2	0.2
3	0.2
4	0.1

Table 7.19

15. Not a probability distribution; probabilities add to 0.9, not 1.

16. Use **Table 7.20.**

Value of Random Variable	Probability
10	1/3
20	2/5
30	1/4

Table 7.20

16. Not a probability distribution; 1/3 + 2/5 + 1/4 = 59/60, not 1.

For 17–18, complete the probability distribution.

17. See **Table 7.21.**

Value of Random Variable	Probability
5	0.2
10	0.25
15	0.1
20	?

Table 7.21

17. In order to make the total probability equal to 1, the missing value must be 0.45.

18. See **Table 7.22.**

Value of Random Variable	Probability
2	1/16
4	1/8
6	1/4
8	1/2
10	?

Table 7.22

18. In order to make the total probability equal to 1, the missing value must be 1/16.

19. A project scheduler for a construction company has kept data on the number of days it takes to excavate the site for a certain type of building, as summarized in **Table 7.23.**

Number of Days	Probability
2	0.2
3	0.4
4	0.3
5	0.1

Table 7.23

What is the probability that site excavation takes longer than three days?

19. 0.3 + 0.1 = 0.4.

20. **Table 7.24** shows a probability distribution for the number of computers in a computer laboratory that are connected to the Internet at any instant.

Number of Computers Online	Probability
0	0.07
1	0.13
2	0.16
3	0.17
4	0.21
5	0.15
6	0.08
7	0.03

Table 7.24

a) What is the probability that at least five of the computers are connected at any one time?

20. a) **0.15 + 0.08 + 0.03 = 0.26.**

b) What is the probability that fewer than three computers are connected at any one time?

b) **0.07 + 0.13 + 0.16 = 0.36.**

21. **Table 7.25** shows a probability distribution for heights of American men.

Height (inches)	Probability
Below 63	0.01
63	0.01
64	0.03
65	0.04
66	0.07
67	0.09
68	0.12
69	0.13
70	0.13
71	0.12
72	0.09
73	0.07
74	0.04
75	0.03
76	0.01
Above 76	0.01

Table 7.25

If a doorway in an office building is 75 inches high, what is the probability that a randomly selected man will be able to walk through it without ducking?

21. **Probability of not having to duck = 1 − (probability of having to duck) = 1 − (probability of being taller than 75 inches) = 1 − (0.01 + 0.01) = 0.98.**

22. **Table 7.26** shows a probability distribution for the quantity of a certain model of air conditioner in a department store's inventory at any time.

Quantity	Probability
0	0.02
1	0.08
2	0.15
3	0.21
4	0.32
5	0.16
6	0.06

Table 7.26

A new order for this model is placed whenever inventory drops below three units. What is the probability that inventory is below three units?

22. **Probability that inventory is below 3 = probability that inventory is either 0, 1, or 2 = 0.02 + 0.08 + 0.15 = 0.25.**

For 23–29, find the expected value of the probability distribution.

23. The lifetimes before recharging of batteries made by a company for portable CD players (see **Table 7.27**).

Battery Life (hours)	Probability
1	0.09
2	0.20
3	0.32
4	0.25
5	0.14

Table 7.27

23. **1(0.09) + 2(0.20) + 3(0.32) + 4(0.25) + 5(0.14) = 3.15 hours.**

24. The distribution of site excavation times in Table 7.23 (Exercise 19)

24. **2(0.2) + 3(0.4) + 4(0.3) + 5(0.1) = 3.3 days.**

25. The number of online computers in Table 7.24 (Exercise 20)

25. **0(0.07) + 1(0.13) + 2(0.16) + 3(0.17) + 4(0.21) + 5(0.15) + 6(0.08) + 7(0.03) = 3.24 computers.**

26. The probability distribution for air conditioner inventory in Table 7.26 (Exercise 22)

26. 0(0.02) + 1(0.08) + 2(0.15) + 3(0.21) + 4(0.32) + 5(0.16) + 6(0.06) = **3.45 air conditioners.**

27. The number of weeks to complete the product development project in Example 10 of Section 7.4

27. 3(0.10) + 4(0.30) + 5(0.40) + 6(0.20) = **4.7 weeks.**

28. The number of full terms served by U.S. presidents (see Table 7.15 of Exercise 9)

28. 0(10/41) + 1(19/41) + 2(11/41) + 3(1/41) = 54/41 ≈ **1.32 terms.**

29. Use **Table 7.28.**

Value of Random Variable	Probability
1	0.5
2	0.3
3	0.1
4	0.1

Table 7.28

29. 1(0.5) + 2(0.3) + 3(0.1) + 4(0.1) = **1.8.**

Section 7.6 Drawing Conclusions from Data

What You Need To Know

- How to find a sample mean
- How to draw a frequency histogram

What You Will Learn

- To use sample measurements to estimate values of unknown quantities
- To understand the meaning of sampling error

Materials

- Coffee stirrers or straws
- Scissors
- Bag or other container
- Metric ruler
- Coin

Sampling

There are many reasons for collecting data, and there are many kinds of data that can be collected. Frequently, some value of a numerical property is sought without prior knowledge of it. For example, an environmental scientist might wish to determine the concentration of a toxin in the soil near a factory. Or a communications company might want to know the average weekly Internet usage in a city. Or an opinion pollster might seek the approval rating of an official. In such situations, it may be impossible, or sometimes just too costly, to measure all of the data to allow exact determination of the value of interest.

Usually, a sample of data will be collected in order to try to determine the value as accurately as possible. The sample size may be as small as one or as large as several thousand or more. As you might expect, larger samples give more accurate information. But at best a sample can only give an approximate value, or estimate, of the value sought. The process of sampling, that is, collecting data on relatively small samples from a larger population, will be explored in Discovery 7.5.

Discovery 7.5 Quality Control and Estimating *(SE pages 269–271)*

Materials: coffee stirrers or straws, scissors, bag or other container, metric ruler

For this exploration, you will be given a bag containing a large number of pieces of straws of varying lengths. Imagine that you are a company's quality control inspector, whose job is to determine the average size of parts produced on a product assembly line. It is essential that a company know accurately the properties of its products so it can be assured of meeting clients' requirements. Because all manufacturing processes have inherent variability, sampling is necessary in order to estimate the property of interest. In this case, that property is the average length of a piece. (The variability in this Discovery is exaggerated for ease of measurement, compared to what would be typical in a manufacturing environment.)

1. Shake the bag to ensure a random mixture of the different size pieces. Then pull one piece from the bag and measure its length to the nearest millimeter. Do you think this value would be a good one to use for an estimate of the average length of all the pieces in the bag?

1. **Answers will vary from one to several centimeters depending on the selection. Measuring the length of one item would not be a very good way to estimate the average length of the population of pieces of various lengths.**

2. Now pull four more pieces from the bag, for a total sample size of five. Measure the length of each piece. Compute and record the sample mean length by finding the sum of the five lengths and dividing by 5. Return the five pieces to the bag.

2. **One possible sample: 7.7 cm, 6.0 cm, 4.6 cm, 3.2 cm, 2.1 cm, for a sample mean length of 4.72 cm.**

3. Are you more confident about using the mean length of your sample of five pieces to estimate the average length of all pieces in the bag than you were after measuring only one? Explain.

3. **A larger sample should probably inspire greater confidence. A single randomly selected piece is liable to be either considerably larger or considerably smaller than average, whereas means of larger samples tend to smooth out the extremes.**

4. Repeat the sampling and measuring process many times (as many as a hundred or so if you have several people in your group and enough available time). Return your measured sample of five pieces to the bag each time, and shake the bag to randomize the samples. When you have finished collecting data, summarize your findings by completing **Table 7.29**. This **grouped frequency table** sorts the sample mean data into larger groups for comparison.

Sample Mean Length (cm)	Frequency (Number of Samples in Each Group)
2 to just under 2.5	
2.5 to just under 3.0	
3.0 to just under 3.5	
3.5 to just under 4.0	
4.0 to just under 4.5	
4.5 to just under 5.0	
5.0 to just under 5.5	
5.5 to just under 6.0	
6.0 to just under 6.5	
6.5 to just under 7.0	
7.0 to just under 7.5	
7.5 to just under 8.0	

Table 7.29

4. **Sample data are shown.**

Sample Mean Length (cm)	Frequency (Number of Samples in Each Group)
2 to just under 2.5	0
2.5 to just under 3.0	2
3.0 to just under 3.5	6
3.5 to just under 4.0	12
4.0 to just under 4.5	13
4.5 to just under 5.0	18
5.0 to just under 5.5	17
5.5 to just under 6.0	15
6.0 to just under 6.5	6
6.5 to just under 7.0	6
7.0 to just under 7.5	4
7.5 to just under 8.0	1

5. Because you have made a large number of measurements of piece length, you could actually compute a very accurate estimate of average length for this large sample (of perhaps as many as 500 pieces). But in this Discovery, we are investigating the accuracy of individual samples of size five, so that value will not be computed at this time. The center of the histogram should provide a very good visual estimate of the true mean, however.

Construct a frequency histogram of your grouped frequency data for the sample means from item 4.

5.

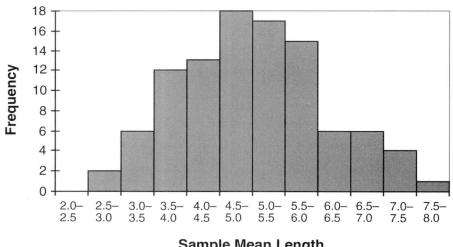

6. Compute a number that is 5% of the total number of samples you collected in item 2.

6. 5 samples (out of a total of 100 samples)

7. On your histogram, find a sample mean length for which 5% of the sample means are to its left, and draw in a vertical line at this value. Draw a second vertical line at a value for which 5% of the sample means are to its right. Calculate half the width of the interval between the two vertical lines, and call it E.

7.

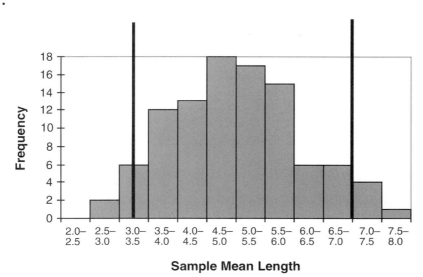

Sample Mean Length

$E = (7.0{-}3.25)/2 \approx 1.9$ cm.

The actual average length of all the pieces in the bag is probably very near the center of the interval bounded by your two vertical lines. If so, then only 10% of the samples you measured have means that are farther than E from the true average length: E is the maximum error for 90% of the samples.

What if you had measured the mean length of only one sample of five pieces? There would be a 10% probability that the sample mean was further than E from the true average. Another way of looking at this is to say that there would be a 90% probability that your sample mean was within E of the true average. Or you could say that you are 90% confident that your sample mean is within E of the true average. In statistical language, we say that the interval from E units below a sample mean to E units above it is a 90% **confidence interval** for the true average value.

8. Write a 90% confidence interval based on your first quality control sample. For example, if your first sample had a mean length of 3.8 cm and your value for E was 2.3, then the 90% confidence interval for the true average length of all pieces would be between 1.5 cm and 6.1 cm.

8. **Sample data give limits of (4.72 − 1.9 cm) and (4.72 + 1.9 cm). A 90% confidence interval for the true average length of all the pieces is from 2.82 cm to 6.62 cm.**

9. Finally, calculate the best estimate of the true average length, from the overall mean of all the samples. Does it fall within your confidence interval? There's a 90% chance that it will.

9. Overall mean is 4.86 cm. This does fall within the confidence interval. About 10% of students' first samples will not provide confidence intervals that contain the true mean.

Confidence Interval

A confidence interval is an estimate of the value of an unknown quantity based on a sample measurement. If the sample mean is $\bar{x}$ and the maximum error of the estimate (which depends on the desired confidence) is E, then the confidence interval is from $(\bar{x} - E)$ to $(\bar{x} + E)$.

Example 14

A town is concerned over reports of lead in water supplies. State environmental regulations require that average lead concentration in each town's drinking water be less than 15 parts per million (ppm). Tests are performed on a sample of water in twelve randomly selected households. The measured lead concentrations (in ppm) are 16, 12, 15, 16, 18, 17, 19, 17, 13, 14, 11, and 18. If the error E for 90% confidence for this sample is known to be 1.3, find a 90% confidence interval for the true average lead concentration for all of the town's households.

Solution:

The sample mean lead concentration is
$$\bar{x} = \frac{16 + 12 + 15 + 16 + 18 + 17 + 19 + 17 + 13 + 14 + 11 + 18}{12} = 15.5 \text{ ppm. A } 90\%$$
confidence interval is from $(\bar{x} - E)$ to $(\bar{x} + E)$, or $(15.5 - 1.3)$ to $(15.5 + 1.3)$. Therefore, we say the town's true average lead concentration is between 14.2 and 16.8 ppm. And when we make this statement, we are 90% sure that it is correct.

Confidence intervals are found in a large variety of applications. One common use is in opinion polling. A *Newsweek* poll taken just after the 2000 Democratic convention reported that 48% of the people surveyed planned on voting for Al Gore in the general election. In the same poll, 42% planned on voting for George W. Bush. The **margin of error** was given as 4%. Because only a sample (806 registered voters) of the voting population was polled, these figures are only estimates of the true percentages of people that planned on voting for either Gore or Bush. The 4% margin of error is equivalent to the *E*-value in your analysis of Discovery 7.5. So another way to state the poll results is to use confidence intervals. Most opinion polls are based on 95% confidence, so we might say that a 95% confidence interval for the percentage of voters preferring Gore is from 44% to 52%. A complete statement of the poll result would be, "We are 95% sure that the percentage of the population that intends to vote for Gore is between 44% and 52%." Similarly, we could say, "We are 95% sure that the percentage of the population that intends to vote for Bush is between 38% and 46%." Gore's apparent 6% lead is therefore not as solid as the published results make it appear. And although it is rarely stated, based on our work in Discovery 7.5, we know that a 95% confidence interval implies that there is a 5% chance that the pollsters are wrong.

Confidence Level and Sample Variability

When we measure a sample in order to find information about the value of an unknown quantity, we know that the sample cannot provide the true value of the quantity. This is why a confidence interval is called an estimate. **Sampling error** refers to the error that results from not measuring the entire population that we are studying. It does not mean mistakes were made or that measurements were not accurate. Sampling error is a known amount that we have called *E*, the maximum error of the estimate, sometimes called the margin of error. Stating a confidence interval in terms of *E* makes it possible to know how much we can trust the result. The **confidence level** of 90%, 95%, or sometimes 99% is a measure of that trust.

The confidence level for a given sample can be increased, but only if we are willing to accept a larger margin of error. Consider again your histogram from item 5 of Discovery 7.5. If the vertical lines you drew separated only 2.5% of the data on each end of the graph instead of 5%, you would have identified the size of a 95% confidence interval. The tradeoff would be that the error *E* was larger. However, it is always possible to reduce the margin of error by measuring more data. The pollsters for *Newsweek* knew that their Gore/Bush poll's margin of error would be close to 4% if they sampled about 800 voters (see Exercise 1).

Exercises 7.6

I. Investigations *(SE pages 273–276)*

1. When polling organizations tell us about the accuracy of their polls by giving a margin of error, we can never be absolutely certain that the sample results are as close to the truth about the population as the margin of error implies. We only know that the method used gives results that are at least as close to the truth as the margin of error 95% of the time. But how is the margin of error found? For most polls with a 95% confidence level, the margin of error is approximately equal to $\dfrac{100\%}{\sqrt{n}}$, where n is the number of people interviewed. For the Gore/Bush *Newsweek* poll, $n = 806$ and $\dfrac{100\%}{\sqrt{806}} \approx 3.5$, so this suggests an approximate margin of error of 3.5% in the poll. The $\dfrac{100\%}{\sqrt{n}}$ approximation is only good if the poll results are close to 50%. When they are not, a more detailed analysis is necessary in order to determine the margin of error.

   a) If a polling organization interviews 2000 voters to determine the percentage favoring either candidate in a close presidential race, what is the approximate margin of error in the poll results?

   1. a) $\dfrac{100\%}{\sqrt{2000}} \approx 2.2\%$.

   b) If a campus organization polls student opinion on whether a new library should be built and interviews only 100 students, how accurate are the results?

   b) Margin of error is $\dfrac{100\%}{\sqrt{100}} = 10\%$.

   c) A poll was conducted by a local radio station and Arizona State University prior to the 2000 Republican primary in that state. Out of 527 likely Republican voters interviewed, 37% said they preferred Senator John McCain in the primary. Find a 95% confidence interval for the true percentage of all Republican voters in Arizona that preferred McCain.

   c) Margin of error is $\dfrac{100\%}{\sqrt{527}} \approx 4.4\%$, so between (37% − 4.4%) = 32.6% and (37% + 4.4%) = 41.4% of all Arizona Republican voters preferred McCain.

   d) During April 21–24, 1995, 51% of a polling sample of 250 people said they approved of the way President Clinton was doing his job. Considering the likely margin of error of this poll, does this provide evidence that a majority of people approved of President Clinton's performance?

   d) Margin of error is $\dfrac{100\%}{\sqrt{250}} \approx 6.3\%$. No, we only know that the true approval rating is somewhere between 45.7% and 57.3%.

2. All the discussion of estimation in this section assumes that samples, whether of measurements on machine parts or of voters' opinions, are taken randomly. The statistical methods on which confidence intervals and other types of conclusions about data are based require that selection of objects or people to be sampled not be biased in any way. Much time and effort is devoted by polling organizations to ensure that samples are randomly selected.

a) A famous example of the consequences of non-random sampling occurred during the first military draft lottery of the Vietnam era, held in 1969. In that lottery, 366 balls with all possible birth dates were mixed in a bin, and the order in which they were drawn out of the bin determined the order of priority for the draft. If every birth date were equally likely to end up in the first 100 dates drawn, what fraction of the first 100 would you expect to be December birth dates?

2. **a) 31/366 ≈ 0.085.**

b) As it turned out, 15 December birth dates were among the first 100 drawn. What fraction does that represent?

b) 15/100 = 0.15.

A statistical analysis done some time after the lottery confirmed that birth dates toward the end of the year were more likely to have been assigned low draft numbers. It has been suggested that the bin containing the 366 balls was not turned enough to completely randomize the selection.

c) In the mid-1980s, Midway Airlines carried 8% of the passenger traffic between New York and Chicago. But they claimed in their ads that "84 percent of frequent business travelers to Chicago prefer Midway Metrolink to American, United, and TWA." The fine print at the bottom of the ad said that the survey was "conducted among Midway Metrolink passengers between New York and Chicago." From the standpoint of sampling, what is wrong with the first statement?

c) The sample is not a random sample of all business travelers between New York and Chicago. Passengers surveyed have already made a choice to fly Midway.

d) During the 1936 presidential election campaign, *The Literary Digest* predicted that the Republican candidate, Alf Landon, would beat Franklin Roosevelt in a landslide. In fact, Roosevelt carried every state but two in winning his second term as president. *The Literary Digest* had conducted its survey by sending questionnaires to 10 million people selected from lists of owners of cars and telephones. The completed questionnaire was returned by 2,376,523 people. How do you think a survey based on such a huge sample could be so wrong in its prediction?

d) In 1936, a smaller fraction of Americans owned cars and telephones. Poorer people tended to be Democratic voters. Also, the sample was self-selecting, as only about 1/5 of people returned questionnaires. They may have been more likely to be dissatisfied with Roosevelt.

3. The margin of error for a measurement of mean value of a continuous variable like length, time, temperature, and strength, can be determined directly from the values of the sample measurements. The maximum error E as examined in Discovery 7.5 depends on how variable the distribution is for all measurements of the quantity being estimated. The key to determining the size of E is in a measure of variability called the **sample standard deviation,** symbolized by s. The value of s is found by first determining a sample mean value $\bar{x}$ from all the separate measurements of x. The deviation of each measurement from the mean $(x - \bar{x})$ is then found. The sample standard deviation is sometimes called the "root-mean-square" of these deviations: The squares of the deviations are added, then divided by $(n - 1)$, where n is the number of measurements in the sample.

The square root of that result is the standard deviation. Expressed as a formula, it is

$$s = \sqrt{\frac{\sum (x - \bar{x})^2}{n - 1}}.$$

For example, for the five time measurements 2.7 hr, 3.4 hr, 3.1 hr, 4.0 hr, and 3.5 hr, the sample mean is 3.34 hr and the standard deviation is:

$$\sqrt{\frac{(2.7 - 3.34)^2 + (3.4 - 3.34)^2 + (3.1 - 3.34)^2 + (4.0 - 3.34)^2 + (3.5 - 3.34)^2}{5 - 1}} \approx 0.48 \text{ hr.}$$

a) Find the sample standard deviation of the four diameter measurements 8.4 cm, 7.2 cm, 7.6 cm, 8.1 cm.

3. **a) Sample mean is 7.825, so sample standard deviation is**

$$\sqrt{\frac{(8.4 - 7.825)^2 + (7.2 - 7.825)^2 + (7.6 - 7.825)^2 + (8.1 - 7.825)^2}{4 - 1}} \approx 0.53 \text{ cm.}$$

For large data sets, calculating a standard deviation in this way can be very tedious. Fortunately, most calculators can compute standard deviations automatically. **Figure 7.18a** shows the List Math menu on a TI-83 calculator. (If another calculator is used, see the user's manual.) **Figure 7.18b** shows the result of using the standard deviation command on list L_1 containing the time measurement data.

Figure 7.18a

Figure 7.18b

b) Use a calculator to compute the sample standard deviation for the diameter measurements in (a).

b)

c) The margin of error E for a 95% confidence interval is approximately given by $E = \dfrac{2s}{\sqrt{n}}$, but this is strictly valid only for larger sample sizes where n is close to 30 or more.

If a sample of 40 bolts has a mean breaking strength of 786 pounds with a standard deviation of 52 pounds, find a 95% confidence interval for the true mean breaking strength of all such bolts.

c) $E = \dfrac{2(52)}{\sqrt{40}} \approx 16$, **so the true mean breaking strength is between 770 lb and 802 lb.**

II. Projects and Group Activities *(SE pages 276–277)*

4. Materials: none

On the Internet, find an opinion poll taken prior to a recent local or national election, or one that discusses the public's approval of the president. Discuss its sampling design, method of analysis, final report, and other features you think are important.

4. Answers will vary but should reflect the mathematics discussed in this section.

5. Materials: coin

As discussed in this section, estimation involves drawing a conclusion about the value of some unknown quantity from measurements that are made on a sample. Another kind of conclusion that can be made using statistical methods involves deciding whether or not some statement is true based on collected data. If a quantity is thought to be a certain size, but measurement shows too large a difference to have reasonably happened by chance, then the difference is called **statistically significant**. A test to look for such a difference is called a **significance test**.

Consider the data on the lead concentration in a town's water supply discussed in Example 14. The mean lead concentration in the measured sample was 15.5 ppm, which is above the limit of 15 ppm set by the state. But is it *significantly* above the limit? Or, to put it another way, could the town's actual lead concentration be less than 15 ppm if the mean of a sample of 12 readings is 15.5? Because the 90% confidence interval was between 14.2 ppm and 16.8 ppm, we know that there is a reasonable chance that the true lead level may be less than 15 ppm. In fact, only if the sample mean were higher than 16.4 would the confidence interval not include 15 ppm. If that were to happen, then we could justifiably pronounce that the lead level in the town's water is significantly greater than 15 ppm. Because we are only concerned with one side of the confidence interval in this situation, we could then also say that there is no more than a 5% chance that the pronouncement is wrong.

When a drug is tested by a pharmaceutical company, it is common to hear a report that the effect of the drug was either significant or not. The results of such tests can be of great importance to the success of a company. For example, a May 2000 study reported that a drug called ancrod, purified from snake venom, was given to half a group of 500 stroke patients. Of the patients receiving ancrod, 42% showed improvement, compared with only 34% of those who didn't receive the drug. This was a statistically significant effect. But in another study, a company's proposed drug for treating traumatic brain injury was shelved when it did not perform significantly better than a placebo in tests on 340 patients. The company's stock dropped 45% as a result.

Is 53% significantly different from 50%? Is 60%? 70%? In order to explore what it means for a measurement to be statistically significant, you can look at something that you know has a true value of 50%, which is the expected percentage of heads for many tosses of a coin.

a) Toss a coin 10 times. How many tosses came up heads?

5. a) Answers will vary. From 3 to 7 heads is typical.

b) Repeat (a) until you have 100 sets of 10 tosses. Use a tally table similar to **Table 7.30** to record the frequency with which each possible number of heads from 0 to 10 occur.

Number of Heads	Frequency
0	
1	
2	
3	
4	
5	
6	
7	
8	
9	
10	

Table 7.30

b) Sample table:

Number of Heads	Frequency
0	0
1	1
2	4
3	13
4	23
5	17
6	22
7	11
8	6
9	3
10	0

c) How often did 8 or more heads occur? How often did 2 or fewer heads occur?

c) Based on the data in (b), 8 or more heads occurred nine times; 2 or fewer occurred five times.

d) Combine your two answers in (c), and determine the probability that at least an 80-20 percentage split occurred between heads and tails. Because you know that the chance of a coin landing heads or tails is 50-50, the probability value you have determined gives you an estimate of your chance of being wrong if you were to say that 80% is significantly different from 50% based on a sample of size 10. This probability is called the *p*-value for the test and is the basis for concluding whether the observed effect is significant.

d) **An imbalance of at least 80%–20% occurred 14 times out of 100 trials. The probability of such a split occurring when a coin is tossed 10 times is therefore about 14%, based on this experiment.**

e) Repeat (a)–(d) for 100 sets of 20 coin tosses.

e) **Sample data:**

Number of Heads	Frequency
0	0
1	0
2	0
3	0
4	1
5	1
6	4
7	7
8	15
9	13
10	18
11	17
12	12
13	10
14	0
15	1
16	1
17	0
18	0
19	0
20	0

f) Compare the probability results for 100 samples of size 10 and size 20. What can you conclude?

f) 80% of 20 is 16, and there is only 1 trial (out of 100) in the sample data from (e) that resulted in 16 or more heads; 20% of 20 is 4, and there is 1 trial that resulted in 4 or fewer heads. So a total of 2 out of 100 samples of size 20 had a heads/tails split of at least 80%–20%. Based on this experiment, we see that when an event has a 50% probability of occurrence, it is still possible to observe a fraction as high as 0.8 or as low as 0.2 about 2% of the time if samples of size 20 are measured. If such deviations from 50% are judged to be significant, there is about a 2% chance of that judgment being wrong. On the other hand, a similar judgment based on samples of size 10 would have about a 14% chance of being wrong.

III. Additional Practice *(SE page 278–279)*

6. The diameters of 30 juice bottle caps from a production run are measured and found to average 4.82 cm. If the margin of error for 90% confidence for this sample is 0.04 cm, find a 90% confidence interval for the true mean diameter of all bottle caps from the production run.

6. **Limits are (4.82 – 0.04) = 4.78 cm and (4.82 + 0.04) = 4.86 cm. The mean diameter of all bottle caps from the production run is between 4.78 cm and 4.86 cm.**

7. The focal lengths of a sample of camera lenses are measured and found to be 50.8, 49.7, 49.3, 50.2, 50.6, 49.6, 51.0, and 50.3 mm. The margin of error for 90% confidence for this sample is 0.41 mm. Find a 90% confidence interval for the true mean focal length of all such lenses.

7. **Sample mean is 50.1875 ≈ 50.19 mm. Limits of confidence interval are (50.19 – 0.41) = 49.78 mm and (50.19 + 0.41) = 50.60 mm. The mean focal length of the lenses is between 49.8 mm and 50.6 mm.**

8. A preliminary sample of 12 patients with high cholesterol showed that a new drug lowered cholesterol levels by the following amounts (in milligrams per deciliter): 20, 35, 32, 29, 45, 23, 32, 22, 30, 35, 36, and 32. The margin of error for 95% confidence is 4.3 mg/dl. Find a 95% confidence interval for the mean amount that the drug lowers cholesterol in all patients who take it.

8. **Sample mean is 30.917 ≈ 30.9 mg/dl. Limits of confidence interval are (30.9 – 4.3) = 26.6 mg/dl and (30.9 + 4.3) = 35.2 mg/dl. The drug lowers cholesterol by an average amount between 26.6 mg/dl and 35.2 mg/dl.**

9. A 1996 poll of 812 Minnesota voters asked the question "Should drivers under the age of 18 be allowed to have only one other teenager in the car with them?" Of those surveyed, 22% answered Yes to this question. The margin of error in the poll was 3.5%, based on a 95% confidence level. State a 95% confidence interval for the true percentage of voters who agreed with the question.

9. Limits are (22% – 3.5%) = 18.5% and (22% + 3.5%) = 25.5%. Between 18.5% and 25.5% of all Minnesota voters agreed that drivers under the age of 18 should be allowed to have only one other teenager in the car with them.

10. A Gallup poll conducted prior to the 1996 presidential election reported that voters preferred President Clinton 57% to Bob Dole 40%. The results were based on 1010 adults and had a margin of error of ±3%. If the confidence level for the poll was 95%, state confidence intervals for the true percentage of voters that preferred either of the two candidates.

10. Limits for Clinton are (57% – 3%) = 54% and (57% + 3%) = 60%. Limits for Dole are (40% – 3%) = 37% and (40% + 3%) = 43%. Between 54% and 60% of all voters preferred Clinton, whereas between 37% and 43% preferred Dole.

11. The Bureau of Labor Statistics announces monthly unemployment figures based on a 90% confidence level, after a survey of 60,000 American households. The resulting margin of error is 0.2 percentage points. In a month for which the national unemployment rate is given as 6.4%, find a 90% confidence interval for the true national unemployment rate.

11. Limits are (6.4% – 0.2%) = 6.2% and (6.4% + 0.2%) = 6.6%. The true national unemployment rate for the month is between 6.2% and 6.6%.

12. For the same sample size, which should be wider—a 95% confidence interval or a 99% confidence interval? Why?

12. The 99% confidence interval would be wider. In order to increase the probability that a confidence interval contains the true mean, it is necessary to widen the interval.

13. The most commonly used confidence intervals are based on 90%, 95%, or 99% confidence levels.

    a) Suppose you are a production manager for a company producing bolts to be used in the assembly of space shuttles, and it is extremely important to know the true average strength of the bolts. What type of confidence interval would you be likely to use? Why?

13. a) A 99% confidence interval gives a greater chance of knowing that the mean strength lies within it.

    b) What could you do to ensure that the desired confidence interval has a small margin of error?

b) Use a larger sample size.

14. If the Gallup poll in Exercise 10 had sampled 3000 adults, do you think the margin of error would be higher or lower? Why?

14. The margin of error should be lower because the larger sample size would give a better estimate of the true voter preferences.

15. If the Gallup poll in Exercise 10 had sampled only 500 adults, how would the margin of error be affected?

15. The margin of error would be larger.

16. Why do you think the Bureau of Labor Statistics uses such a large sample size of 60,000 in determining its estimate of the U.S. unemployment rate (see Exercise 11)?

16. Many government and business financial decisions, like interest rates, can be affected by the unemployment rate. A large sample size helps ensure that the measured rate has a small margin of error.

Section 7.7 Modeling through Simulations

What You Need To Know

- How to construct a probability distribution
- How to find the expected value of a probability distribution

What You Will Learn

- To construct a table of random number assignments for a simulation
- To conduct a simulation of an experiment using random numbers
- To generate random numbers using calculators and/or computers

Materials

- Handout 7.5
- Cardboard
- Paper clips
- Scissors
- Glue
- Protractor

Simulation provides us with a method for imitating a real situation in a different way. We often think of a simulation as acting out the details of a situation we are modeling. For example, a re-creation of a historical event with modern actors is a simulation of the event. A chair may be covered with a plastic that simulates leather. Car crashes are simulated in order to provide information on the performance of passenger restraints. And computers are used to simulate many things, from changes in world climate to exploding a nuclear warhead.

Most computer-based simulations are constructed using probability. We have already conducted probabilistic simulations in this chapter. In Discovery 7.1, a spinner was used to simulate positions of a control stick for an amusement park ride. Exercise 5 of Section 7.2 used selection of labeled cards to simulate a genetics experiment. In this section, we will explore how we can create models using different types of simulators that are based on probability considerations.

Discovery 7.6 Simulating a Free Throw Contest

(SE pages 281–282)

Materials: Handout 7.5, cardboard, paper clips, glue

The free throw percentage for a professional basketball player is usually relatively constant over the course of a season, or even a career. Reggie Miller of the National Basketball Association's Indiana Pacers has had a consistent average of 90% success from the free throw line over a long career. Of course, there are games in which he makes fewer than 90% and others in which he makes all of his free throws. But it is reasonable to say that there is a probability of 0.90 that he will make any particular free throw.

Lisa Leslie of the Los Angeles Sparks in the Women's National Basketball Association has a career free throw percentage of 70% since the league began operation in 1997. Could she ever beat Reggie Miller in a free throw shooting contest? Although we may never see that particular match-up, we can use probability to create a simulation that models such a contest.

1. Handout 7.5 contains one kind of simulator that can be used to gather data on a possible free throw contest between Reggie and Lisa. Cut out the circular figures, mount them on thin cardboard, and use pencils and paper clips to create spinners from them. Examine the one labeled "Reggie Miller." Why might this spinner work to simulate a Reggie Miller free throw attempt?

1. **The region marked "MAKE FREE THROW" on the Reggie Miller spinner occupies 90% of the area of the spinner, and we would expect the spinner to land in this region 90% of the time.**

2. Have one person spin the Reggie Miller simulator and another spin the Lisa Leslie simulator. Simulate 10 free throws by each player, alternating shots, and record the results. On the average, Reggie will make 9 of 10 free throws and Lisa will make 7. Was the score 9 to 7? If not, how can you explain it?

2. **Sample answers: In the table, Yes corresponds to a made free throw, No to a miss.**

FT No.	1	2	3	4	5	6	7	8	9	10
Reggie	No	No	Yes	Yes	Yes	Yes	Yes	Yes	No	Yes
Lisa	No	No	Yes	Yes	Yes	No	Yes	Yes	Yes	No

Reggie wins, 7 to 6. Reggie's 90% and Lisa's 70% are only averages. In any particular trial of 10 free throws, the outcome is uncertain. Only over many trials will the observed averages approach 90% and 70%.

3. Simulate a dozen or so identical contests, and summarize the results. Are they what you expected?

3. Sample results for 12 more simulations (winner and score given):

   Reggie 8–5; Lisa 9–8; tie 8–8; Reggie 9–6; Reggie 10–6; Reggie 9–7; Reggie 10–6; tie 8–8; tie 9–9; Reggie 9–6; Reggie 8–4; Reggie 8–6. Overall, Reggie wins 8, Lisa wins 1, with 3 ties. Reggie made between 8 and 10 free throws throughout the trials, whereas Lisa's totals varied more, from 4 to 9. Reggie won the most contests, as expected.

4. How could you use the spinner in **Figure 7.19** to simulate a Reggie Miller free throw? Is there more than one way to set up the simulation?

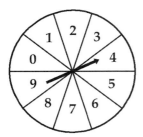

Figure 7.19

4. The numbers 0 through 8 could be assigned to a made free throw, with 9 being a miss. Or any other group of nine digits could be assigned to a made free throw.

5. The key to simulating anything (like Reggie's free throws) that occurs with 90% probability is to have a simulator that can be counted on to do one thing 90% of the time and something else 10% of the time. What is another possible way, besides using a spinner, to simulate Reggie's free throws?

5. Sample answer: Chips or marbles that are 90% one color and 10% another could be mixed in a bag.

6. A computer simulation of Reggie's free throws can be constructed in a manner similar to the use of the spinner in Figure 7.19. We could say that the computer can "simulate the spinner" and can thus replace it as the contest simulator. This can be done by having the computer generate random integers ranging from 0 to 9. If you are using a TI-83 calculator, enter the command **randInt(0,9)** from the Math PRB menu. (For any other calculator, consult the user's manual to find a similar command.) Repeat this entry numerous times. What do you see?

6. A series of integers, varying randomly between 0 and 9, should appear.

7. Choose a range of numbers to represent Reggie's successful free throw attempts. Remember to include 90% of the integers from 0 through 9. What will represent an unsuccessful attempt?

7. For example, let 0–8 represent a made free throw. Then 9 represents a miss.

8. Now set up appropriate simulation ranges for Lisa's successful and unsuccessful free throw attempts.

8. **Let 0–6 represent a made free throw, with 7–9 representing a miss.**

9. In order to simulate the free throw contest using random number generators, it will be more realistic to have two people (one representing Lisa and the other Reggie) use separate calculators. Have each person "shoot" a free throw using the **randInt(0,9)** command. Compare the resulting output integers to the simulator ranges you constructed in items 7 and 8. What happened? Did Reggie make his free throw? Did Lisa?

9. **Sample results: For Reggie, RN is 0, free throw was made. For Lisa, RN is 3, free throw was made.**

10. Simulate a complete 10-shot contest. Who won?

10. **Sample answer:**

Shot No.	Reggie RN	Reggie Makes FT?	Lisa RN	Lisa Makes FT?
1	6	Yes	5	Yes
2	8	Yes	5	Yes
3	6	Yes	7	No
4	0	Yes	3	Yes
5	3	Yes	6	Yes
6	4	Yes	5	Yes
7	2	Yes	0	Yes
8	1	Yes	9	No
9	8	Yes	5	Yes
10	8	Yes	3	Yes

Reggie wins 10–8.

11. Simulate many such contests. How often did Lisa win? How often did Reggie win? How many ties were there?

11. **A sample of 20 contests yielded the following results:**

    Reggie won 14 times, by scores of 10–8, 9–6, 10–8, 10–7, 9–8, 9–6, 9–8, 9–8, 10–5, 8–6, 10–8, 9–7, 9–5, and 10–8.

    Lisa won twice, by scores of 9–8 and 8–7.

    There were 4 ties, all 8–8.

12. Based on your simulations, write an answer to the question "How often could Lisa Leslie beat Reggie Miller in a 10-shot free throw contest?"

12. **In a total of 34 simulated contests, Lisa beat Reggie 4 times, so she should be able to beat him about 10% of the time (with about 20% ties).**

In business and industry, simulation is often used in situations for which it would be difficult or expensive to perform the actual activity being studied. Computer modeling of possible engineering designs is used extensively to avoid the expense of building many prototypes. As with all models, however, it is essential that such simulations be tested against reality to ensure that results obtained from them are meaningful.

Designing a Simulation

A simulation is one type of model. Therefore, simulations are generally constructed using the modeling guidelines we have discussed previously.

Simulations based on probability have particular requirements related to their chance nature. They are summarized here and then illustrated with an example.

1. Identify the situation to be simulated.

Models are usually constructed because someone needs to solve a problem or to gather information about a phenomenon of interest. Whether it is predicting sales for a business or analyzing the response of an airplane to wind turbulence, there should be a clearly stated objective for which the simulation will be designed.

2. Create a simplified and idealized version of the situation.

Most real-world problems have a variety of factors influencing their behavior. In order to get usable results from a simulation (or any model, for that matter), it is often necessary to ignore some of those factors, at least for the first trial. For example, the question "Are men better basketball players than women?" could provoke endless discussion without resolving anything. The free throw simulation in Discovery 7.6 addresses only one very specific aspect of the larger question. It also assumes that the probability of making a free throw is constant for a given individual, which may or not be true. This is because it ignores such things as the stress associated with close game situations and the health of the individual. But making simplifying assumptions is part of the modeling process. It makes getting usable results easier at the expense of possibly ignoring significant aspects of the problem.

3. Determine the relevant probabilities.

Simulations require information about probabilities for various possible outcomes. For most real-world problems, those probabilities are based on historical data. They are therefore usually experimental probabilities rather than theoretical probabilities. In Discovery 7.6, the free throw probabilities were based on career data for the two athletes.

4. Decide on an appropriate simulator.

We have seen that there is more than one way to conduct a simulation. But modern simulation methods rely to a great extent on computer-generated random numbers. There are a variety of special-purpose simulation programs that are commercially available. Some are particularly well-suited for business forecasting, others for construction project scheduling, and still others for electronic design. Regardless of purpose, such programs usually rely on probabilities associated with randomly generated numbers.

5. Identify appropriate random number assignments.

Once a simulator has been chosen, possible events must be matched with random number ranges that reflect their probabilities of occurrence. For instance, to model an 80% free throw shooter, we might choose the whole numbers 0 through 7 to represent a made free throw, with 8 and 9 representing a missed free throw. Other assignments are possible, such as 1 through 8 for a success and 9 and 0 for a miss. However, because of the way that computers generate random numbers, it is customary to begin the first range with 0.

6. Conduct the simulation.

Once probabilities and random number intervals have been determined, all that remains is to run the simulation for the desired number of repetitions. Results can be tabulated and treated much the same as actual experimental data.

7. Interpret and test results.

Any model is only as good as the results it produces. If the output doesn't make sense or doesn't accurately reflect known experimental results, the model must be refined. Crucial aspects of the problem may have been ignored in idealizing the model.

Example 15

A city hospital emergency room is experiencing unexpected delays, and the administrators would like to analyze the problem. The ER size and staffing level were designed to handle 12 patients per hour. Data on the number of arrivals in the ER during any 10-minute period has been accumulated, with the information summarized in **Table 7.31.**

Arrivals in 10 Minutes	Probability
0	0.1
1	0.3
2	0.4
3	0.2

Table 7.31

The expected value of the number of arrivals in any 10-minute period is $0(0.1) + 1(0.3) + 2(0.4) + 3(0.2) = 1.7$. This translates into $6(1.7) = 10.2$ expected arrivals per hour. This arrival rate is within the capacity of the ER. Simulate emergency room arrivals during a 24-hour period to see if the number of arrivals per hour ever exceeds 12.

Solution:

An hour of arrivals can be treated as six consecutive 10-minute periods. Because all of the probability values are single-digit numbers, random integers from 0 to 9 can be used for the simulation. A TI-83 calculator using the **randInt** command will be used to generate the random integers. The assignments in **Table 7.32** will produce the correct probabilities.

Arrivals in 10 Minutes	Probability	Random Numbers
0	0.1	0
1	0.3	1–3
2	0.4	4–7
3	0.2	8–9

Table 7.32

Figure 7.20 shows a simulation of the first hour. The random numbers 9, 7, 0, 5, 7, and 1 correspond to arrivals of 3, 2, 0, 2, 2, and 1 patients in the six 10-minute intervals of the hour. The total number of arrivals is therefore 10. **Table 7.33** shows results of simulating 24 hours. In this case, notice that there are 5 hours during which arrivals exceeded 12. Therefore, even though the emergency room was equipped to handle the average expected arrivals, the chances of exceeding capacity appear to be significant.

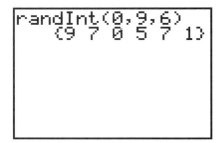

Figure 7.20

Hour	Random Numbers	Arrivals
1	970571	$3 + 2 + 0 + 2 + 2 + 1 = 10$
2	616803	$2 + 1 + 2 + 3 + 0 + 1 = 9$
3	906747	$3 + 0 + 2 + 2 + 2 + 2 = 11$
4	892396	$3 + 3 + 1 + 1 + 3 + 2 = 13$
5	568321	$2 + 2 + 3 + 1 + 1 + 1 = 10$
6	773385	$2 + 2 + 1 + 1 + 3 + 2 = 11$
7	576635	$2 + 2 + 2 + 2 + 1 + 2 = 11$
8	922908	$3 + 1 + 1 + 3 + 0 + 3 = 11$
9	886472	$3 + 3 + 2 + 2 + 2 + 1 = 13$
10	794143	$2 + 3 + 2 + 1 + 2 + 1 = 11$
11	337901	$1 + 1 + 2 + 3 + 0 + 1 = 8$
12	905943	$3 + 0 + 2 + 3 + 2 + 1 = 11$
13	775107	$2 + 2 + 2 + 1 + 0 + 2 = 9$
14	405187	$2 + 0 + 2 + 1 + 3 + 2 = 10$
15	238978	$1 + 1 + 3 + 3 + 2 + 3 = 13$
16	958648	$3 + 2 + 3 + 2 + 2 + 3 = 15$
17	194036	$1 + 3 + 2 + 0 + 1 + 2 = 9$
18	127492	$1 + 1 + 2 + 2 + 3 + 1 = 10$
19	400580	$2 + 0 + 0 + 2 + 3 + 0 = 7$
20	718131	$2 + 1 + 3 + 1 + 1 + 1 = 9$
21	446693	$2 + 2 + 2 + 2 + 3 + 1 = 12$
22	454730	$2 + 2 + 2 + 2 + 1 + 0 = 9$
23	771402	$2 + 2 + 1 + 2 + 0 + 1 = 8$
24	605888	$2 + 0 + 2 + 3 + 3 + 3 = 13$

Table 7.33

The simulation in Example 15 could be programmed to run hundreds, or even thousands, of times in order to produce more data. The fraction of the time that the ER's capacity is exceeded could then be estimated. Appropriate action, which might include altered staffing patterns or a redesign of the waiting room, could then be taken.

It is important to understand that the result of Example 15 is not unique. If two people conduct the identical simulation, they may get entirely different results for any trial of the experiment. Each execution of the simulation provides a "for instance" example of what might occur under specified conditions. For this reason, research of this kind is sometimes referred to as "What If?... Analysis."

Exercises 7.7

I. Investigations *(SE pages 288–291)*

1. The spinner shown in Figure 7.19 is a good simulator for producing random numbers from 0 through 9. The TI-83 **randInt(0,9)** command works equally well for the same purpose.

   a) How could the spinner in Figure 7.19 be used to simulate a coin toss? How could the **randInt(** command be used to simulate a coin toss?

   1. a) Sample answer: Assign five of the digits (say 0–4) to heads and the other five (5–9) to tails. One spin of the spinner, or one execution of the randInt(command, would simulate one toss of the coin.

   b) How could you design a spinner that would correctly simulate (on a single spin) the number of heads that result when two coins are tossed (see Example 1 in Section 7.2)?

   b) Sample answer: Two heads or two tails will each occur 1/4 of the time, while one head and one tail will occur 1/2 of the time. Let a half circle on the spinner represent the event "one heads out of two tosses," and assign a quarter circle to each of the events "zero heads out of two tosses" and "two heads out of two tosses."

   c) What physical device (other than a spinner) could you use as a simulator to generate random integers between 0 and 5?

   c) Sample answer: One die could be used, with the side having six dots representing the number 0.

   d) What physical device (other than a spinner) could you use as a simulator to simulate a 75% free throw shooter? How could the **randInt(** command be used?

   d) Sample answer: A tetrahedral (four-sided) die could be used, with three of the sides assigned to a made free throw. The command randInt(0,3) could be used, with the numbers 0–2 representing a made free throw.

   e) How could the **randInt(** command be used to simulate an 83% free throw shooter?

   e) Sample answer: randInt(0,99) could be used, with the numbers 0–82 (or 1–83) representing a made free throw.

2. There are many games that are based on simulations of sports. Suppose that a game company wants to design a non-computerized simulation of a football game. In order to randomly select 1 player from among the 11 members of a football lineup, it would be necessary to use a simulator with 11 possible outcomes. A member of the design team suggests, "Let's use a pair of dice as the simulator. When dice are rolled, there are 11 possible sums, and each sum could be assigned to a different player." Is there anything wrong with this suggestion? If so, what?

2. **The 11 dice outcomes are not equally likely. As shown in Exercise 5 of Section 7.4, the probability of rolling a 7 is six times as great as the probability of rolling a 2 or a 12, etc. So the player assigned to number 7 would be more likely to be selected than any other player.**

3. The **randInt(** command is useful for generating randomized integers. Including any two numbers as inputs of the command will result in generation of a random integer chosen from a uniform distribution of integers within a specified range. For example, **randInt(3,20)** will return an integer from 3 to 20 (and include the 3 and the 20 as possibilities). An optional third input can be used to specify how many times the command is to be repeated. The command **randInt(3,20,5)** results in a list of five random numbers from 3 to 20 (see **Figure 7.21**).

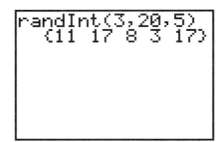

Figure 7.21

   a) Use the **randInt(** command to generate a single random integer between 10 and 100, including the 10 and 100 as possibilities. (A more precise way of stating this interval is "between 10 and 100 **inclusive**," meaning that the boundaries of the interval are included.)

3. **a) randInt(10,100)**

   b) Use the **randInt(** command to generate a single random number between 10 and 100, without including the 10 or the 100 as possibilities.

b) randInt(11,99)

   c) Use the **randInt(** command to generate a list of six random numbers between 0 and 100 inclusive.

c) randInt(0,100,6) produces a sample result of {90 27 42 91 69 46}.

   d) Use the **randInt(** command to generate a list of 10 random numbers between 0 and 99 inclusive.

d) randInt(0,99,10) produces a sample result of {14 94 83 52 9 79 79 38 84 36}.

The **randInt(** command can be used in programs, in conjunction with various program control commands, to create a variety of user-friendly simulations.

4. A computer spreadsheet program can be used to create simulations that can be easily run hundreds or even thousands of times. (The instructions in this exercise are for Excel. For any spreadsheet program other than Excel, consult the user's manual.)

a) Excel contains a **RAND()** command that is similar to the TI-83 **rand** command. Its output is a decimal number greater than or equal to 0 and less than 1. This command can be modified so that it produces random numbers for specific intervals in the following way: Entering **=RAND()*(b–a)+a** results in a random decimal number greater than or equal to **a** and less than **b**. For example, **=RAND()*5+3** returns a random decimal number greater than or equal to 3 and less than 8. Use the **RAND()** command in this way to generate a random number greater than or equal to 20 and less than 30.

4. a) RAND()*10+20

b) In order to produce random integers, a separate Excel command called **INT** can be used. The **INT** command rounds any number down to the nearest integer. Thus **=INT(3.7894)** will return the number 3, and **=INT(0.28)** will return a 0. Use the **INT** command in combination with the **RAND()** command to create a combined command that will function exactly like the TI-83's **randInt(0,9)** command.

b) =INT(RAND()*10+0)

c) To create multiple random numbers, make multiple copies of the cell that contains the command you wrote in (b). For example, if you placed that command in cell A1, highlight that cell and click on the Copy icon. Then click on cell A2 and drag down to highlight a range of cells, say from A2 to A10. Now if you click on the Paste icon, the command in A1 will be copied to the highlighted cells. Try this yourself.

c) Cells A1 through A10 should contain 10 random integers in the interval 0–9 inclusive.

Notice that if you then make any entry into another cell, all the random numbers in column A change. Spreadsheets are set up to automatically recalculate formulas whenever something changes in the spreadsheet. Because **RAND()** does not calculate the same number every time, cells containing that command will almost always change whenever anything in the spreadsheet changes.

d) The entire simulation process can be automated using a spreadsheet feature called a "lookup table." This essentially involves telling the computer to first compare a number (in our case, a random number) to a list of other numbers. Then, depending on where the number falls in the list, it will "lookup" a replacement value in an adjoining list and replace the original number with it. For example, if a lookup table like that in **Figure 7.22** is created, any number greater than or equal to 0 but less than 1 will be replaced with X, a number greater than or equal to 1 but less than 2 will be replaced with Y, and so on.

Enter a table exactly like Figure 7.22 into cells A1 through B3 of your spreadsheet. Then enter the command **VLOOKUP(1.4,$A$1:$B$3,2)** into any other cell and see what happens.

0	X
1	Y
2	Z

Figure 7.22

d) The cell with the VLOOKUP command returns a Y.

The V in the **VLOOKUP** command stands for "Vertical" because the table is organized in vertical columns. The 1.4 is the value that is compared to the first row of the lookup table, in cells A1:B3. (Note: The $ symbols are needed to make sure the program always looks in the same place for the table, even when the command is written in other cells.) Because 1.4 is greater than or equal to 1 but less than 2, the program looks in the second column next to 1 and finds Y, which it then displays instead of the 1.4. (The last input to the **VLOOKUP** command tells the program the number of the column, in this case the second one, where the output value is to be found.)

e) If you enter **RAND()*3+0** in place of the 1.4 in your **VLOOKUP** command, the program will generate a random number between 0 and 3 (including 0), compare it to the first column of the lookup table, and return the corresponding letter from the second column. Try it and see what happens. Notice that you don't see the random number, only the final output letter.

e) VLOOKUP(RAND()*3+0,$A$1:$B$3,2) randomly returns one of the letters X, Y, or Z.

f) Copy this command into another range of cells and see what happens.

f) The cells should show a random mix of the letters X, Y, and Z.

g) Finally, use what you have learned to create a computer simulation (using Excel) of the Reggie Miller/Lisa Leslie free throw contest from Discovery 7.6. To simulate a Reggie Miller free throw, you will need a lookup table in which the first column contains only the numbers 0 and 9 (where the 0 corresponds to the random number interval 0–8). The second column of the lookup table can contain Yes and No for made and missed free throws. The Lisa Leslie lookup table is constructed in a similar manner but using her random number intervals.

By extending your simulation, you can "shoot" as many free throws as desired and get an improved answer to the question "How often could Lisa Leslie beat Reggie Miller in a 10-shot free throw contest?" Another option is to use the F9 key to recalculate the spreadsheet for successive contests.

g) A sample Excel output is shown.

	A	B	C	D	E	F	G
1							
2					Trial No.	Reggie	Lisa
3		Reggie Miller			1	Yes	Yes
4		0	Yes		2	Yes	Yes
5		9	No		3	Yes	Yes
6					4	Yes	Yes
7					5	Yes	Yes
8		Lisa Leslie			6	Yes	No
9		0	Yes		7	No	Yes
10		7	No		8	Yes	No
11					9	Yes	Yes
12					10	Yes	No

Cell F3 contains the command
VLOOKUP(INT(RAND()*10+0),$B$4:$C$5,2), which was then copied
into cells F4 through F12. Cell G3 contains the command
VLOOKUP(INT(RAND()*10+0),$B$9:$C$10,2), which was then copied
into cells G4 through G12. This particular result shows Reggie beating
Lisa by a score of 9 to 7. Pressing the F9 key will generate a new set of
random numbers in the cells F3 through G12, resulting in a new
simulation. This can be repeated indefinitely, generating a series of
contests.

5. Write a calculator program that will conduct a complete 24-hour simulation
of the problem in Example 15. The output should be the number of hours
during a 24-hour period that the room's capacity is exceeded.

5. One possibility, for a TI-83 calculator, is shown here.

ClrHome

Disp "SIMULATION OF"

Disp "EMERGENCY ROOM"

Disp "ARRIVALS IN"

Disp "24 HR PERIOD"

$0 \rightarrow X$

For(H,1,24)

$0 \rightarrow A$

```
For(N,1,6)
randInt(0,9) → R
If R=0
A + 0 → A
If (R ≥ 1 and R ≤ 3)
A + 1 → A
If (R ≥ 4 and R ≤ 7)
A + 2 → A
If R ≥ 8
A + 3 → A
End
If A > 12
X + 1 → X
End
ClrHome
Disp "HOURS CAPACITY"
Disp "EXCEEDED=",X
```

II. Projects and Group Activities *(SE pages 291–292)*

6. Materials: a simulator of choice

   A current family-planning policy in China provides incentives for married couples to stop having children once they have had a son. How would such a policy affect average family size and the birth rates for boys and girls?

   Design a simulation to test the effects of this policy. Assume that all married couples strictly adhere to the policy and have children until a boy is born, then stop. Further assume that the probabilities of having a girl or boy baby are equal. Whatever you choose for a simulator should reflect this equal probability. Then use your simulation to find how many girls will be born in a family before the first boy is born. Repeat the simulation many times in order to determine the overall fraction of children that will be girls under the policy. Collect enough data so that you are reasonably confident of your results. As we have seen, a graph of "fraction girls" vs. "number of families" may be helpful.

6. **Flipping a penny is an adequate simulator, although some students may want to use a calculator or spreadsheet. If heads represents girls and tails represents boys, then a possible sequence of 20 "families" is given in the accompanying table.**

Family No.	1	2	3	4	5	6
Simulation	HHT	T	HT	HHT	HHHHHT	HHHT
No. of girls	2	0	1	2	5	3

7	8	9	10	11	12	13
HT	T	HT	T	T	T	T
1	0	1	0	0	0	0

14	15	16	17	18	19	20
T	HHHT	T	T	T	T	HT
0	3	0	0	0	0	1

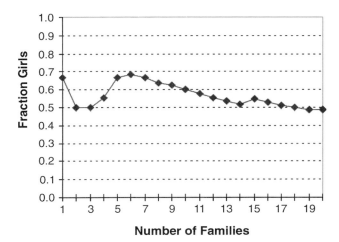

In these 20 families, there are 19 girls and 20 boys. It can be shown that the "have children until a boy is born" policy will result in equal numbers of boys and girls being born.

7. Materials: cardboard, scissors, glue, paper clip, protractor

Design a simulator to simulate the batting ability of a baseball player. Look up complete statistics on a player's career batting record, including number of at bats, singles, doubles, triples, home runs, strikeouts, walks, outs, and so on. Then determine the fraction of at bats in which each type of event occurred. Divide the 360 degrees of the circle proportionally to correctly represent the fractions you calculated. You can then spin the spinner to simulate what the player might do on any at bat. By repeating the process, you could create an entire simulated team that could be matched against someone else's team in a game. Discuss the limitations of this simulation.

7. **Sample statistics, converted to angular degrees for spinner construction, are shown for the first 11 years of the career of Ken Griffey, Jr. Data are taken from the Web site www.baseball-reference.com. Total number of "at bats" of 6579 was found by adding number of walks to official at bats. (Number of outs is not listed but is assumed to be the difference between the number of total at bats and the sum of all other categories.)**

Statistic	Frequency	Percentage	Angle (degrees)
Singles	994	15.1%	54
Doubles	320	4.9%	18
Triples	30	0.5%	2
Home runs	398	6.0%	22
Walks	747	11.4%	41
Strikeouts	984	15.0%	54
Other outs	3106	47.2%	170

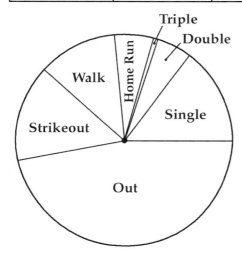

Because the simulation assumes constant fractions for singles, home runs, etc., it makes no allowance for different pitchers, different-sized ballparks, etc.

8. When managers of large projects use modern quantitative techniques to predict how long a project will last, they sometimes try to estimate the length of a particular project activity using probabilities. Consider a (simplified) project to test market a new razor, consisting of only three tasks that must be performed in order: (1) design prototype, (2) produce prototype, and (3) do market research. Suppose that you, as project manager, have made probability estimates of the times required for the three tasks as shown in **Table 7.34,** based on past experience.

Activity	Time Estimate	Probability
Design prototype	3 months	0.3
	5 months	0.5
	9 months	0.2
Produce prototype	2 months	0.1
	3 months	0.5
	4 months	0.4
Market research	2 months	0.3
	4 months	0.4
	7 months	0.3

Table 7.34

Use a computer spreadsheet program to design a simulation of the three-activity project (see Exercise 4), and run the simulation a number of times to determine how long the project might take.

8. One possibility, using Excel:

	A	B	C	D	E	F	G	H
1								
2		Design prototype			Project simulation			
3		RN limit	Months					
4		0	3		Design	Produce	Mar. res.	Months
5		3	5		3	3	4	10
6		8	9		5	3	7	15
7					3	3	2	8
8		Produce prototype			5	3	4	12
9		RN limit	Months		3	3	4	10
10		0	2		3	3	2	8
11		1	3		9	3	7	19
12		6	4		5	3	4	12
13					5	4	2	11
14		Market research			3	3	4	10
15		RN limit	Months					
16		0	2					
17		3	4					
18		7	7					

Columns B and C contain lookup tables for the three projects.
Cells E5 through E14 contain the command
=VLOOKUP(INT(RAND()*10+0),$B$4:$C$6,2).
Cells F5 through F14 contain the command
=VLOOKUP(INT(RAND()*10+0),$B$10:$C$12,2).
Cells G5 through G14 contain the command
=VLOOKUP(INT(RAND()*10+0),$B$16:$C$18,2).
Cell H5 contains the command =SUM(F6:H6), which was copied down
through cell H14 in order to sum the total time for the three activities in
each of the 10 simulated completions of the project. The average project
completion time for the 10 simulations is 11.5 days.

III. Additional Practice *(SE pages 292–293)*

For 9–16, use a spinner like that in Figure 7.19 or a calculator command like **randInt(** as a simulator.

9. A copying machine has a history of breakdowns requiring service as shown in **Table 7.35.**

Time Before Breakdown	Probability	Random Numbers
1 day	0.1	
2 days	0.2	
3 days	0.5	
4 days	0.2	

Table 7.35

Complete Table 7.35 to show the random number assignments that reflect the probabilities in the table.

9. One possibility:

Time Before Breakdown	Probability	Random Numbers
1 day	**0.1**	**0**
2 days	**0.2**	**1–2**
3 days	**0.5**	**3–7**
4 days	**0.2**	**8–9**

Other assignments are possible, as long as they produce the correct relative probabilities for the chosen simulator.

10. Suppose that the Seattle Mariners and the Chicago Cubs are playing in baseball's World Series. Assume that the probability of Seattle winning any single game is 60%, while the probability of Chicago winning is 40%. Further assume that these probabilities are constant throughout the series. (This is not usually a good assumption, because the pitchers can strongly affect the probabilities.) Construct a table of random number assignments that reflect the probabilities of a Seattle or Chicago win in any one game of the series.

10. One possibility:

Team	Probability of Winning	Random Numbers
Seattle Mariners	**0.6**	**0–5**
Chicago Cubs	**0.4**	**6–9**

Other assignments are possible, as long as they produce the correct relative probabilities for the chosen simulator.

11. The durations of calls to a computer company's customer service telephone line follows the distribution in **Table 7.36.**

Duration of Call (min)	Probability	Random Numbers
1	0.1	
2	0.2	
3	0.3	
4	0.2	
5	0.1	
6	0.1	

Table 7.36

Complete Table 7.36 to show the random number assignments that reflect the probabilities in the table.

11. One possibility:

Duration of Call (min)	Probability	Random Numbers
1	0.1	0
2	0.2	1–2
3	0.3	3–5
4	0.2	6–7
5	0.1	8
6	0.1	9

Other assignments are possible, as long as they produce the correct relative probabilities for the chosen simulator.

12. Simulate the next 30 days of use of the copying machine whose performance is summarized in Table 7.35. That is, run a simulation using the random number assignments from Exercise 9 until 30 days has elapsed.

12. The table shows the results of a typical simulation.

Random Number	Time Before Breakdown
8	4 days
8	4 days
6	3 days
9	4 days
6	3 days
1	2 days
9	4 days
4	3 days
1	2 days
4	3 days

13. Use the random number assignments from Exercise 10 to simulate a World Series between the Seattle Mariners and the Chicago Cubs. The series ends when either team has won a total of seven games.

13. The table shows the results of a typical simulation.

Game Number	Random Number	Winning Team
1	1	Cubs
2	4	Cubs
3	6	Mariners
4	6	Mariners
5	3	Cubs
6	3	Cubs

The Cubs win the series in six games.

14. Assume you have called the customer service line of the computer company in Exercise 11. You are greeted by a voice that says, "There are 10 calls ahead of you, and your call will be handled in order." Simulate the durations of 10 calls using the random number assignments from Exercise 11 to find out how long you might have to wait for service.

14. The table shows the results of a typical simulation.

Call Number	Random Number	Duration of Call
1	3	3 minutes
2	2	2 minutes
3	5	3 minutes
4	5	3 minutes
5	2	2 minutes
6	6	4 minutes
7	4	3 minutes
8	4	3 minutes
9	5	3 minutes
10	2	2 minutes

The total wait time is 28 minutes.

15. Simulate a 10-shot free throw contest between WNBA players Sheryl Swoopes, an 80% free throw shooter, and Jennifer Rizzotti, a 60% shooter.

15. One possibility for random number assignments:

	Probability for Sheryl	Random Numbers	Probability for Jennifer	Random Numbers
Make FT	0.8	0–7	0.6	0–5
Miss FT	0.2	8–9	0.4	6–9

A possible simulation using these assignments:

Random Number	Sheryl Makes FT?	Random Number	Jennifer Makes FT?
2	Yes	7	No
3	Yes	0	Yes
3	Yes	2	Yes
8	No	0	Yes
1	Yes	3	Yes
6	Yes	8	No
9	No	4	Yes
2	Yes	8	No
1	Yes	7	No
4	Yes	3	Yes

Sheryl wins 8–6.

16. Craig Hodges made several appearances in the NBA's annual three-point contest. He retired from the NBA with a 40% career shooting percentage for three-point field goals. Isiah Thomas averaged only 30% for three-point field goals. Simulate a 10-shot three-pointer contest between Hodges and Thomas.

16. One possibility for random number assignments:

	Probability for Craig	Random Numbers	Probability for Isiah	Random Numbers
Make 3	0.4	0–3	0.3	0–2
Miss 3	0.6	4–9	0.7	3–9

A possible simulation using these assignments:

Random Number	Craig Makes 3-Pointer?	Random Number	Isiah Makes 3-Pointer?
0	Yes	3	No
3	Yes	4	No
2	Yes	4	No
3	Yes	7	No
6	No	6	No
6	No	6	No
5	No	6	No
6	No	3	No
1	Yes	3	No
8	No	5	No

Craig wins 5–0.

Chapter 7 Summary

Definition of probability

Theoretical probabilities

Probabilistic models

Basic Multiplication Principle for counting

Experimental probabilities

Complement of an event

Multiplication Rule for Probabilities for independent events

Addition Rule for Probabilities for mutually exclusive events

Probability distributions

Expected value

Confidence intervals

Sampling error

Simulation

Chapter 7 Review *(SE pages 295–298)*

1. Write a summary of the important mathematical ideas found in Chapter 7.

1. Answers will vary. Following are some of the important ideas that should be listed:

The probability of an event can be found by calculating the ratio of
$$\frac{\text{Total number of ways the event can occur}}{\text{Total number of equally likely outcomes}}.$$

Probabilities can be expressed as ratios, decimals, and percents.

Mathematical models can be either deterministic or probabilistic.

The Basic Multiplication Principle can be used to find the number of ways an event can occur.

The experimental probability that a randomly selected member of a group has a certain characteristic is equal to the fraction of the group that has that characteristic.

The probabilities of an event and its complement add to 1.

The probability that two or more independent events occur is equal to the product of the probabilities of the individual events.

The probability that any of two or more mutually exclusive events occurs is equal to the sum of the probabilities of the individual events.

A probability distribution includes all possible values of a random variable, along with their probabilities; the sum of the probabilities in the distribution must be equal to 1.

The expected value of a probability distribution is an average that is found by adding the products formed by multiplying each value of the random variable by its corresponding probability.

Sample measurements can be used to estimate values of unknown quantities with certain confidence levels.

Real situations can sometimes be simulated using probabilities.

For 2–4, give answers as fractions, decimals, and percents.

2. What is the probability of guessing the correct answer to a multiple-choice question with four choices given?

2. 1/4 = 0.25 = 25%.

3. A bag of nails contains 70 eight-penny nails, 80 ten-penny nails, and 50 sixteen-penny nails, all mixed together. If a carpenter reaches into the bag and pulls out the first nail he touches, what is the probability that it will be a ten-penny nail?

3. $\dfrac{80}{70+80+50} = \dfrac{80}{200} = \dfrac{2}{5} = 0.4 = 40\%.$

4. The pair of tetrahedral dice in **Figure 7.22** (each numbered 1 through 4) is tossed. Use an area model to determine the probability that the sum of the two face-down numbers will be 5.

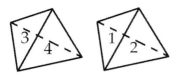

Figure 7.22

4.

First Die

		1	2	3	4
	1	2	3	4	5
Second Die	2	3	4	5	6
	3	4	5	6	7
	4	5	6	7	8

(sums are inside squares)

Probability of a 5 is 4/16 = 1/4 = 0.25 = 25%.

5. If a four-digit password for an Internet retail account can be any combination of letters and numbers (that is, any "alphanumeric" characters), how many different passwords are possible?

5. **There are 26 + 10 = 36 alphanumeric characters. Using the Basic Multiplication Principle, the number of possible passwords is $36^4 = 1,679,616$.**

6. The human population of Earth is over 6 billion. How many digits would be needed if a numerical code were to be assigned to every currently living person?

6. **With 9 digits, the number of codes is $10^9 = 1$ billion. With 10 digits, the number of codes is $10^{10} = 10$ billion. 10 digits are needed.**

For 7–8, express answers as fractions, decimals, and percents.

7. During the mid-1990s, 1149 of every 100,000 men aged 70–74 years had prostate cancer. What is the probability that a man aged 70–74 had cancer at that time?

7. $\dfrac{1149}{100,000} = 0.01149 = 1.149\%.$

8. At the start of the 1999–2000 season, 9 out of 308 players in the National Basketball Association were less than six feet tall. What was the probability that a randomly selected NBA player was at least six feet tall?

8. The probability of being less than six feet was $\dfrac{9}{308}$. Being at least six feet is the complement of being less than six feet, so the probability of being at least six feet was $1 - \dfrac{9}{308} = \dfrac{299}{308} \approx 0.971 = 97.1\%$.

9. A one-year 2000 study of children aged 2–5 involved in vehicle crashes found that children were either unprotected, protected by a seat belt, or protected by a special car seat. The probability of a child being protected by a seat belt was 0.40, and the probability of a child being protected by a car seat was 0.58. What was the probability that a child involved in a crash was unprotected?

9. Probabilities for all possible events must have a sum of 1, so the required probability is 1 – (0.40 + 0.58) = 0.02.

10. In 1997, 25% of American 12th-grade students smoked daily. What is the probability that a randomly selected American 12th grader in 1997 did not smoke daily?

10. The complement of 25% is 100% – 25% = 75%.

11. Quality control standards for a juice-bottling company allow 1 defective bottle cap out of every 10,000. What is the probability that a randomly selected bottle cap from this company will not be defective?

11. The probability of a defective cap is $\dfrac{1}{10,000}$. Not being defective is the complement of being defective, so the required probability is $1 - \dfrac{1}{10,000} = \dfrac{9999}{10,000}$, or 0.9999.

12. On average, 25% of flights arriving at a certain airport are delayed. What is the probability that two independent flights scheduled to arrive at the airport are delayed?

12. Use the Multiplication Rule for Probability: (0.25)(0.25) = 0.0625.

13. What is the probability of rolling a number greater than 4 on a single six-sided die?

13. Use the Addition Rule for Probability: Probability of a number greater than 4 = probability of a 5 + probability of a 6 = 1/6 + 1/6 = 1/3.

14. The U.S. Department of Defense had a 30% defective rate in its stocks of fasteners (screws, bolts, washers, etc.) for military hardware at one time during the 1980s because of problems with unscrupulous suppliers. If a particular tank part was attached with four bolts, what is the probability that all four bolts were defective.

14. Use the Multiplication Rule for Probability: $(0.30)^4 = 0.0081$.

15. A 50-year flood is a flood of such magnitude that its probability of occurrence during a single year is 1/50. What is the probability that there will be at least one flood as large as a 50-year flood during the 30-year design lifetime of a town's sewerage treatment plant?

15. **Probability of at least one flood = 1 – probability of no floods in 30 years = $1 - (49/50)^{30} \approx 0.45$.**

16. The probability distribution for the number of heads when four coins are tossed is shown in **Table 7.37**.

Number of Heads	Probability
0	0.0625
1	0.25
2	0.375
3	0.25
4	0.0625

Table 7.37

What is the probability of more than one coin showing heads when four coins are tossed?

16. **Probability of more than one heads = sum of probabilities of 2, 3, or 4 heads (by the Addition Rule for Probability) = 0.375 + 0.25 + 0.0625 = 0.6875.**

17. A sailboat rental business on a lake has accumulated data, shown in **Table 7.38**, on the demand for rentals of small sailboats on summer afternoons.

Number of Rentals in an Hour	Frequency
0	25
1	78
2	102
3	114
4	41

Table 7.38

Construct a probability distribution for the number of small sailboat rentals per hour.

17.

Number of Rentals in an Hour	Probability
0	0.069
1	0.217
2	0.283
3	0.317
4	0.114

18. Complete the probability distribution in **Table 7.39.**

Number	Probability
1	0.26
2	0.35
3	0.31
4	?

Table 7.39

18. Probabilities must sum to 1, so the last entry is 1 − (0.26 + 0.35 + 0.31) = 0.08.

19. Find the expected value of the probability distribution in **Table 7.40.**

Number	Probability
0	0.19
1	0.45
2	0.36

Table 7.40

19. (0)(0.19) + (1)(0.45) + (2)(0.36) = 1.17

20. A consumer survey of 10 households showed the following amounts spent on groceries per week: $132, $78, $111, $107, $163, $92, $122, $103, $141, $118. If the margin of error at the 90% confidence level is $14, find a 90% confidence interval for the true mean weekly grocery bill for all such households. (Round amounts to the nearest dollar.)

20. Mean amount is $117, so the confidence interval is from ($117 − $14) to ($117 + $14). The true mean grocery bill is between $103 and $131.

21. A *Newsweek* poll in August 2000 after the Republican convention showed George W. Bush leading Al Gore among registered voters by 49% to 38%. The poll surveyed 817 registered voters and had a margin of error, with 95% confidence, of 4%. Find a 95% confidence interval for the true percentage of voters favoring each candidate.

21. **Bush was favored by between 45% and 53% of voters. Gore was favored by between 34% and 42% of voters.**

22. The probability distribution for colors of peanut M&Ms candies is shown in **Table 7.41.**

Color	Probability	Random Numbers
Blue	0.3	
Brown	0.2	
Green	0.1	
Orange	0.1	
Red	0.1	
Yellow	0.2	

Table 7.41

Complete Table 7.41 to show the random number assignments that reflect the probabilities in the table.

22.

Color	Probability	Random Numbers
Blue	**0.3**	0–2
Brown	**0.2**	3–4
Green	**0.1**	5
Orange	**0.1**	6
Red	**0.1**	7
Yellow	**0.2**	8–9

Other assignments are possible, as long as they produce the correct relative probabilities for the chosen simulator.

23. Simulate the colors of six M&Ms drawn consecutively (and then replaced) from a bag of peanut M&Ms.

23. **One possibility is shown using the randInt(command of a TI-83. A spinner numbered from 0 to 9 could also be used. Based on the assignments in Exercise 22, the colors are 2-blue, 0-blue, 5-green, 7-red, 2-blue, and 5-green. There are 3 blues, 2 greens, and 1 red M&M.**

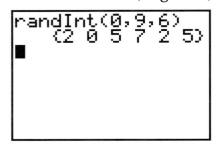

Chapter 8—Trigonometry

Goals of the Chapter

- To investigate similar triangles and use their properties to solve real-world problems

- To relate angles and their trigonometric ratios

- To apply trigonometric ratios to determine unknown distances and angles in a variety of contexts

Preparation Reading

When two sides of a right triangle are known, we can use the Pythagorean theorem to find the length of the third side. Unfortunately, there are times that we don't know the lengths of two of the sides. For example, we might need to find the height of an object, say a flagpole or a mountain, where it would be impractical to try to measure its height directly. **Figure 8.1** shows a diagram of this situation. It would be easy to measure the length of the side a of the triangle because it is on the ground. But the other two sides, b and c, are not so easy to measure directly.

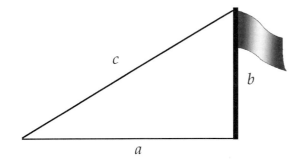

Figure 8.1

Problems such as this are posed daily by engineers, architects, artists, surveyors, astronomers, and pilots. In order to solve problems such as these, we need to know more about the angles and sides of triangles and their relationships to each other. In this chapter, we will investigate similar triangles and the relationships between their sides. Our study will then turn to the branch of mathematics known as **trigonometry,** the study of triangles. We will develop the tools necessary to solve the flagpole problem and others like it.

Reflect and Discuss *(SE page 300)*

1. If you want to estimate the height of the flagpole in Figure 8.1, how might you do so?

1. **Stand someone of known height near the flagpole and guess how many people high it is. Or students might compare it to a height with which they are familiar. For example, it's about twice as tall as our room, which is 10 feet tall. Some students might think of looking at the shadow it casts and using something of known height to see how long the known height's shadow is. If this is brought up, have the student draw a picture so others can "see" what they are saying.**

2. Name several other situations where you might need to know a distance or length and are unable to measure it directly.

2. **Distance from Earth to the Moon or Sun, height of a tall building, height of a fountain or geyser such as Old Faithful, height of a kite in the air, distance across a river or canyon, and so on.**

Section 8.1 Similar Triangles

What You Need to Know

- How to find the ratio of two numbers
- How to solve proportions
- How to use a protractor to measure an angle
- How to find lengths of line segments

What You Will Learn

- To identify similar triangles
- To use the properties of similar triangles to solve problems

Materials

- Protractor
- Ruler
- Handout 8.1

Figure 8.2

Old buildings and monuments such as the pyramids in Egypt and Mexico, the Parthenon in Greece, the Taj Mahal in India, and the Eiffel Tower in France have always held the interest of young and old alike. Another structure that has recently captured the attention of the public is the 800-year-old Leaning Tower of Pisa shown in **Figure 8.2.**

In 1990, engineers predicted that if something was not done to stop the ever-increasing southward leaning of the tower, it would fall over by the year 2050. After experimenting with various options, engineers and architects agreed on a soil extraction method as a solution.

Many questions similar to the following had to be answered before the process could begin:

- How could the angle of lean of the tower be determined at any given time in the restoration process?

- How much would the north end of the tower have to be lowered if the lean of the tower needed to be decreased by 0.5 degrees?

- If a pipe used to extract the soil is designed to enter the ground at a 25-degree angle, how long must it be in order to reach the desired location under the tower?

In the following activity, we will explore similar triangles and find out how they can be used to answers questions such as these.

Construction of the Tower of Pisa began in 1173. It was built to honor the naval forces in the city of Pisa, Italy. Shortly after construction started, the tower began to lean to the south due to the weak soil beneath it. Despite the lean, builders decided to proceed, and in 1370, construction of the 187-foot tower was completed.

In 1838, a walkway was dug around the tower. The digging further weakened the foundation of the structure, and for the first time, the tower was in danger of falling over.

In 1990, architects and engineers decided to search for a solution to the problem before the tower collapsed. Several methods such as steel cables and counterweights were tried in an attempt to save the monument. Nothing seemed to work until, in 1999, soil was slowly removed from under the tower's north side (approximately 5 gallons per day). By June 2001, the tilt of the tower had decreased one-half degree, from 5.5 degrees to 5.0 degrees. Engineers felt that this movement was enough to stabilize the tower's position; the drilling ceased and visitors could once again climb the tower.

Activity 8.1 Properties of Similar Triangles *(SE pages 302–306)*

By definition, a **polygon** is a closed figure formed by line segments that meet only at their endpoints (see **Figure 8.3**).

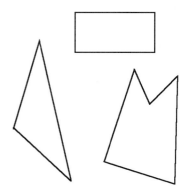

Figures that are polygons   Figures that are not polygons

Figure 8.3

Any polygon with three sides is called a **triangle.** Sometimes the vertices of the triangle are labeled, as in **Figure 8.4.** Triangle ABC has three sides, AB, AC, and BC; three angles, $\angle A$, $\angle B$, and $\angle C$; and three vertices, points A, B, and C.

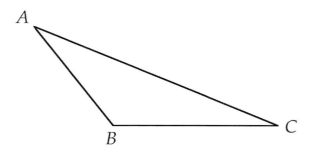

Figure 8.4

Figure 8.5 shows three triangles, $\triangle LMN$, $\triangle LNP$, and $\triangle LMP$. In figures like this, we must be careful how we name the angles. We can no longer refer to $\angle L$, because there are three angles at vertex L. To avoid confusion, we use three letters to name the angles. In this case, the angles with vertex L are $\angle MLN$, $\angle NLP$, and $\angle MLP$.

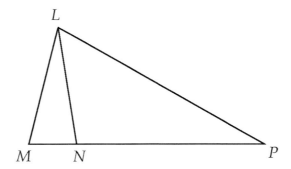

Figure 8.5

Similar triangles are triangles that have the same shape but not necessarily the same size. **Figure 8.6** shows several pairs of similar triangles. Use Handout 8.1 and work with a partner to answer the following questions.

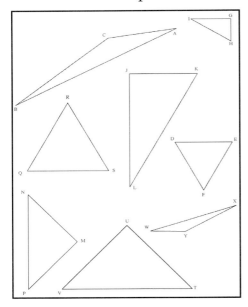

Figure 8.6

1. In the handout, there are four pairs of similar triangles. Look for the triangles that are the same shape but not necessarily the same size. Which triangles are similar? (Note: The mathematical symbol for "is similar to" is ~. To indicate that triangle *ABC* is similar to triangle *XYZ*, we write $\triangle ABC \sim \triangle XYZ$.)

1. $\triangle ABC \sim \triangle WXY; \triangle HGI \sim \triangle KJL; \triangle QRS \sim \triangle DEF; \triangle PMN \sim \triangle TUV.$

2. For each pair of similar triangles, use a protractor to carefully measure the three angles of each triangle. Label the angles on your handout. Use this information to complete **Table 8.1**. (Use the first line as a model.)

	Names	**Angle Measurements**		
Pair #1	$\triangle ABC$	$\angle A = 19°$	$\angle B = 11°$	$\angle C = 150°$
Pair #2				
Pair #3				
Pair #4				

Table 8.1

2. **Answers:**

	Names	Angle Measurements		
Pair #1	$\triangle ABC$	$\angle A = 19°$	$\angle B = 11°$	$\angle C = 150°$
	$\triangle WXY$	$\angle W = 19°$	$\angle X = 11°$	$\angle Y = 150°$
Pair #2	$\triangle GHI$	$\angle G = 90°$	$\angle H = 60°$	$\angle I = 30°$
	$\triangle JKL$	$\angle J = 90°$	$\angle K = 60°$	$\angle L = 30°$
Pair #3	$\triangle QRS$	$\angle Q = 60°$	$\angle R = 60°$	$\angle S = 60°$
	$\triangle DEF$	$\angle E = 60°$	$\angle D = 60°$	$\angle F = 60°$
Pair #4	$\triangle TUV$	$\angle T = 45°$	$\angle U = 90°$	$\angle V = 45°$
	$\triangle PMN$	$\angle P = 45°$	$\angle M = 90°$	$\angle N = 45°$

3. What do you notice about the angles of the triangles that are similar?

3. **For each pair of similar triangles, the corresponding angles are equal in measure.**

4. Choose one pair of similar triangles, measure the lengths of the sides of each triangle to the nearest 0.1 centimeter, and label the figures on your handout. Compute the ratios of the lengths of corresponding sides and complete **Table 8.2**. (For example, choose the longest side of each triangle and find the ratio. Then choose the shortest sides and do the same.)

Names of Triangles in the Pair	Ratio of Side 1 of the First Triangle to Side 1 of the Second Triangle	Ratio of Side 2 of the First Triangle to Side 2 of the Second Triangle	Ratio of Side 3 of the First Triangle to Side 3 of the Second Triangle

Table 8.2

4. **See item 5.**

5. Repeat item 4 for your other pairs of similar triangles.

5. **Side lengths:** ΔABC **(15.5 cm, 10.0 cm, 6.0 cm) and** ΔWXY **(7.8 cm, 5.0 cm, 3.0 cm);**

ΔDEF **(5.0 cm, 5.0 cm, 5.0 cm) and** ΔQRS **(7.0 cm, 7.0 cm, 7.0 cm);**

ΔGHI **(2.0 cm, 4.0 cm, 3.5 cm) and** ΔJKL **(6.0 cm, 12.0 cm, 10.4);**

ΔTUV**(11.3 cm, 8.0 cm, 8.0 cm) and** ΔPMN **(8.5 cm, 6.0 cm, 6.0 cm)**

Names of Triangles in the Pair	Ratio of Side 1 of the First Triangle to Side 1 of the Second Triangle	Ratio of Side 2 of the First Triangle to Side 2 of the Second Triangle	Ratio of Side 3 of the First Triangle to Side 3 of the Second Triangle
ΔABC; ΔWXY	2:1 or 1:2	2:1 or 1:2	2:1 or 1:2
ΔDEF; ΔQRS	7:5 or 5:7	7:5 or 5:7	7:5 or 5:7
ΔGHI; ΔJKL	1:3 or 3:1	1:3 or 3:1	1:3 or 3:1
ΔTUV; ΔPMN	3:4 or 4:3	3:4 or 4:3	3:4 or 4:3

6. If two triangles are similar, a side of one triangle *corresponds* to a side of the other if the sides are opposite equal angles. These pairs of sides are called **corresponding sides.** What do you notice about the ratios of the corresponding sides of similar triangles?

6. **For each pair of similar triangles, the ratios of the corresponding sides are equal.**

Our investigations in items 1–6 suggest that for two triangles to be similar, the corresponding angles are equal in measure. They also suggest that if two triangles are similar, the ratios of their corresponding sides are equal.

We can now return to one of the questions posed at the beginning of this section and use these properties of similar triangles to find a solution.

Figure 8.7 shows a drilling rig similar to the one used to extract the soil from under the tower. The aboveground end of the rig is approximately 24 feet long and stands 10 feet above the ground. It enters the ground at an angle of 25°.

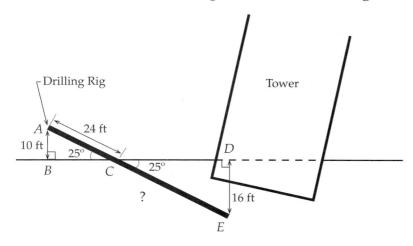

Figure 8.7

7. Use the information shown in the figure to explain why $\triangle ABC \sim \triangle EDC$.

7. **$\angle ACB = \angle ECD$ because both are 25°. $\angle ABC = \angle EDC$ because both are 90°. And because the sum of the angles of any triangle is 180°, the third angles of the triangles, $\angle BAC$ and $\angle DEC$, must be equal (65°). If the corresponding angles of the two triangles are equal, the two triangles are similar.**

Because the two triangles are similar, the ratios of their corresponding sides are equal. Another way to say that the ratios of the corresponding sides are equal is to say that the corresponding sides are proportional. We can write this in symbols as $\dfrac{AB}{ED} = \dfrac{AC}{EC} = \dfrac{BC}{DC}$. Notice that the corresponding sides are those opposite the angles of equal measure.

8. If we want to find the length of the in-ground portion of the rig, we can write a proportion using two of the equal ratios. (Hint: The proportion we write must include three known lengths and the unknown length EC.)
    Write a proportion and solve it to find the length of the in-ground portion of the rig EC.

8. $\dfrac{AB}{ED} = \dfrac{AC}{EC}$; $\dfrac{10}{16} = \dfrac{24}{EC}$; $EC = \dfrac{(24)(16)}{10}$; **$EC = 38.4$ ft.**

Extend the Activity *(SE pages 307–308)*

9. In **Figure 8.8,** $\triangle MNP \sim \triangle RST$ because three angles of $\triangle MNP$ are equal in measure to three angles of $\triangle RST$. And because the triangles are similar, we know that the corresponding sides are proportional. Complete the proportion $\dfrac{MN}{RS} = \dfrac{?}{ST} = \dfrac{?}{?}.$

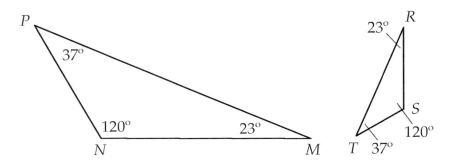

Figure 8.8

9. $\dfrac{MN}{RS} = \dfrac{NP}{ST} = \dfrac{MP}{RT}.$

10. We can also show that two triangles are similar by examining the ratios of the sides. In other words, if corresponding sides of two triangles are proportional, then the two triangles are similar. Show that the two triangles in **Figure 8.9** are similar by showing that the corresponding sides are proportional.

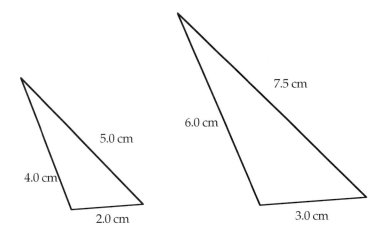

Figure 8.9

10. **Does** $\dfrac{2.0}{3.0} = \dfrac{4.0}{6.0} = \dfrac{5.0}{7.5}?$ **The two triangles are similar because each ratio is equivalent to 2/3.**

11. Return to our original flagpole problem in Preparation Reading. One way to find the unknown height of a flagpole is to use the properties of similar triangles.

Suppose that at 1:00 P.M., a flagpole casts a shadow of 3 feet 6 inches, while a nearby pole 3 feet tall casts a shadow of 5 inches (see **Figure 8.10**).

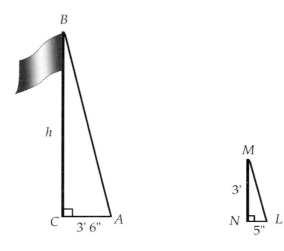

Figure 8.10

a) Explain why $\triangle ABC \sim \triangle LMN$. (Assume that the angle between the sun's rays and the ground is the same for each shadow.)

11. a) In $\triangle ABC$ and $\triangle LMN$, the flagpole and nearby pole are both perpendicular to the ground. So $\angle C = \angle N$. Because the angle of elevation of the sun is the same for each shadow, $\angle A = \angle L$. And because the sum of the angles of any triangle is 180°, the third angles of the triangles, $\angle B$ and $\angle M$, must be equal. If the corresponding angles of the two triangles are equal, the two triangles are similar.

b) Because the triangles are similar, the corresponding sides are proportional. We can write the proportion using ratios of the lengths of the sides of the small triangle to the lengths of the sides of the large triangle or the lengths of the sides of the large triangle to the lengths of the sides of the small triangle. For example, we could write $\dfrac{36 \text{ in}}{h \text{ ft}} = \dfrac{5 \text{ in}}{3.5 \text{ ft}}\left[\dfrac{\text{small triangle}}{\text{large triangle}}\right]$, or we could write $\dfrac{h \text{ ft}}{36 \text{ in}} = \dfrac{3.5 \text{ ft}}{5 \text{ in}}\left[\dfrac{\text{large triangle}}{\text{small triangle}}\right]$. Find the height h of the flagpole.

b)

$$\dfrac{h \text{ ft}}{36 \text{ in}} = \dfrac{3.5 \text{ ft}}{5 \text{ in}}\left[\dfrac{\text{large triangle}}{\text{small triangle}}\right]$$

$$h = \dfrac{(36 \text{ in})(3.5 \text{ ft})}{5 \text{ in}}$$

$$h = 25.2 \text{ ft.}$$

The height of the flagpole is approximately 25 feet.

Section 8.2 Properties of Triangles

What You Need to Know

- How to use the Pythagorean theorem
- How to identify similar triangles

What You Will Learn

- To classify triangles by their sides and by their angles
- To use the converse of the Pythagorean theorem to determine whether a triangle is a right triangle
- To identify similar polygons
- To use the properties of similar polygons to solve problems

Materials

- Geometric drawing utility such as The Geometer's Sketchpad (optional)

A triangle, the simplest of the polygons, is one of the most useful. Look around. Triangles are everywhere. They can be seen in art, engineering designs, and construction. If the lengths of the sides of a triangle are fixed, the triangle is rigid. Its size and shape cannot change. In fact, it is the only polygon with this property. When bracing and strength are needed, triangles are very important.

Note the marks on the sides of triangle A in **Figure 8.11.** These are called "tick marks" and are used to indicate segments of equal lengths. Hence, the two sides with one tick mark are equal in length. The tick marks in rectangle B indicate that the sides with two tick marks are equal in length and the two sides with three tick marks are equal in length. The markings in polygon C indicate that the pentagon has two angles of equal measure.

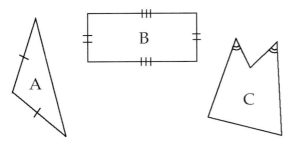

Figure 8.11

Classifying Triangles

Triangles can be classified by their sides and also by their angles. As shown in **Figure 8.12,** we can classify a triangle by its sides as either **scalene** (no sides equal), **isosceles** (two sides equal), or **equilateral** (all three sides equal in length).

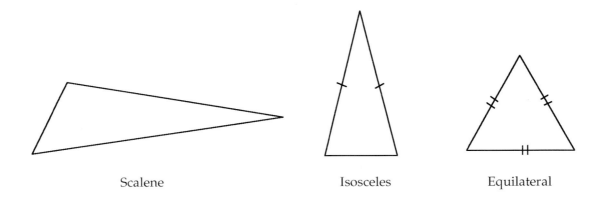

Scalene   Isosceles   Equilateral

Figure 8.12

One unit used to measure angles is the **degree.** The angle formed by rotating one side of an angle one complete turn (one revolution) is said to have a measure of 360 degrees. Hence, 1 degree is 1/360 of a revolution.

The use of the term *degree* to measure angles dates back to the Babylonians, whose number system was based on 60. Why they chose 1/360 of a full rotation as their basic measure is unknown. Some believe that it has to do with the calendar, whereas others trace it back to the time it took to travel a certain unit of distance.

Figure 8.13 shows how triangles can be classified by their angles. If all three of the angles of a triangle are acute (less than 90°), then we say it is an **acute triangle.** If the triangle has a right angle (equal to 90°), then it is a **right triangle,** and if it has an obtuse angle (an angle greater than 90°), then it is an **obtuse triangle.**

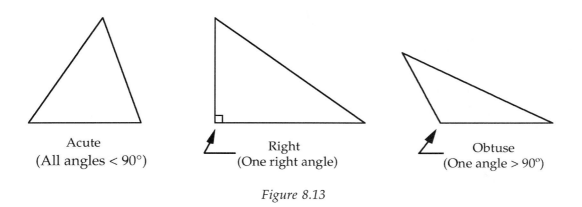

Acute
(All angles < 90°)

Right
(One right angle)

Obtuse
(One angle > 90°)

Figure 8.13

Example 1

Classify **Figures 8.14a** and **8.14b** by their sides and by their angles.

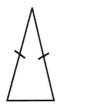

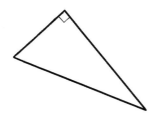

Figure 8.14a

Figure 8.14b

Solution:

Figure 8.14a can be classified as both an acute triangle and an isosceles triangle because all of its angles are less than 90° and it has two equal sides.

Figure 8.14b is a scalene right triangle. It has a right angle, as indicated, and no sides equal in measure.

Right Triangles

Recall that one of the most famous theorems in mathematics, the Pythagorean theorem, describes the relationships between the three sides of a right triangle. We can use it to find the length of one side of a right triangle if the lengths of two of the other sides are known.

The Pythagorean Theorem

In a right triangle, the square of the length of the hypotenuse is equal to the sum of the squares of the lengths of the two legs.

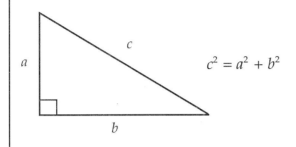

$$c^2 = a^2 + b^2$$

But what if we know all three of the sides of a triangle and want to know if the triangle is a right triangle? Can the Pythagorean theorem help us? The answer is yes, but only because the **converse of the Pythagorean theorem** is true.

Converse of the Pythagorean Theorem

If the square of the length of the longest side of a triangle is equal to the sum of the squares of the lengths of the other two sides, then the triangle is a right triangle.

If-then statements are known as conditional statements. They are made up of two parts, the hypothesis p and the conclusion q. For example, consider the following statement:

"If today is Saturday, then I can stay home"
$\quad$ p, hypothesis $\qquad$ q, conclusion

We often write a conditional as "If p, then q" or $p \Rightarrow q$.

The converse of a conditional $p \Rightarrow q$ is $q \Rightarrow p$ or "If q, then p." The converse of the statement above is, "If I can stay home, then today is Saturday." The converse of a true conditional statement may be either true or false.

Example 2

Is the triangle in **Figure 8.15** a right triangle?

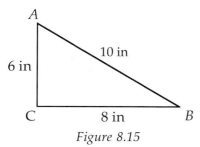

Figure 8.15

Solution:

Does $(10 \text{ in})^2 = (8 \text{ in})^2 + (6 \text{ in})^2$? $100 \text{ in}^2 = 64 \text{ in}^2 + 36 \text{ in}^2$, so the triangle is a right triangle and the angle opposite the longest side, $\angle C$, is a right angle.

Similar Polygons

Our investigations in Activity 8.1 suggested that for two triangles to be similar, the corresponding angles must be equal in measure. We also discovered that if the triangles are similar, their corresponding sides are proportional.

In addition to triangles, other kinds of polygons can be similar. Two polygons are similar if their corresponding angles are equal. And if they are similar, their corresponding sides are proportional.

There are many applications of similar figures in scientific and technical contexts. For example, a photograph is similar to the actual original image, a floor plan for a house is a scale model of the house itself, and an image in a microscope is an enlargement of the object being observed. In each case, the dimensions of the image are proportional to the original.

Example 3

In **Figure 8.16**, pentagon $ABCDE$ ~ pentagon $RSTUV$. Find the lengths of sides ST and CD.

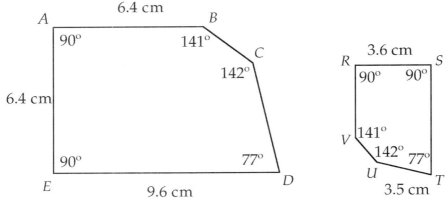

Figure 8.16

Solution:

Because the pentagons are similar, the corresponding sides are proportional. Side AB corresponds to RV, BC to UV, CD to TU, DE to ST, and AE to RS. So

$$\frac{AB}{RV} = \frac{BC}{UV} = \frac{CD}{UT} = \frac{DE}{ST} = \frac{AE}{RS}.$$

To find the length of ST:

$$\frac{AE}{RS} = \frac{DE}{ST}$$

$$\frac{6.4}{3.6} = \frac{9.6}{ST}$$

$$ST = \frac{(9.6)(3.6)}{6.4}$$

$$ST = 5.4 \text{ cm}$$

To find the length of CD:

$$\frac{AE}{RS} = \frac{CD}{UT}$$

$$\frac{6.4}{3.6} = \frac{CD}{3.5}$$

$$CD = \frac{(6.4)(3.5)}{3.6}$$

$$CD = 6.222... \approx 6.2 \text{ cm}$$

Exercises 8.2

I. Investigations *(SE pages 315–321)*

1. **Figure 8.17** shows a template with dimensions in millimeters.

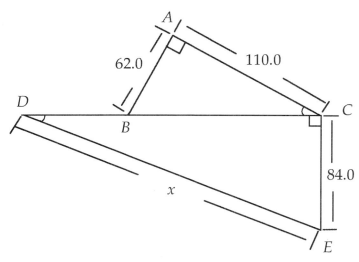

Figure 8.17

a) Find the length of BC.

a) Because $\triangle ABC$ is a right triangle, the Pythagorean theorem can be used to find the hypotenuse BC: $BC^2 = 110.0^2 + 62^2$; $BC^2 = 15{,}944$; $BC \approx$ 126.3 mm.

b) Explain why $\triangle ABC \sim \triangle CED$?

b) $\angle D = \angle ACB$ and $\angle DCE = \angle A$. Two angles of one triangle are equal to two angles of the other, so the third angles are equal and the triangles are similar.

c) Determine length x to the nearest tenth of a millimeter.

c) Because $\triangle ABC \sim \triangle CED$, the corresponding sides are proportional. Hence, $\dfrac{AB}{CE} = \dfrac{BC}{x}$. Then $\dfrac{62.0}{84.0} = \dfrac{126.3}{x}$; $x \approx 171.1$ mm.

2. a) Some ancient Egyptian tombs have drawings that show people with ropes with equally spaced knots on them. It is speculated that these ropes were used in surveying and possibly constructing the pyramids. By having one person hold the first knot and the thirteenth knot together on a rope, a second person hold the fourth knot, and a third person hold the eighth knot, a
right triangle can be formed when the rope is stretched tight. Explain why this is true.

2. a) Answer:

Knots 1 and 13

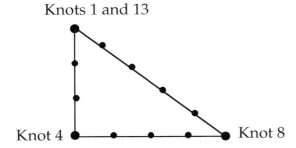

Knot 4   Knot 8

When the knots are held as suggested, a triangle with sides of lengths 3, 4, and 5 units is formed. Using the converse of the Pythagorean theorem, $3^2 + 4^2 = 5^2$, we know the triangle is a right triangle.

b) A method similar to the one described in (a) is still used today in agricultural surveying. It is often referred to as the 3-4-5 method of laying out a right angle. This method requires that the same units and the same multiples of 3, 4, and 5 be used on all three sides. It works best if at least three people are working together.

The process begins by determining the three lengths (multiples of 3, 4, and 5) that are to be used. Then a baseline is established using either of the two shorter lengths. The corners of the baseline are marked. One person stands at each of the corners and holds a tape measure. The third person then holds the other ends of the two tapes together at the correct lengths and moves the third corner until the tapes are tight.

Make a sketch of the 3-4-5 method using a 15-20-25-foot triangle. Indicate what tape marks should be held together by each person and where the right angle will be located.

b) Answers may vary depending on whether each of the three people holds the 0 mark of his or her tape. One possible sketch is shown:

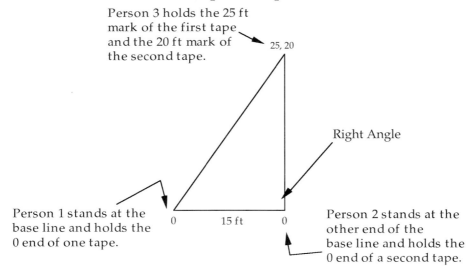

3. **Figure 8.18** shows a picture frame and its outside dimensions.

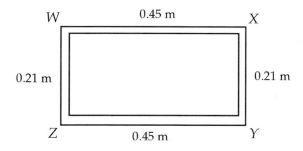

Figure 8.18

a) Suppose the diagonal *WY* measures 0.51 meters. Is ∠*Z* a right angle, an acute angle, or an obtuse angle? Explain.

3. a) Using the converse of the Pythagorean theorem, check to see if $(0.51)^2 = (0.21)^2 + (0.45)^2$. It does not. 0.2601 > 0.0441 + 0.2025. So △*WYZ* is not a right triangle and ∠*Z* is not a right angle. Because the square of the hypotenuse is greater than the sum of the squares of the two legs, $(0.51)^2 > (0.21)^2 + (0.45)^2$, the triangle must be obtuse. So ∠*Z* is an obtuse angle.

b) Generalize your findings by completing the following statements: In △*ABC* with sides *a*, *b*, and *c*,

if $c^2 = a^2 + b^2$, then △*ABC* is a(n) _____.

if $c^2 < a^2 + b^2$, then △*ABC* is a(n) _____.

if $c^2 > a^2 + b^2$, then △*ABC* is a(n) _____.

b) Right triangle; acute triangle; obtuse triangle

4. A pinhole box camera can be made out of any closed box, such as a shoebox.
A pinhole is made in one end of the box, and a piece of unexposed film is attached to the other end. When light enters the pinhole, an inverted image is produced on the film (see **Figure 8.19**).

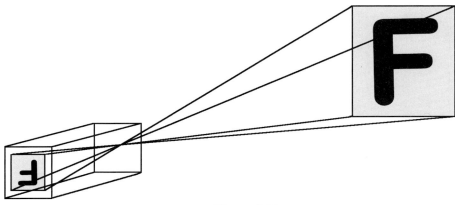

Figure 8.19

Suppose you are taking a picture of a square object 36 cm on each side, your pinhole camera has a depth of 20 cm, and your film is 5 cm square. How far should you place the object in front of the box if you want the image to fill the square piece of film?

4. **Using similar figures, we know that**

$$\frac{\text{side of the image}}{\text{distance from the image to the pinhole}} = \frac{\text{side of the object}}{\text{distance from the pinhole to the object}}.$$

So $\dfrac{5}{20} = \dfrac{36}{d}$ and $d = 144$ cm.

5. If we use reasoning similar to that used for the pinhole box camera in Exercise 4, it is possible to experimentally determine the diameter of the sun. To do so, place a pinhole in a piece of thin cardboard such as an index card. Around noon on a sunny day, take the cardboard, a piece of paper, and a ruler outside. Place the paper on a flat surface such as a sidewalk, and hold the cardboard approximately 30 cm above the paper.

Measure the diameter of the image of the sun on the paper. Also measure the distance from the pinhole to the image. Using those two distances and the distance from the cardboard to the sun (approximately 93,000,000 miles), write a proportion and solve it to find the diameter of the sun.

5. **Sample answer:**

$$\frac{\text{diameter of the image}}{\text{distance from image to cardboard}} = \frac{\text{diameter of the sun}}{\text{distance to the sun}}$$

$$\frac{3 \text{ mm}}{310 \text{ mm}} = \frac{d}{93{,}000{,}000 \text{ mi}}$$

$$d \approx 900{,}000 \text{ mi}$$

This answer is significant to one digit. The actual diameter of the sun is 865,000 miles.

6. Another way to find the height of an object indirectly is to use a mirror. In this method, you place a mirror with a dot on it on the ground between you and the object you want to measure. Look into the mirror and walk back and forth until you see the top of the object on the dot on the mirror. Knowing your height, your distance from the mirror, and the distance between the mirror and the object, you can use similar triangles to find the height of the object (see **Figure 8.20**).

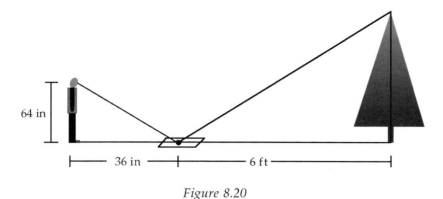

64 in

36 in

6 ft

Figure 8.20

Suppose your eye is 64 inches above the ground and you are 36 inches from the mirror. How tall is the treetop if the mirror is 6 feet away from the tree?

6. **Using similar triangles, $\dfrac{64 \text{ in}}{36 \text{ in}} = \dfrac{x \text{ ft}}{6 \text{ ft}}$. Solving for x, $x \approx 10.7$ feet.**

7. In Section 6.4, Exercise 7, we used a magnifying glass, a light source, and a screen to explore the relationship $\dfrac{1}{f} = \dfrac{1}{d} + \dfrac{1}{i}$, where d is the distance from the magnifying glass to the object, i is the distance from the magnifying glass and the image of the object, and f is the focal length (see **Figure 8.21**). In this exercise, we will use similar triangles to verify this relationship.

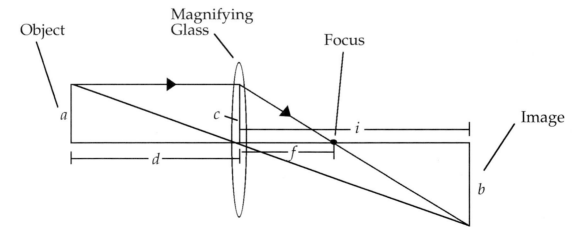

Object

Magnifying
Glass

Focus

a

c

i

d

f

Image

b

Figure 8.21

a) **Figure 8.22** shows Figure 8.21 redrawn and labeled for the purpose of examining the similar triangles.

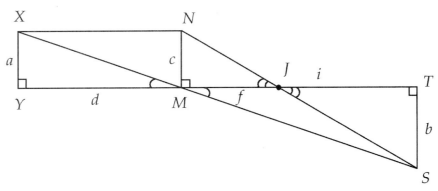

Figure 8.22

Examine $\triangle XYM$ and $\triangle STM$. Explain why the two triangles are similar. Because the triangles are similar, complete the proportion $\dfrac{a}{b} = \dfrac{?}{i}$.

7. a) $\angle XMY = \angle TMS$ and $\angle Y = \angle T$, so $\angle MXY = \angle MST$. **And if the angles of one triangle are equal to the angles of another triangle, the triangles are similar;** $\dfrac{a}{b} = \dfrac{d}{i}$.

b) Examine $\triangle NMJ$ and $\triangle STJ$. We know the two triangles are similar because their angles are equal in measure. Complete the proportion $\dfrac{c}{b} = \dfrac{?}{i-f}$. And because $a = c$, we can rewrite this proportion as $\dfrac{a}{b} = \dfrac{?}{i-f}$.

b) $\dfrac{c}{b} = \dfrac{f}{i-f}$; $\dfrac{a}{b} = \dfrac{f}{i-f}$.

c) From (a) and (b), we know that $\frac{a}{b}$ equals $\frac{d}{i}$ and $\frac{f}{i-f}$. So $\frac{d}{i} = \frac{f}{i-f}$. This proportion relates the desired variables d, f, and i.

Multiply both sides of the proportion by the common denominator. Then divide both sides of the equation by ifd to show that this proportion is equal

to $\frac{1}{f} = \frac{1}{d} + \frac{1}{i}$.

c)
$$\frac{d}{i} = \frac{f}{i-f}$$
$$d(i-f) = fi$$
$$di - df = fi$$
$$\frac{di}{fid} - \frac{df}{fid} = \frac{fi}{fid}$$
$$\frac{1}{f} - \frac{1}{i} = \frac{1}{d}$$
$$\frac{1}{f} = \frac{1}{d} + \frac{1}{i}.$$

8. When we measure angles, there are two ways of expressing fractional parts of degrees. They can be expressed as degrees, minutes, seconds ($14°12'45''$) or as decimal degrees ($14.2125°$). We rarely need to convert from degrees, minutes, seconds (DMS) to decimal degrees (DD), or visa versa, but it is still important to know what each of the forms means because there are times, such as the tables in many trade manuals, that it is necessary to convert between the two forms.

In order to assist with the conversions, the following relationships are important:

1 degree (°) = 60 minutes (')

1 minute (') = 60 seconds ('')

For example, if we want to change 35.127 degrees to degrees, minutes, and seconds, we multiply the decimal part of the value by 60 to obtain minutes: $0.127 \text{ degrees} \cdot \frac{60 \text{ minutes}}{1 \text{ degree}} = 7.62$ minutes. Multiply the decimal part of this value by 60 to obtain seconds:

$0.62 \text{ minutes} \cdot \frac{60 \text{ seconds}}{1 \text{ minute}} = 37.2$ seconds. Rounding 37.2 to 37 seconds, $35.127° = 35°7'37''$.

If we want to change $27°48'$ to decimal degrees, we divide 48 by 60 to obtain the fractional part of a degree. Hence, $27°48' = (27 + 48/60)° = 27.8°$.

a) Convert $27.375°$ to degrees, minutes, seconds.

8. a) $27°22'30''$

b) Convert 6°12′ to decimal degrees.

b) (6 + 12/60)° = 6.2°.

c) Convert 120°36′15″ to decimal degrees.

c) Begin by changing seconds to decimal minutes: (36 + 15/60)′ = 36.25′. Then change minutes to decimal degrees: (120 + 36.25/60) ≈ 120.60417″.

d) Which angle is larger, 63°27′ or 63.47°?

d) 63.47° > 63°27′.

9. Because of the numerical relationships that exist between their sides, some right triangles are considered "special." In this investigation, we will explore one such special triangle, the 45°-45°-90°, also known as the isosceles right triangle.

a) Construct an isosceles right triangle with legs 5 cm in length. Measure the hypotenuse. Then calculate the length of the hypotenuse. Compare your measurement and your calculation.

9. a) The measurement and the calculation should be close to the same value, $\sqrt{50}$, or about 7 cm.

b) If the length of one leg in an isosceles right triangle is 1 unit, find the length of the hypotenuse.

b) $\sqrt{2}$ units

c) In an isosceles right triangle, what are the ratios of leg:leg:hypotenuse?

c) 1:1:$\sqrt{2}$

d) If you know the ratios of the three sides of an isosceles right triangle, you can quickly find a missing side. For example, if the leg is 8 cm, the hypotenuse is $8\sqrt{2}$ cm. Explain how to find the length of one of the legs if the hypotenuse is given.

d) To find the length of a leg when the hypotenuse is known, divide the length of the hypotenuse by $\sqrt{2}$.

e) Find x in **Figure 8.23**.

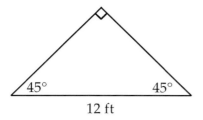

12 ft

Figure 8.23

e) $x = \dfrac{12}{\sqrt{2}} = 6\sqrt{2} \approx 8.5\,\text{ft.}$

f) Use your knowledge of isosceles right triangles to find the length of the diagonal of a square whose side is 10 meters. Explain.

f) The diagonal of a square is the hypotenuse of a 45°-45°-90° triangle. Its length is the product of the length of one side of the square and $\sqrt{2}$, or $10\sqrt{2}$, which is approximately 14.1 meters.

> If the length of the leg of a 45°-45°-90° triangle is x, the length of the hypotenuse is $x\sqrt{2}$.

10. A second special triangle is the 30°-60°-90° triangle. As in Exercise 9, a simple relationship exists between the sides of these triangles.

   a) Construct an equilateral triangle ABC. Draw a line from vertex A to the midpoint (M) of side BC (see **Figure 8.24**). Explain why $\triangle ABM$ and $\triangle ACM$ are each 30°-60°-90° triangles.

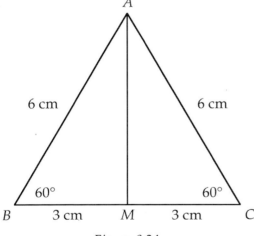

Figure 8.24

10. a) In Figure 8.24, the two triangles are congruent (same size and shape) because the lengths of their three sides are equal; hence, their angles must be equal. We know one angle is 60° because $\triangle ABC$ is equilateral. Angle BMC is a straight angle (180°), so $\angle BMA$ and $\angle CMA$ must each be 90°. Therefore, $\angle BAM$ and $\angle CAM$ must each be 30°.

b) Consider $\triangle ABM$. The lengths of the hypotenuse and the short leg are already known: 6 cm and 3 cm, respectively. Measure the length of the long leg AM. Then calculate its length using the Pythagorean theorem. Compare your calculation and your measurement.

b) The measurement and the calculation should be close, approximately 5.2 cm.

c) If the length of the short leg of the 30°-60°-90° triangle is 1 unit, what is the length of the hypotenuse and the long leg?

c) Hypotenuse: 2 units; using the Pythagorean theorem to find the long leg: $\sqrt{2^2 - 1^2} = \sqrt{3}$ units.

d) In a 30°-60°-90° triangle, what are the ratios of short leg:long leg:hypotenuse?

d) $1:\sqrt{3}:2$

e) If the length of the short leg of a 30°-60°-90° triangle is 5 inches, find the lengths of the other two sides.

e) Long leg = (short leg)($\sqrt{3}$) = $5\sqrt{3} \approx 8.7$ inches; hypotenuse = (2)(short leg) = 10 inches.

f) If the length of the short leg of a 30°-60°-90° triangle is a units, what are the lengths of the long leg and of the hypotenuse?

f) Long leg: $a\sqrt{3}$ units; hypotenuse: $2a$ units

g) If the length of the long leg of a 30°-60°-90° triangle is 12 cm, find the lengths of the other two sides.

g) Short leg = $\dfrac{\text{long leg}}{\sqrt{3}} = \dfrac{12}{\sqrt{3}} = 4\sqrt{3} \approx 6.9$ cm; hypotenuse = (2)(short leg) = $8\sqrt{3} \approx 13.8$ cm.

> If the length of the short leg of a 30°-60°-90° triangle is x, the length of the long leg is $x\sqrt{3}$ and the length of the hypotenuse is $2x$.

11. We can use the Pythagorean theorem to show that a triangle with sides 3, 4, 5 is a right triangle. Any set of three numbers that can be sides of a right triangle is known as a **Pythagorean triple.**

   a) Is the set 5, 12, 13 a Pythagorean triple? Is the set 24, 25, 7? Is the set 8, 13, 15?

11. a) Yes; yes; no

b) If you multiply each of the numbers in the Pythagorean triple 3, 4, 5 by 2, do you get another Pythagorean triple? Explain.

b) Yes, 6, 8, 10 satisfies the Pythagorean theorem.

c) Prove that if the set a, b, c is a Pythagorean triple, then the set ka, kb, kc, where $k > 0$, is also a Pythagorean triple.

c) From the Pythagorean theorem, you know that $a^2 + b^2 = c^2$. Multiplying both sides by k^2, we get $k^2(a^2 + b^2) = k^2c^2$. Using the distributive property, we get $k^2a^2 + k^2b^2 = k^2c^2$, which can be rewritten as $(ka)^2 + (kb)^2 = (kc)^2$. Hence, the set ka, kb, kc is a Pythagorean triple.

II. Projects and Group Activities *(SE page 321)*

12. According to the Guinness Book of Records, the Pythagorean theorem has more published proofs than any other theorem in mathematics. Either try to prove the theorem yourself or use a source such as the Internet to find a proof. Write a short paper explaining the proof you found. Be prepared to explain your proof to your instructor or classmates.

12. Answers will depend on proofs found.

13. Use a geometric drawing utility such as The Geometer's Sketchpad to explore similar triangles.

   a) Construct a triangle ABC.

   b) Construct a second triangle XYZ that is smaller than triangle ABC and with $\angle A = \angle X$ and $\angle B = \angle Y$. Then $\angle C = \angle Z$. Why?

   c) Have the Sketchpad measure the lengths of the sides of each triangle and then find the ratios of the corresponding sides. Does this confirm what you know about the sides being proportional?

13. Sample Sketchpad drawing with angles and sides measured and ratios calculated.

m∠BAC = 33°   m $\overline{AB}$ = 3.17 cm   m∠YXZ = 33°   m $\overline{XY}$ = 2.05 cm
m∠ABC = 120°   m $\overline{BC}$ = 3.80 cm   m∠XYZ = 120°   m $\overline{YZ}$ = 2.41 cm
m∠ACB = 27°   m $\overline{CA}$ = 6.04 cm   m∠XZY = 27°   m $\overline{XZ}$ = 3.87 cm

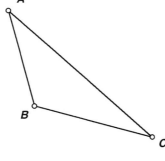

$\dfrac{\text{m } \overline{AB}}{\text{m } \overline{XY}} = 1.55$

$\dfrac{\text{m } \overline{BC}}{\text{m } \overline{YZ}} = 1.58$

$\dfrac{\text{m } \overline{CA}}{\text{m } \overline{XZ}} = 1.56$

14. Use either the shadow or the mirror method to find the height of some unknown object. Write a paper explaining your method. Be sure to include drawings.

14. For a sample solution, refer to Exercises 6 of this section and item 11 of Activity 8.1.

III. Additional Practice *(SE pages 322–327)*

15. In **Figure 8.25**, find the measure of $\angle ACD$.

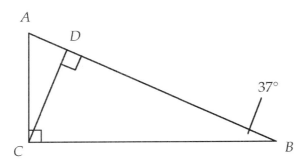

Figure 8.25

15. In right triangle *BCD*, $\angle DCB = 180 - (90 + 37) = 53°$. And because $\angle ACB = 90°$, $\angle ACD = 90 - 53 = 37°$.

16. In **Figure 8.26**, if $\angle XWZ = 35°$, $\angle XYW = 42°$, and $\angle Z = 31°$, find the measure of $\angle XWY$.

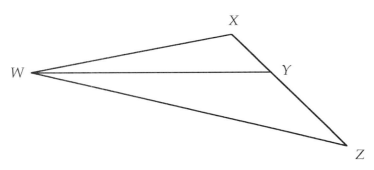

Figure 8.26

16. In $\triangle WXZ$, we know $\angle X = 180 - (31 + 35) = 114°$. Then in $\triangle WXY$, $\angle XWY = 180 - (114 + 42) = 24°$.

For 17–20, solve each proportion for the variable.

17. $\dfrac{4}{5} = \dfrac{x}{15}$.

17. $5x = 60$; $x = 12$.

18. $\dfrac{3}{2y} = \dfrac{6}{18}$.

18. $12y = 54$; $y = 9/2$, or $4\dfrac{1}{2}$.

19. $\dfrac{x-3}{2} = \dfrac{9}{12}$.

19. $12x - 36 = 18$; $x = 9/2$, or $4\dfrac{1}{2}$.

20. $\dfrac{3}{y-2} = \dfrac{4}{2y}$.

20. $4y - 8 = 6y$; $2y = -8$; $y = -4$.

21. In **Figure 8.27,** the triangles are similar. Find the value of x.

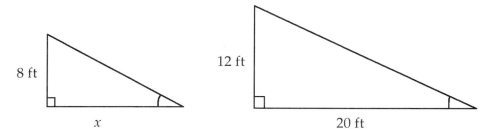

Figure 8.27

21. $\dfrac{8}{12} = \dfrac{x}{20}$; $12x = 160$; $x = 40/3$, or $13\dfrac{1}{3}$ ft.

22. In **Figure 8.28,** the triangles are similar. Find the value of y.

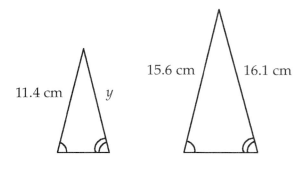

Figure 8.28

22. $\dfrac{11.4}{15.6} = \dfrac{y}{16.1}$; $15.6y = (11.4)(16.1)$; $y \approx 11.8$ cm.

23. The sides of ∆*ABC* are 4.2 m, 7.6 m, and 8.0 m. The sides of ∆*XYZ* are 19.0 m, 20.0 m, and 10.5 m. Are the two triangles similar? Explain.

23. Yes, the two triangles are similar because the sides are proportional:
$$\frac{4.2}{10.5} = \frac{7.6}{19.0} = \frac{8.0}{20}.$$

24. In **Figure 8.29**, ∆*ABC* ~ ∆*XYZ*. Find the measures of the angles in ∆*XYZ*.

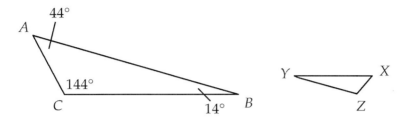

Figure 8.29

24. Because the two triangles are similar, the corresponding angles are equal in measure. So ∠X = 44°, ∠Y = 14°, and ∠Z = 144°.

25. The two triangles in **Figure 8.30** are **congruent,** that is, they have the same size and shape. We write ∆*ABC* ≅ ∆*DEF*.

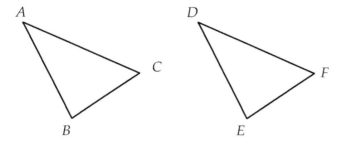

Figure 8.30

    a) What do you know about the angles of the two triangles?

25. a) Their corresponding angles are equal in measure.

    b) What is the ratio of the corresponding sides?

b) 1:1

    c) Are the two triangles similar? Explain.

c) Yes, they are similar because the corresponding angles are equal and corresponding sides are proportional. All congruent figures are similar.

26. Goniometers are a group of instruments designed to measure joint angles. They are used to measure the range of motion of joints such as wrists, elbows, shoulders, backs, ankles, and necks. The typical elbow has a range of motion of about 140°, while the normal knee has a range of motion of about 135°.

Suppose after knee replacement, a person has a range of motion of only 90°, and the therapist wants to increase his range by 20° to 25°. What will be the final range of motion for the patient?

26. Between 110° and 115°

27. In **Figure 8.31,** classify each triangle (a–d) by its sides and angles.

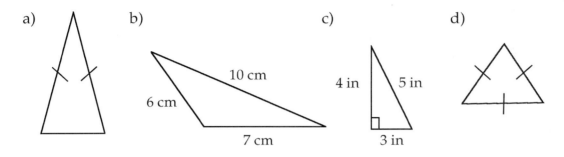

Figure 8.31

27. a) Isosceles, acute

b) Scalene, obtuse

c) Scalene, right

d) Equilateral, acute

28. Complete each statement:

a) 1/4 revolution = _____°

b) 1/2 revolution = _____°

c) 3/4 revolution = _____°

d) 1/12 revolution = _____°

28. a) 90°; b) 180°; c) 270°; d) 30°

29. In ΔDEF, if $\angle D = 38°$ and $\angle F = 47°$, find $\angle E$.

29. $\angle E = 180 - (38 + 47) = 95°$.

30. Triangle *GHI* is an isosceles right triangle. If $\angle H$ is the right angle, find $\angle G$ and $\angle I$.

30. $\angle G = \angle I = 45°$.

31. In **Figure 8.32,** solve for x.

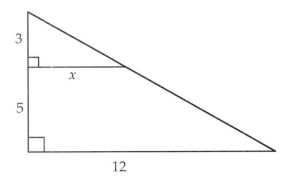

Figure 8.32

31. Using similar triangles, we know $\dfrac{3}{x} = \dfrac{8}{12}$. **Hence, $8x = 36$ and $x = 4.5$.**

32. In **Figure 8.33,** solve for x and y.

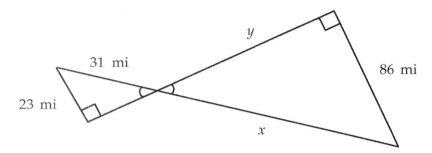

Figure 8.33

32. $\dfrac{23}{86} = \dfrac{31}{x}$; $x \approx 115.9$ **mi. Solve for y using the Pythagorean theorem:**
$y^2 = 115.9^2 - 86.0^2$; $y^2 = 6036.8$; $y \approx 77.7$ **mi.**

For 33–36, if the numbers represent the lengths of the sides of a triangle, classify the triangle as right, acute, or obtuse.

33. 3, 10, 12

33. $3^2 + 10^2 < 12^2$, **so the triangle is obtuse.**

34. 12, 16, 20

34. $12^2 + 16^2 = 20^2$, **so the triangle is a right triangle.**

35. 7, 8, 10

35. $7^2 + 8^2 > 10^2$, **so the triangle is acute.**

36. 2, 5, 8

36. This is not a triangle, because the sum of the lengths of the two shortest sides is less than the length of the third side.

37. In **Figure 8.34,** is $\triangle MNP$ right, obtuse or acute?

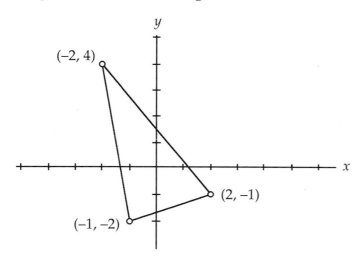

Figure 8.34

37. **Find the lengths of the sides of the triangle using the distance formula: $\sqrt{37}$, $\sqrt{10}$, and $\sqrt{41}$. Then use the converse of the Pythagorean theorem: $\sqrt{37}^2 + \sqrt{10}^2 > \sqrt{41}^2$. The triangle is acute.**

38. If a triangle with sides 15 cm, 21 cm, and 14 cm is similar to a triangle whose shortest side is 6 cm, find the other two sides of the smaller triangle.

38. $\dfrac{14 \text{ cm}}{6 \text{ cm}} = \dfrac{15 \text{ cm}}{x \text{ cm}} = \dfrac{21 \text{ cm}}{y \text{ cm}}$; $x = 45/7$, or 6.4 cm and $y = 9.0$ cm.

39. If a triangle with sides 4.0 cm, 8.0 cm, and 10.0 cm is similar to a triangle whose longest side is 23.0 cm, find the perimeter of the larger triangle.

39. $\dfrac{10 \text{ cm}}{23 \text{ cm}} = \dfrac{4 \text{ cm}}{x \text{ cm}} = \dfrac{8 \text{ cm}}{y \text{ cm}}$; $x = 9.2$ cm and $y = 18.4$ cm; perimeter: 9.2 + 18.4 + 23.0 = 50.6 cm.

40. The image of a microprocessor chip under a magnifying glass is a rectangle 3.1 cm wide by 6.2 cm long. If the width of the actual chip is 0.4 mm, what is the length of the actual chip?

40. $\dfrac{0.4 \text{ mm}}{3.1 \text{ cm}} = \dfrac{l \text{ mm}}{6.2 \text{ cm}}$; $l = 0.8$ mm.

41. On a blueprint for a new home, 1 inch corresponds to 5 feet. Find the length of the garage if it is represented on the blueprint by a line that is $3\dfrac{1}{8}$ inches long.

41. $\dfrac{1 \text{ in}}{5 \text{ ft}} = \dfrac{3\frac{1}{8} \text{ in}}{l \text{ ft}}$; $l = 5(3.125) = 15.625$ ft.

42. To estimate the height of a building, a 5-foot-9-inch-tall person measured her shadow and the shadow of the building at approximately the same time. The building cast a shadow of 39.6 feet, while her shadow was 33 inches. How tall is the building?

42. $\dfrac{33 \text{ in}}{39.6 \text{ ft}} = \dfrac{69 \text{ in}}{h \text{ ft}}$, **$h = 82.8$ ft.**

43. Students were told that triangles I and II in **Figure 8.35** are similar.

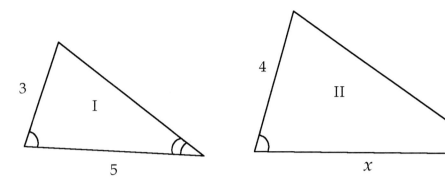

Figure 8.35

Three of the students set up the following proportions for the similar triangles:

A. $\dfrac{3}{5} = \dfrac{4}{x}$ B. $\dfrac{3}{4} = \dfrac{5}{x}$ C. $\dfrac{x}{5} = \dfrac{4}{3}$

The instructor said that all three were correct. Explain each student's thinking.

43. **Student A: Shorter side of triangle I is to the longer side of triangle I as the shorter side of triangle II is to the longer side of triangle II.**

Student B: Shorter side of triangle I is to its corresponding side of triangle II as the longer side of triangle I is to its corresponding side of triangle II.

Student C: Longer side of triangle II is to its corresponding side of triangle I as the shorter side of triangle II is to its corresponding side of triangle I.

Section 8.3 Trigonometric Ratios

What You Need to Know

• How to write proportions for the corresponding sides of similar triangles

• How to solve proportions

What You Will Learn

• To find the sine, cosine, and tangent of an acute angle

• To use the sine, cosine, and tangent of an angle to find a missing side of a right triangle when given one side and an acute angle

Materials

• Meter or yardsticks

• Protractor

• Geometric utility such as Geometer's Sketchpad or Cabri (optional)

• Handout 8.2

• Ruler (metric)

• Masking tape

• String with a weight on the end (plumb bob)

In Section 8.1, we investigated the properties of similar triangles and found that the relationships between corresponding sides could be used to help find heights and distances that could not be measured directly. In this lesson, we will explore the relationships between the sides and angles of right triangles. The study of these relationships is known as right triangle **trigonometry.**

The word *trigonometry* comes from Latin and means "the study or measurement of trigons (triangles)."

Discovery 8.1 The Tangent of an Angle *(SE pages 329–331)*

Have each group of students place a point *A* on the wall approximately 50–60 cm above the floor. (Do not measure the distance above the floor exactly.) Place a meter stick at a diagonal from point *A* to the floor at point *B* as shown in **Figure 8.36.**

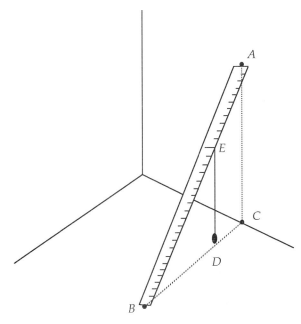

Figure 8.36

Hang a plumb bob (made of a piece of string and a weight) at some point *E* along the meter stick. Call the point where the plumb bob would touch the floor point *D*.

1. What kind of triangle is △*EBD*? Why?

1. **A right triangle; the floor, which is horizontal, is perpendicular to the plumb bob, which is vertical.**

2. Use a protractor to measure ∠*EBD* as precisely as possible. (see **Figure 8.37**). Also measure the lengths of *ED* and *BD* to the nearest 0.1 cm. Fill in the first line of **Table 8.3.**

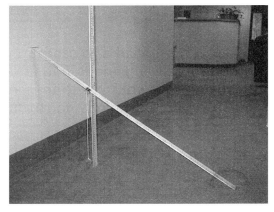

Figure 8.37

Move the plumb bob to a different location on the meter stick and fill in the second line of Table 8.3. Repeat this for three additional locations of the plumb bob. Complete Table 8.3.

Plumb Bob Location	Measure of $\angle EBD$	Length of ED	Length of BD	Ratio of ED to BD
#1				
#2				
#3				
#4				
#5				

Table 8.3

2. Answers will depend on where the meter stick is located in relationship to the wall and to the floor.

3. What did you notice about the measure of $\angle EBD$ for each of the different locations of the plumb bob?

3. **$\angle EBD$ remained unchanged.**

4. What did you notice about the lengths of ED and BD as the plumb bob was moved to a new location? What about $\dfrac{ED}{BD}$? Explain.

4. **The lengths of ED and BD changed with the change in location of the plumb bob, but the ratio of ED to BD remained unchanged. The ratio stayed constant because the triangles are similar and similar triangles have sides that are proportional.**

5. Measure BC. Use the length of BC and the information in your table to find AC indirectly.

5. **To find AC, solve the proportion $\dfrac{ED}{BD} = \dfrac{AC}{BC}$ for AC.**

6. Measure AC to see how accurate your indirect measurement was in item 5. How close were your indirect and direct measurements? If they were not exactly the same, explain why you think this happened.

6. **Most likely, the measurements will not be exactly the same. Differences can be explained by errors in measurement.**

Long ago, mathematicians and astronomers discovered the same relationship that you just did. They noticed that for a given acute angle in a right triangle, the length of the leg opposite the acute angle divided by the length of the leg adjacent to the angle is the same no matter the lengths of the sides. That is, the value of the ratio depends only on the measure of the angle, not the size of the triangle.

A **trigonometric ratio** is a ratio of the lengths of any two sides of a right triangle. The ratio consisting of the length of the leg opposite the acute angle divided by the length of the leg adjacent to the angle is one of the trigonometric ratios and is called the **tangent** of the angle.

In **Figure 8.38**, the tangent of $\angle M = \dfrac{\text{length of the leg opposite } \angle M}{\text{length of the leg adjacent to } \angle M} = \dfrac{NP}{MP}$.

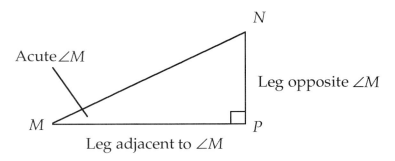

Figure 8.38

In addition to the tangent, there are two other basic trigonometric ratios: the **sine** and the **cosine**. The sine and cosine of the acute angle M in Figure 8.38 are defined as follows:

sine of $\angle M = \dfrac{\text{length of the leg opposite } \angle M}{\text{length of the hypotenuse}} = \dfrac{NP}{MN}$

cosine of $\angle M = \dfrac{\text{length of the leg adjacent } \angle M}{\text{length of the hypotenuse}} = \dfrac{MP}{MN}$

These ratios are abbreviated as $\tan M$, $\sin M$, and $\cos M$.

Example 4

Consider the triangle in **Figure 8.39.** Find the sine, cosine, and tangent of $\angle A$.

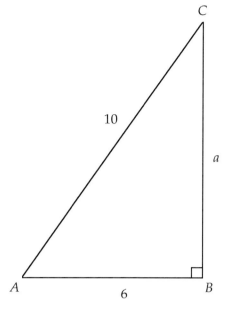

Figure 8.39

Triangles are often labeled as shown in Figure 8.39, where capital letters are used to indicate the vertices of the angles and lowercase letters are used to refer to the sides. Note that side a is opposite $\angle A$.

Solution:

In $\triangle ABC$, the leg adjacent to $\angle A$ is equal to 6 and the hypotenuse is equal to 10. The leg opposite $\angle A$, a, can be found by using the Pythagorean theorem.

$$a^2 + b^2 = c^2$$
$$a^2 + 6^2 = 10^2$$
$$a^2 = 100 - 36$$
$$a^2 = 64$$
$$a = 8$$

The sine, cosine, and tangent of $\angle A$ are as follows:

$$\sin A = \frac{\text{opp.}}{\text{hyp.}} = \frac{8}{10} = \frac{4}{5}; \quad \cos A = \frac{\text{adj.}}{\text{hyp.}} = \frac{6}{10} = \frac{3}{5}; \quad \tan A = \frac{\text{opp.}}{\text{adj.}} = \frac{8}{6} = \frac{4}{3}.$$

Use Handout 8.2.

1. For each of the right triangles on the handout, measure the acute angles to the nearest degree and the lengths of the sides to the nearest 0.1 cm. Use your measurements to complete the trigonometry table in **Table 8.4**.

Angle Measure (degrees)	Tangent of the Angle	Sine of the Angle	Cosine of the Angle
10			
20			
30			
40			
50			
60			
70			
80			

Table 8.4

1.

Angle Measure (degrees)	Tangent of the Angle	Sine of the Angle	Cosine of the Angle
10	1.7/10 = 0.17	1.7/10.1 = 0.17	10/10.1 = 0.99
20	3.7/10 = 0.37	3.7/10.7 = 0.35	10/10.7 = 0.93
30	5.8/10 = 0.58	5.8/11.5 = 0.50	10/11.5 = 0.87
40	8.3/10 = 0.83	8.3/13 = 0.64	10/13 = 0.77
50	10/8.3 = 1.2	10/13 = 0.77	8.3/13 = 0.64
60	10/5.8 = 1.7	10/11.5 = 0.87	5.8/11.5 = 0.50
70	10/3.7 = 2.7	10/10.7 = 0.93	3.7/10.7 = 0.35
80	10/1.7 = 5.9	10/10.1 = 0.99	1.7/10.1 = 0.17

2. Examine the patterns in your table and consider a right triangle whose angle is so small that it is almost 0 degrees. What do you think cos(0°) is? What are sin(0°) and tan(0°)?

2. **cos(0°) = 1; sin(0°) = 0; tan(0°) = 0.**

3. Again, examine the patterns in your table. This time think about a right triangle whose angle is very close to 90 degrees. What do you think cos(90°) is? What are sin(90°) and tan(90°)?

3. cos(90°) = 0; sin(90°) = 1; tan(90°) is undefined (students may say that it is a big number).

4. To complete Table 8.4, why did you need only four triangles rather than eight?

4. In a right triangle, if one of the acute angles is *x*, the other is 90 – *x*. So if you have a triangle with a 10-degree angle, then that same triangle has an 80-degree angle. Hence, two of the rows in the table could be calculated from one triangle.

5. Two angles whose sum is 90° are called **complementary angles**. In a right triangle, the acute angles are complementary. Explain why.

5. The sum of the three angles in a triangle is 180°. One of the angles is 90°, so the sum of the other two must be 90°. Hence, they are complementary angles.

6. Use your table to find the length of leg *x* in **Figure 8.40.**

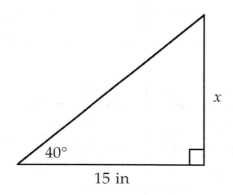

Figure 8.40

6. tan(40°) = *x*/15 in. From Table 8.4, tan(40°) = 0.83. So *x* = (15.0)(0.83); *x* ≈ 12.5 in.

7. You can use your calculator to check your values in Table 8.4. (Hint: Make sure your calculator is in degree mode.) To find sin(10°), try one of the following:

10 | sin | or | sin | 10 | enter

If neither of these seems to work, you may need to refer to your calculator manual.

Another unit used to measure angles is called the **radian.** In one revolution (or 360°), there are 2π radians. One radian is about 57°, which is approximately the angle formed by your middle and index fingers when spread apart (see **Figure 8.41**). For additional information about radian measure, see Exercise 8.

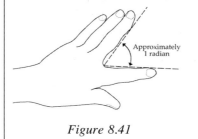

Figure 8.41

How accurate is your table in item 1? Explain why it may not agree exactly with the calculator. What is the advantage of using a calculator over measuring right triangles?

7. **Values may differ due to the inaccuracy of measuring. The advantage of the calculator is that it is more accurate and much quicker than making tables.**

TABLE SETUP
TblStart=10
ΔTbl=10
Indpnt: **Auto** Ask
Depend: **Auto** Ask

X	Y1
10	.17633
20	.36397
30	.57735
40	.8391
50	1.1918
60	1.7321
70	2.7475
Y1=tan(X)	

X	Y1
10	.17365
20	.34202
30	.5
40	.64279
50	.76604
60	.86603
70	.93969
Y1=sin(X)	

X	Y1
10	.98481
20	.93969
30	.86603
40	.76604
50	.64279
60	.5
70	.34202
Y1=cos(X)	

8. Use your calculator to find the distance across the lake in **Figure 8.42**.

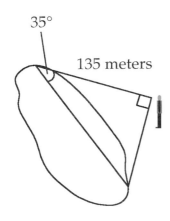

35°

135 meters

Figure 8.42

8. $\cos(35°) = 135 \text{ m}/x$; $0.819x = 135$; $x \approx 165$ m.

Once you know the value of one trigonometric ratio, you can determine the values of the others.

Example 5

In a given right triangle, $\cos A = \dfrac{1}{4}$. Find the exact value of $\sin A$ and $\tan A$.

Solution:

Draw a right triangle showing $\angle A$. Based on the value of $\cos A$, two sides of the triangle can be labeled. In **Figure 8.43,** the side adjacent to $\angle A$ and the hypotenuse are labeled 1 and 4, respectively. They could be labeled with any lengths as long as the ratio of the side adjacent to $\angle A$ to the hypotenuse is 1:4. For example, the sides could be labeled 2 and 8 or 1.03 and 4.12.

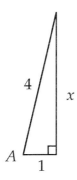

Figure 8.43

From the Pythagorean theorem, $4^2 = 1^2 + x^2$. Hence, $x = \sqrt{15}$. Then the exact value of $\sin A$ is $\dfrac{\sqrt{15}}{4}$, and $\tan A = \dfrac{\sqrt{15}}{1} = \sqrt{15}$.

Sine, Cosine, and Tangent Summary

In a right triangle, the three trigonometric ratios that we have investigated—the sine, cosine, and tangent ratios for any acute angle (refer to **Figure 8.44**)—are defined as follows:

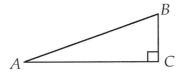

Figure 8.44

$$\sin A = \frac{\text{length of the side opposite } \angle A}{\text{length of the hypotenuse}} = \frac{BC}{AB}.$$

$$\cos A = \frac{\text{length of the side adjacent to } \angle A}{\text{length of the hypotenuse}} = \frac{AC}{AB}.$$

$$\tan A = \frac{\text{length of the side opposite } \angle A}{\text{length of the side adjacent to } \angle A} = \frac{BC}{AC}.$$

$$\sin B = \frac{\text{length of the side opposite } \angle B}{\text{length of the hypotenuse}} = \frac{AC}{AB}.$$

$$\cos B = \frac{\text{length of the side adjacent to } \angle B}{\text{length of the hypotenuse}} = \frac{BC}{AB}.$$

$$\tan B = \frac{\text{length of the side opposite } \angle B}{\text{length of the side adjacent to } \angle B} = \frac{AC}{BC}.$$

Exercises 8.3

I. Investigations *(SE pages 336–339)*

1. In Section 8.1, you found the height of a flagpole using similar triangles. Reconsider the problem when different information is known (see **Figure 8.45**). Find the height of the flagpole.

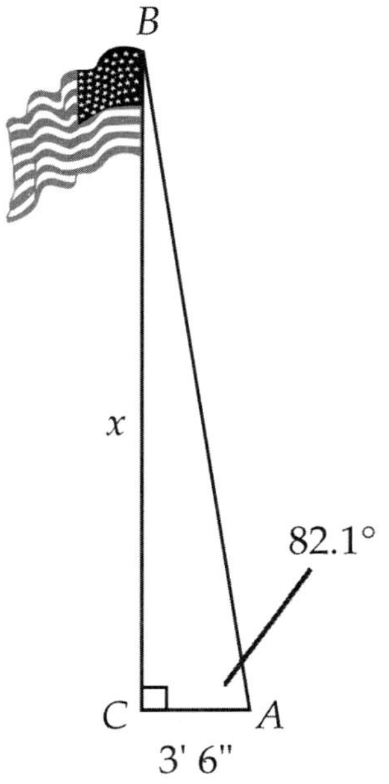

Figure 8.45

1. $\tan(82.1°) = \dfrac{x}{3.5}$; $x \approx 25.2$ **ft.**

2. As mentioned in Section 8.1, the angle of lean of the Leaning Tower of Pisa was decreased from approximately 5.5° to 5.0° by removing soil from under the north end of the building.

   Figure 8.46 shows a portion of the tower before the soil was removed. **Figure 8.47** shows a portion of the tower after soil removal. (Note: The figures are not drawn to scale.)

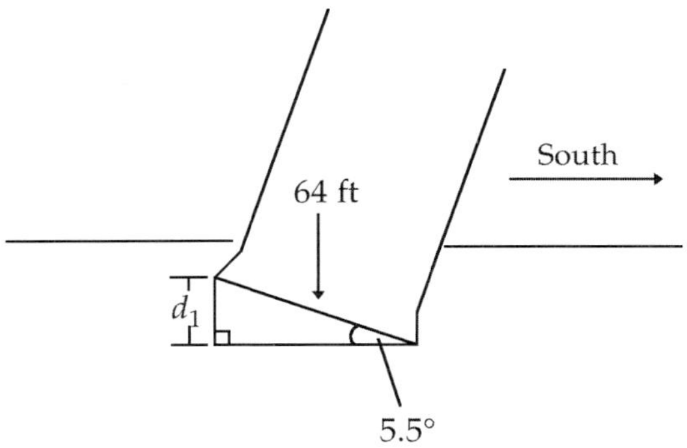

Figure 8.46

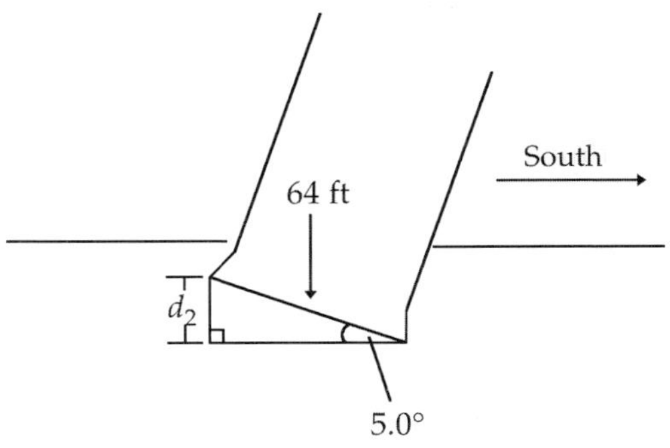

Figure 8.47

a) Use Figure 8.46 to determine approximately how much higher the north end of the building was than the south end before the soil was removed. Assume the diameter of the foundation is approximately 64 feet. Express the distance d_1 in feet.

2. a) $\sin(5.5°) = \dfrac{d_1}{64}$; $d_1 \approx 6.13$ ft.

   b) Use Figure 8.47 to determine approximately how much higher the north end of the building is than the south end after the soil was removed. Express this distance d_2 in feet.

   b) $\sin(5.0°) = \dfrac{d_2}{64}$; $d_2 \approx 5.58$ ft.

   c) How much was the north end of the tower lowered after the soil was removed. Express your answer in inches.

   c) 6.32 ft – 5.58 ft = 0.55 ft ≈ 7 in.

3. In addition to the sine, cosine, and tangent ratios, there are three other trigonometric ratios: the cosecant (csc), the secant (sec), and the cotangent (cot). The additional three ratios are defined as follows (refer to **Figure 8.48**):

Figure 8.48

$$csc A = \frac{\text{length of the hypotenuse}}{\text{length of the side opposite } \angle A} = \frac{AB}{BC}.$$

$$sec A = \frac{\text{length of the hypotenuse}}{\text{length of the side adjacent to } \angle A} = \frac{AB}{AC}.$$

$$cot A = \frac{\text{length of the side adjacent to } \angle A}{\text{length of the side opposite } \angle A} = \frac{AC}{BC}.$$

a) How do $csc A$, $sec A$, and $cot A$ relate to $\sin A$, $\cos A$, and $\tan A$?

3. a) $csc A = \dfrac{1}{\sin A}$; $sec A = \dfrac{1}{\cos A}$; $cot A = \dfrac{1}{\tan A}$, **or you can say that the cosecant of an angle is the reciprocal of the sine, the secant of an angle is the reciprocal of the cosine, and the cotangent of an angle is the reciprocal of the tangent.**

b) Find the value of each of the six trigonometric ratios of $\angle M$ in **Figure 8.49.**

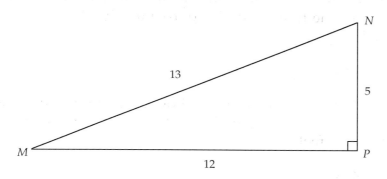

Figure 8.49

b) $\sin M = 5/13$; $\cos M = 12/13$; $\tan M = 5/12$; $\csc M = 13/5$; $\sec M = 13/12$; $\cot M = 12/5$

c) If the sine of $\angle X$ is $\dfrac{2}{3}$, find the value of the other five trigonometric ratios for the acute $\angle X$.

c)

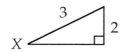

Using the Pythagorean theorem, the leg adjacent to $\angle X$ is $\sqrt{5}$. Hence,
$$\cos X = \frac{\sqrt{5}}{3}; \tan X = \frac{2}{\sqrt{5}} \text{ or } \frac{2\sqrt{5}}{5}; \csc X = \frac{3}{2}; \sec X = \frac{3}{\sqrt{5}} \text{ or } \frac{3\sqrt{5}}{5}; \cot X = \frac{\sqrt{5}}{2}.$$

4. In Section 8.2, Exercises 9 and 10, we explored the relationships between the sides of a 45°-45°-90° triangle and a 30°-60°-90° triangle, respectively. Use those relationships to complete **Table 8.5.** (Hint: Begin by drawing a 45°-45°-90° triangle and a 30°-60°-90° triangle and labeling the sides.)

$\angle A$ (degrees)	$\sin A$	$\cos A$	$\tan A$
45			
30			
60			

Table 8.5

4.

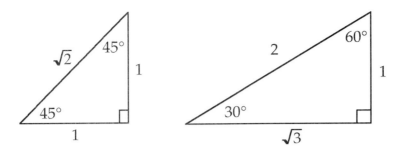

∠A (degrees)	sinA	cosA	tanA
45	$\dfrac{1}{\sqrt{2}}$ or $\dfrac{\sqrt{2}}{2}$	$\dfrac{1}{\sqrt{2}}$ or $\dfrac{\sqrt{2}}{2}$	1
30	$\dfrac{1}{2}$	$\dfrac{\sqrt{3}}{2}$	$\dfrac{1}{\sqrt{3}}$ or $\dfrac{\sqrt{3}}{3}$
60	$\dfrac{\sqrt{3}}{2}$	$\dfrac{1}{2}$	$\sqrt{3}$

5. If an object is located above the horizontal, the angle from the horizontal to the line of sight is called the **angle of elevation.** In **Figure 8.50,** the horizontal is represented by the line *d* and the line of sight by line *l*.

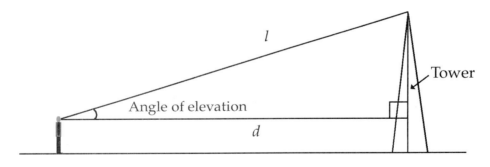

Figure 8.50

a) Assume the angle of elevation in Figure 8.50 is 10.6°, the distance from the person to the tower is 47.3 meters, and eye level for the person is 2.1 meters above the ground. Find the height of the tower.

5. a) **tan(10.6°) = height of the tower above eye level (*t*)/47.3 meters; *t* ≈ 8.9 meters. So the height of the tower is 8.9 + 2.1 ≈ 11.0 meters tall.**

b) Assume that the angle of elevation from a ship to the top of a 153-foot lighthouse on a rocky coast is 35 degrees. Find the distance from the ship to the lighthouse.

b)

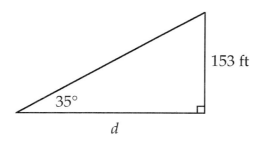

$$\tan(35°) = \frac{153}{d}; \ d \approx 219 \text{ ft.}$$

6. Sine bars, sine plates, and compound sine plates are used to establish or measure angles. These small devices are made of extremely hard metal, milled to tight tolerances, and used by machinists to set machines for processing cuts and measuring angles of cuts other than 90 degrees. To set a desired angle, one end of a sine bar or plate is raised with gage blocks (see **Figure 8.51**). The most common sizes of sine bars and plates are 5 inches and 10 inches between rolls. When using a sine bar, the height setting is limited by the gage block division available (often 0.0001 inch).

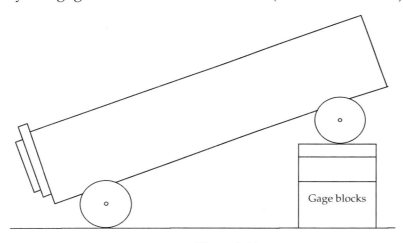

Figure 8.51

a) Suppose a milling machine operator wants to set a 9°10′ angle with a 10-inch sine bar. Determine the gage block height.

6. a)

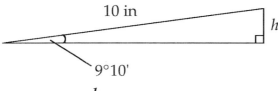

$$\sin(9°10') = \frac{h}{10}; \ h = 1.5931 \text{ in.}$$

b) Determine the gage block height needed to set an angle of 12.73° with a 5-inch sine plate.

b)

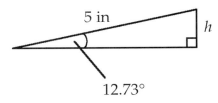

$$\sin(12.73°) = \frac{h}{5}; \; h = 1.1018 \text{ in.}$$

7. Assume $0° \le \angle A \le 90°$. Is it possible for $\sin A$ to be greater than 1 for some $\angle A$? Can $\cos A > 1$? Can $\tan A > 1$? Explain.

7. **The sine and cosine of an angle cannot be greater than 1. In the following figure, $\sin A = CB/AC$. As $\angle A$ gets closer to 90°, CB and AC become almost equal in length.**

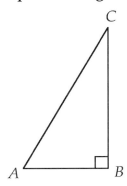

Hence, $\sin A$ gets close to 1, but it can't get any greater than 1 because CB cannot be greater than AC. $\cos A = AB/AC$, and as $\angle A$ gets close to 0°, AB and AC become almost equal in length. So $\cos A$ gets close to 1, but it can't get any greater than 1 because AB cannot be greater than AC. $\tan A = CB/AB$, and as $\angle A$ gets close to 90°, CB gets larger, or AB gets smaller, and the quotient CB/AB gets larger. So $\tan A$ gets very large as the angle approaches 90°.

II. Projects and Group Activities *(SE pages 339–342)*

8. The term *radian* has been in use for a little over 100 years. The need for radian measure evolved from the need for a standard measure for angles radiating from the center of a circle.

Materials: Flattened paper plate, strip of paper longer than the circumference of the plate, scissors

a) Wrap the strip of paper around the edge of the paper plate. Cut the paper so that the length of the strip is equal to the length of the circumference of the circle. How does the length of your paper strip compare with the diameter of the plate?

8. a) The circumference is a little over three times as long as the diameter.

b) Fold your paper plate in half and crease. Fold it in half again, forming creases that divide the plate into four equal sections.

Using one of the radii of your plate as a guide, begin at one end of the paper strip and mark off the strip in radius units (see **Figure 8.52**).

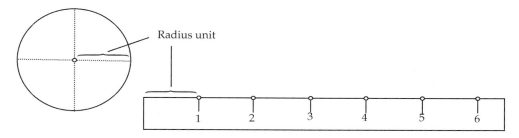

Figure 8.52

If the length of the radius of the plate is *r*, *approximately* how long is your strip of paper? Using prior knowledge about the relationship between the radius of a circle and the circumference, *exactly* how long is the strip?

b) It is a little over 6r long; it is exactly 2πr long.

c) On your plate, mark the end of one radius 0 (zero). Then use your paper strip to mark off 1 radius unit along the circumference of the plate. The central angle whose arc measures 1 radius has a measure of 1 radian. So in **Figure 8.53**, ∠θ = 1 radian. Note: Greek letters, in this case theta (θ), are often used to identify specific angles.

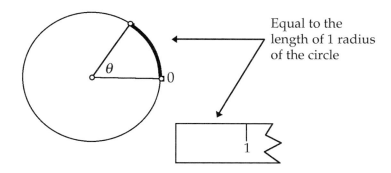

Equal to the length of 1 radius of the circle

Figure 8.53

How many radians are in a 360° angle? (Hint: Recall from (b) that the paper strip is exactly 2π radii (2πr) long.)

c) 360° = 2π radians

d) How many radians are in an angle of 180°? 90°? 270°? 60°?

d) π radians; π/2 radians; 3π/2 radians; π/3 radians

e) Explain why the following proportion can be used to convert degrees to radians or radians to degrees:

$$\frac{x \text{ degrees}}{360°} = \frac{y \text{ radians}}{2\pi \text{ radians}}.$$

e) If you know the number of degrees in an angle and want to find the measure of the angle in radians, you know that the ratio of the given angle in degrees is to 360 degrees as the measure of the angle in radians is to 2π radians because one full rotation = 360 degrees = 2π radians.

f) Change 130° to radians.

f) $\dfrac{130°}{360°} = \dfrac{x \text{ radians}}{2\pi \text{ radians}} = x = 130(2\pi)/360 \approx 2.27$ **radians**

9. A geometric drawing utility can be used to explore the sine, cosine, and tangent of acute angles of right triangles. Begin this exploration by creating a right triangle.

Use a geometric drawing utility such as Geometer's Sketchpad or Cabri to construct a right triangle.

Step 1. Construct a line segment AB and a perpendicular to segment AB through B.

Step 2. Construct a point C on the line perpendicular to segment AB. Construct segments BC and AC to form a right triangle (see **Figure 8.54**).

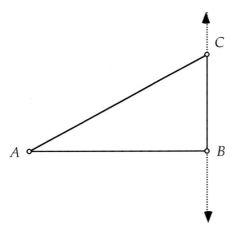

Figure 8.54

Step 3. Hide the perpendicular line.

Step 4. Use the software Measure feature to measure $\angle CAB$ and the ratios $BC{:}AB$, $BC{:}AC$, and $AB{:}AC$ as shown in **Figure 8.55.**

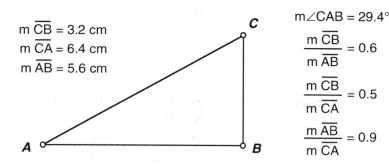

m $\overline{CB}$ = 3.2 cm

m $\overline{CA}$ = 6.4 cm

m $\overline{AB}$ = 5.6 cm

m∠CAB = 29.4°

$$\frac{m \overline{CB}}{m \overline{AB}} = 0.6$$

$$\frac{m \overline{CB}}{m \overline{CA}} = 0.5$$

$$\frac{m \overline{AB}}{m \overline{CA}} = 0.9$$

Figure 8.55

a) Which of the ratios in Figure 8.55 represent tan*A*, sin*A*, and cos*A*?

9. a) $\tan A = \dfrac{CB}{AB}$; $\sin A = \dfrac{CB}{CA}$; $\cos A = \dfrac{AB}{CA}$.

b) Move point C in such a manner that ∠*CAB* gets smaller and smaller. Observe what happens to the tangent, sine, and cosine ratios. From your observations, what do you think cos(0°) is? What are sin(0°) and tan(0°)?

b) cos(0°) = 1; sin(0°) = 0; tan(0°) = 0.

c) Now move point C in such a manner that ∠*CAB* gets very close to 90 degrees. What do you think cos(90°) is? What are sin(90°) and tan(90°)?

c) cos(90°) = 0; sin(90°) = 1; tan(90°) is undefined (students may say that it is a big number).

d) Two angles whose sum is 90 degrees are called **complementary angles.** In a right triangle, the acute angles are complementary. Explain why.

d) The sum of the three angles in a triangle is 180°. One of the angles is 90°, so the sum of the other two must be 90°. Hence, they are complementary angles.

e) Now measure ∠*ACB* and the tangent, sine, and cosine ratios for that angle. As you move point C, what do you notice about the relationship between ∠*ACB* and ∠*CAB*? What do you notice about the sine and cosine ratios of the two angles?

e) The sum of ∠*ACB* and ∠*CAB* is always 90 degrees. The sine of ∠*ACB* = cosine of ∠*CAB*, and the cosine of ∠*ACB* = sine of ∠*CAB*.

f) What can you conclude about the sine and cosine of two angles that are complementary?

f) The sine of an angle is equal to the cosine of its complementary angle.

III. Additional Practice *(SE pages 342–344)*

10. Consider **Figure 8.56.**

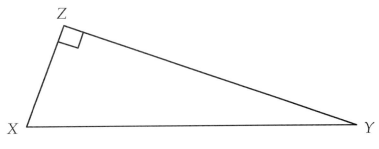

Figure 8.56

    a) Write the side opposite ∠X, the side adjacent to ∠X, and the hypotenuse.

10. a) Opposite, *YZ*; adjacent, *XZ*; hypotenuse, *XY*

    b) Write the side opposite ∠Y, the side adjacent to ∠Y, and the hypotenuse.

b) Opposite, *XZ*; adjacent, *YZ*; hypotenuse, *XY*

For 11–17, use trigonometric ratios to find the indicated side(s) of the right triangle.

11. See **Figure 8.57.**

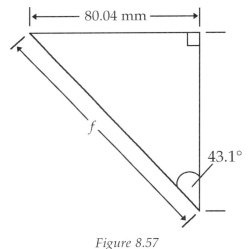

Figure 8.57

11. sin(43.1°) = 80.04/*f*; *f* ≈ 117.1 mm.

12. See **Figure 8.58.**

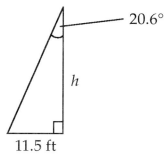

Figure 8.58

12. tan(20.6°) = 11.5/*h*; *h* ≈ 30.6 ft.

13. One acute angle is 57°, and the hypotenuse is 41.4 miles. Find the lengths of the two legs.

13. sin(57°) = *x*/41.4; *x* ≈ 34.7 miles; cos(57°) = *y*/41.4; *y* ≈ 22.5 miles.

14. One acute angle is 32°, and the leg opposite it is 13.2 cm. Find the lengths of the other leg and the hypotenuse.

14. sin(32°) = 13.2/*x*; *x* ≈ 24.9 cm; tan(32°) = 13.2/*y*; *y* ≈ 21.1 cm.

15. See **Figure 8.59.**

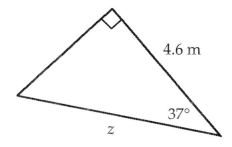

Figure 8.59

15. cos(37°) = 4.6/*z*; *z* ≈ 5.8 m.

16. One acute angle is 41.5°, and the leg adjacent to it is 135 ft. Find the lengths of the other leg and the hypotenuse.

16. tan(41.5°) = *x*/135; *x* ≈ 119 ft; cos(41.5°) = 135/*y*; *y* ≈ 180 ft.

17. One acute angle is 10.1°, and the hypotenuse is 0.138 mm. Find the lengths of the two legs.

17. sin(10.1°) – *x*/0.138; *x* ≈ 0.024 mm; cos(10.1°) = *y*/0.138; *y* ≈ 0.136 mm.

18. Find dimension *d* in **Figure 8.60.**

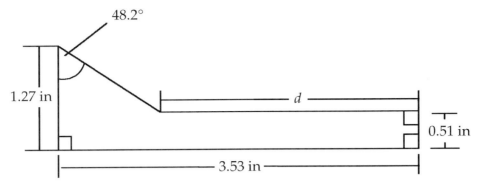

Figure 8.60

18.

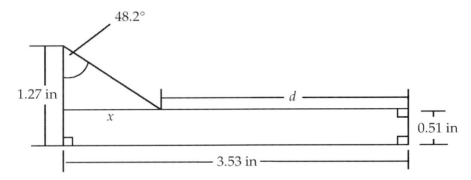

To find *x*: tan(48.2°) = *x*/(1.27 − 0.51); *x* ≈ 0.85 in. Then *d* = 3.53 − 0.85 ≈ 2.68 inches.

For 19–23, find the value of *x* that makes each statement true.

19. sin(10°) = cos*x*.

19. *x* = 80°.

20. cos(35°) = sin*x*.

20. *x* = 55°.

21. sin(36°10′) = cos*x*.

21. *x* = 53°50′.

22. cos(83.7°) = sin*x*.

22. *x* = 6.3°.

23. cos(*y* − 20)° = sin*x*.

23. *x* = (90 − (*y* − 20))° = (110 − *y*)°.

24. Consider the four triangles in **Figure 8.61.** Name the triangles in which the angles θ are equal in measure.

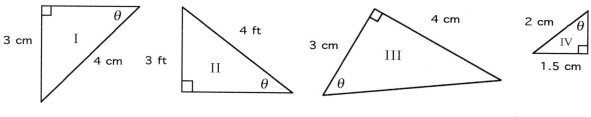

Figure 8.61

24. In triangles I, II, and IV, $\sin\theta = 3/4$, so the angles θ in each of these triangles are equal in measure.

Section 8.4 Modeling with Right Triangles

What You Need to Know

- How to find the sine, cosine, and tangent of a given angle

- How to use the sine, cosine, and tangent of an angle to find a missing side of a right triangle when given one side

What You Will Learn

- To solve a right triangle

- To use right triangle trigonometry to solve real-world problems

Materials

- Protractor

- Straw

- String and weight

- Meter sticks or tape measure

In the first three sections of this chapter, we explored two methods of determining heights of objects that could not be measured directly. In the first method, we used the properties of similar triangles, and in the second, we used the trigonometric ratios. As we examined the two methods carefully, we noticed how they were alike and how they were different. In this section, we will use the tools we learned in the first two sections to "solve right triangles." This technique will be used to solve a variety of real-world problems.

We will begin this section by making a device called a clinometer. We will then use the clinometer to find the height of something we are unable to measure directly such as a tree, a building, or a tall pole.

Discovery 8.3 Build and Use a Clinometer *(SE pages 346–347)*

Materials: Protractor, straw, string, weight

1. With a partner, begin constructing your clinometer by taping a weighted string to the vertex point of a protractor. Then tape a drinking straw to the straight edge of the protractor as shown in **Figure 8.62.**

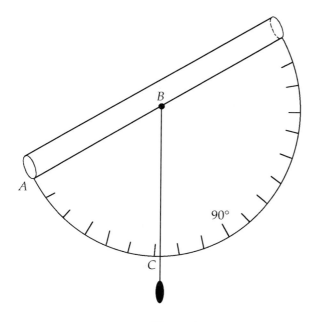

Figure 8.62

To use this device, hold it so that when you look through the straw you see the top of the object you want to measure. The angle formed by the weighted string and the straw (∠ABC) is the complement of the desired angle. Use **Figure 8.63** to help you explain why this is so.

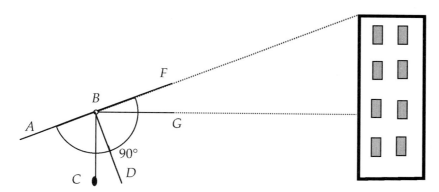

Figure 8.63

1. **∠ABC is the angle measured, ∠FBG is the measurement that is needed. ∠ABC + ∠CBD = 90°; ∠GBD + ∠CBD = 90°. Hence, ∠ABC + ∠CBD = ∠GBD + ∠CBD. Subtracting equals from equals, ∠ABC = ∠GBD. ∠GBD + ∠FBG = 90°. Then by substitution, ∠ABC + ∠FBG = 90°. By definition, ∠GBD and ∠FBG are complementary angles.**

2. Choose a tall object of measure. It should be tall enough so that you can't measure it directly. To find the height of your chosen object,

- measure the height of your eye above the ground
- measure the distance from where you are standing to the object
- use your clinometer to find the desired angle of elevation
- calculate the approximate height of the object

By drawing pictures and showing your calculations, explain how you determined the object's height.

2.

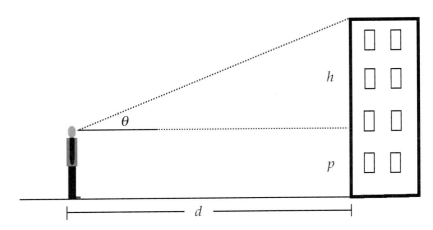

To calculate the height of the object, first use the clinometer to find the complement of θ. Once θ is known, the height of the object (above eye level) h can be calculated.

$\tan\theta = \dfrac{h}{d}$. Then the desired height of the object is $h + p$ (the height of the person's eye above the ground).

Sample calculation:

Height of the eye above the ground: $p = 6$ ft

Distance from the object: $d = 15$ ft

Reading on the clinometer: $20°$

The desired angle of elevation: $\theta = 70°$

To find the height of the object above eye level:

$\tan(70°) = \dfrac{h}{15}$

$h = 41$ ft

The height of the object $= 41$ ft $+ 6$ ft $= 47$ ft.

Finding Acute Angles

At this point, we know how to calculate trigonometric ratios for specific angles and use them to solve problems. We have the tools to find a missing side of a right triangle when one of the sides and one of the acute angles is given. But what if we know two sides and are asked to find one of the acute angles of the triangle?

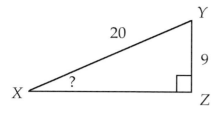

Figure 8.64

Consider the triangle in **Figure 8.64.** Because the leg opposite $\angle X$ and the hypotenuse are known, we can find $\sin X$.

$$\sin X = \frac{9}{20}$$
$$\sin X = 0.45$$

To solve this equation for $\angle X$, we can use the $\sin^{-1}$ key on a calculator (see **Figure 8.65**).

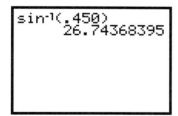

Figure 8.65

To the nearest tenth, $\angle X = \sin^{-1}(0.45) \approx 26.7°$.

We can read $y = \sin^{-1}(0.45)$ as "y is the angle whose sine is 0.45."

Example 6

If $\tan\theta = \dfrac{8}{6}$, find θ.

Solution:

Using the $\tan^{-1}$ key on a calculator,

$$\theta = \tan^{-1}\left(\frac{8}{6}\right) \approx 53.1°.$$

Solving Right Triangles

To **solve a right triangle** means to calculate the measures of each of the sides and each of the angles of the triangle. We now have the background in right triangle trigonometry to find the missing sides and angles of right triangles.

Example 7

Solve the right triangle with a leg of length 2.3 km and hypotenuse of length 9.1 km.

Solution:

Draw a sketch of the information (see **Figure 8.66**).

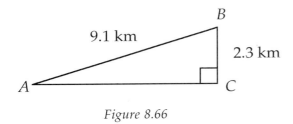

Figure 8.66

In the problem, we are given two sides and the right angle. We need to find the length of AC and the measures of $\angle A$ and $\angle B$.

We know $\sin A = \dfrac{2.3}{9.1}$, so $\angle A = \sin^{-1}\left(\dfrac{2.3}{9.1}\right) \approx 14.6°$. And because angles A and B are complementary, $\angle B = 90° - 14.6° = 75.4°$.

We can find the length of leg AC by using the Pythagorean theorem or trigonometric ratios (once we have found an angle).

Pythagorean theorem:

$(AC)^2 + (BC)^2 = (AB)^2$

$(AC)^2 + (2.3)^2 = (9.1)^2$

$(AC)^2 = 77.52$

$AC \approx 8.8$ km

Trigonometric ratios:

$\tan A = \dfrac{2.3}{AC}$

$AC = \dfrac{2.3}{\tan(14.6)}$

$AC \approx 8.8$ km

Example 8

Solve the right triangle in **Figure 8.67.**

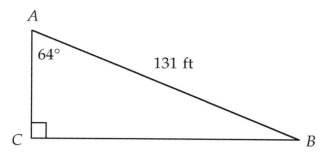

Figure 8.67

Solution:

In the problem, we are given two angles and the length of the hypotenuse. We need to find the measure of $\angle B$ and the lengths of the two legs, AC and BC.

Because we know two angles of $\triangle ABC$, we can find the third angle.

$\angle B = 180 - (90 + 64) = 26°$.

We can find both legs by using trigonometric ratios:

$$\sin(64°) = \frac{BC}{131} \qquad \sin(26°) = \frac{AC}{131}$$
$$BC \approx 117.7 \text{ ft} \qquad AC \approx 57.4 \text{ ft}$$

Example 9

In **Figure 8.68,** find the measures of angles C and B and the distance h between holes A and C.

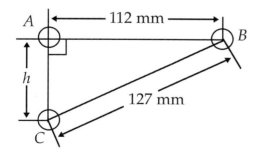

Figure 8.68

Solution:

To find $\angle C$:

$$\sin C = \frac{112}{127}$$
$$\angle C = \sin^{-1}(0.8819) \approx 61.9°$$

To find $\angle B$:

$$\cos B = \frac{112}{127}$$
$$\angle B = \cos^{-1}(0.8819) \approx 28.1°$$

To find the length of h, either trigonometric ratios or the Pythagorean theorem can be used.

Pythagorean theorem:

$h^2 = 127^2 - 112^2$

$h \approx 59.9$ mm

Trigonometric ratios:

$\sin(28.1°) = \dfrac{h}{127}$

$h \approx 59.8$ mm

Exercises 8.4

I. Investigations *(SE pages 351–354)*

1. As the soil was removed from the north end of the Leaning Tower of Pisa, scientists used more than 120 different devices to gauge what was happening to the tower. Included among these devices were three very precise plumb lines.

   The angle of lean of the tower is the angle measured from the vertical to the tower itself (see **Figure 8.69**).

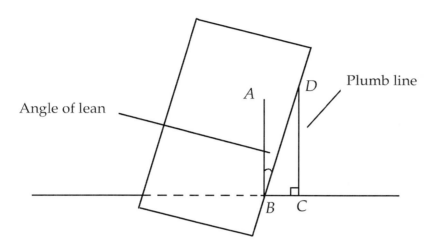

Figure 8.69

   a) Suppose the length of the plumb line from point D on the tower to the ground is 163.7 feet, and it touches the ground 15.25 feet from the ground level of the building. Find the angle of lean of the tower.

1. **a)** Let $\theta = \angle DBC$, the angle the building makes with the ground. Then $\tan(\theta) = 163.7/15.25$; $\theta = \tan^{-1}(10.73) \approx 84.7°$. If $\angle DBC$ is approximately 84.7°, then $\angle ABD = 90° - 84.7° = 5.3°$.

   b) Find the distance along the side of the tower from the ground to where the plumb line is suspended from the building.

   b) The distance from points B to D can be determined in several ways. Using trigonometry: $\sin(84.7°) = 163.7/x$; $x \approx 164.4$ ft. Using the Pythagorean theorem: $x^2 = 163.7^2 + 15.25^2$; $x \approx 164.4$ ft.

2. A tall tower is supported by a cable attached at a point 175 feet above the ground. The other end of the cable is secured to the ground at a point 100 feet from the base of the tower (see **Figure 8.70**).

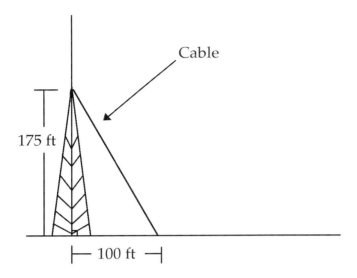

Figure 8.70

a) Assuming the cable is taut, what angle does the cable make with the ground?

2. **a) Let α = the angle the cable makes with the ground. Then tan(α) = 175/100; α = tan^{-1}(1.75) $\approx$ 60.3°.**

b) What angle does the cable make with the tower?

b) 90° – 60.3° = 29.7°

c) How long is the cable?

c) Using the Pythagorean theorem: 175^2 + 100^2 = (length of cable) ^{2}; the length of the cable $\approx$ 202 feet.

3. Hypsometers are instruments used by the U.S. Forestry Service to measure the heights of trees. They also provide information about slopes of roads and pieces of land. There are many different types of hypsometers, but all operate on the principle illustrated in **Figure 8.71.**

In actuality, most of the work is done by the hypsometer. All that the forester has to do is measure a desired horizontal distance from the base of the tree, select an appropriate distance scale on the hypsometer, sight the top of the tree, pull the trigger on the meter, read the height from the scale, site the base of the tree, read the height from the scale, and add the two heights together.

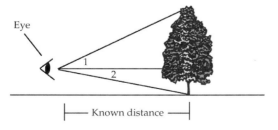

Figure 8.71

Suppose a forester standing 20 meters from a tree finds $\angle 1 = 47.2°$ and $\angle 2 = 5.4°$. Find the height of the tree.

3. **tan(47.2°) = upper part of the tree/known distance = $x/20$; $x \approx 21.6$ meters.**
    tan(5.4°) = lower part of the tree/known distance = $y/20$; $y \approx 1.9$ meters.

    The height of the tree = 21.6 m + 1.9 m = 23.5 meters.

4. In some cities and states, building codes regulate the design of staircases and ramps both inside and outside buildings. These codes are mandated for safety and comfort reasons and represent minimum standards. It's up to the architect and builder to comply with them.

    a) Each step in a typical staircase has an 11-inch run and a 7-inch rise. Find the angle of elevation of the staircase (see **Figure 8.72**).

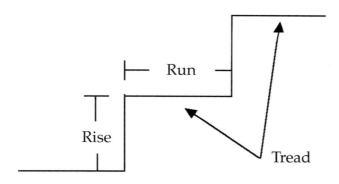

Figure 8.72

4. **a) tan$x$ = 7/11; x = tan^{-1}(7/11); $x \approx 32°$.**

    b) The Americans with Disabilities Act (ADA) of 1990 states that the slope of any ramp must be no greater than 1:12. (This means that for each foot of rise in the ramp, there must be at least 12 feet of run.) What is the maximum angle of elevation for a ramp that meets the specifications for the ADA?

    b)

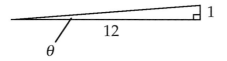

tanθ = 1/12; $\theta \approx 4.76°$.

c) For safety reasons, the minimum depth of a stair tread is often mandated. **Figure 8.73** shows a single stair tread from a spiral staircase. Notice that the tread depth is not constant but increases as the distance from the innermost point of the tread (A) increases.

If code mandates that the tread depth be at least 8 inches at a distance of 15 inches from the innermost point of the tread, what is the minimum angle θ that is required for such a staircase?

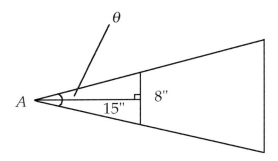

Figure 8.73

c) $\tan\left(\dfrac{\theta}{2}\right) = \dfrac{4}{15}$

$\dfrac{\theta}{2} \approx 14.9°$

$\theta \approx 29.8°$

5. The tape-sine method is used in agricultural surveying to lay out and measure angles. The procedure for measuring an angle is shown in **Figure 8.74.**

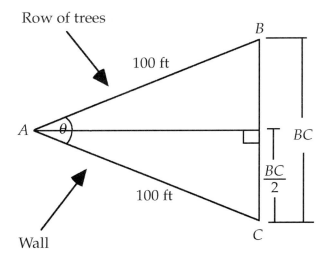

Figure 8.74

Figure 8.74 shows an existing angle (θ) formed by a wall and a line of trees. To find the measure of θ, the tape-sine method says to lay out (measure) an equal distance along each side of the angle (100 feet if possible), then measure the distance BC. The distances $\dfrac{BC}{2}$ and AC form two sides of a right triangle.

a) Suppose $BC = 72.6$ feet. Find θ.

5. a) $\sin(\theta/2) = (BC/2)/100$; $\sin(\theta/2) = 36.3/100$; $\theta/2 \approx 21.3°$; $\theta \approx 42.6°$.

b) Suppose you need to establish a fence at 44° to an existing fence (see **Figure 8.75**). Explain how you would use the tape-sine method to lay out the desired angle.

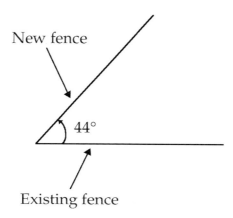

Figure 8.75

b) Using Figure 8.74, find a distance $BC/2$ that when combined with the hypotenuse (say 100 ft) will form the correct ratio for 22° (half the desired 44° angle). To lay out the angle, mark an arc with a radius of 100 feet and lay out an arc with radius BC. A line through the intersection of the two arcs will establish the desired 44° angle.

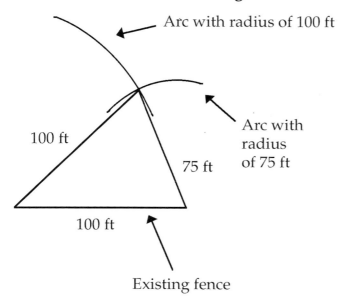

II. Projects and Group Activities *(SE pages 355–356)*

6. Materials: Protractor, cardboard, straw, string, weight, and tape

   Build a simple hypsometer using a straw, a piece of cardboard, a piece of string, and a weight (see **Figure 8.76**).

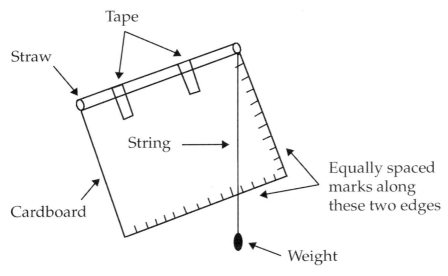

Figure 8.76

   Use your hypsometer to measure the height of a tree or building (see **Figure 8.77**).

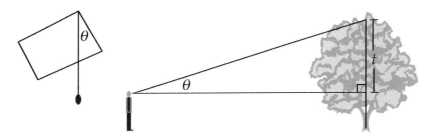

Figure 8.77

   In a paper, describe why your hypsometer works and how you used it to find the height of a tree. Be sure to use mathematics to support your reasoning.

6. **The markings on the hypsometer give two sides (opposite and adjacent) of a right triangle. The triangle on the hypsometer is similar to the right triangle whose leg opposite is part of the unknown height as shown in Figure 8.77.**

   Sample calculation: Suppose the hypsometer reading is shown in the following figure, the person's height is 6 feet, and the distance from the tree is 24 feet. The following proportion can then be used to find the height t of the tree above eye level.

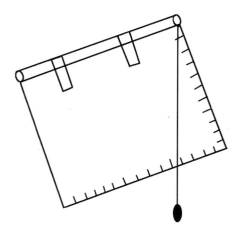

$\dfrac{4.5}{t} = \dfrac{11}{24}$; $t \approx 10$ ft. The height of the tree is then 10 ft + 6 ft = 16 feet.

7. In this activity, you will compare the earth-moon distance to the earth-sun distance just as it was done almost 2000 years ago by the astronomer Aristarchus. Some advance planning is necessary to obtain your calculations, as the activity can only be done when the moon is half-full (first or last quarter) and is visible in the sky along with the sun.

a) Observe both the sun and the moon, determine the angle between them (θ in **Figure 8.78**), then find the ratio of the earth-moon distance to the earth-sun distance.

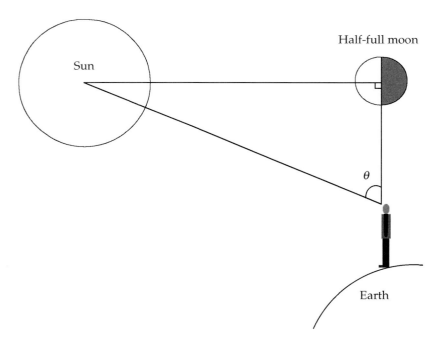

Figure 8.78

7. a) **Answers here will vary widely. The actual angle is 89.83°. The ratio of the two distances is found by finding the cosine of θ and depends on the estimated angle. While the correct ratio is about 0.0026 (1:382), Aristarchus' calculation was about 0.05 (1/19).**

b) Explain how you found this ratio.

b) By finding cos θ.

c) The approximate distance from the earth to the moon is 239,000 miles. Using your ratio from (a), find the distance to the sun.

c) cos θ = 239,000/s; the distance to the sun = 239,000/cos θ.

III. Additional Practice *(SE pages 356–361)*

8. The Monongahela Incline, one of the oldest and steepest incline railways in the region, is located on the south bank of the Monongahela River in Pittsburgh, Pennsylvania. It was built in 1870 and provided transportation from the river valley to the Mount Washington residential area.

   a) The length of the incline is 635 feet, and the height from the station at the bottom of the mountain to the one at the top is 367.39 feet. Find the angle of elevation of the incline (see **Figure 8.79**).

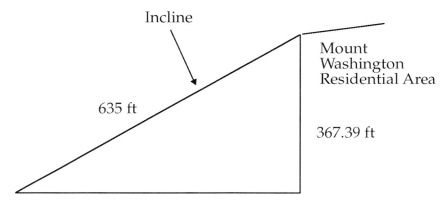

Figure 8.79

8. **a) sin θ = 367.39/635; θ = sin^{-1}(367.39/635); $\theta \approx$ 35.4°.**

   b) The incline travels at a rate of 6 miles per hour. Assuming that it travels at a constant rate (ignoring acceleration and deceleration for starting and stopping), how many minutes does it take for the incline to travel from the river valley station to the station on Mount Washington?

   b) 635 ft $\left(\dfrac{1\ \text{mile}}{5280\ \text{ft}}\right)\left(\dfrac{1\ \text{hr}}{6\ \text{miles}}\right)\left(\dfrac{60\ \text{min}}{1\ \text{hr}}\right)$ = 1.2 minutes.

9. If an object is located below the horizontal, the angle from the horizontal down to the line of sight is called the **angle of depression.** In **Figure 8.80,** the horizontal is represented by the line d and the line of sight by line l.

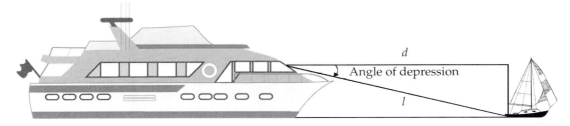

Figure 8.80

a) The angle of depression of the waterline of a sailboat from the bridge of a ship 325 feet above sea level is 23°. How far is the sailboat from the ship?

9. a) tan(23°) = 325/d; $d \approx$ 766 feet.

b) A surveyor is standing at the top of a 68-meter-high cliff. She sights an object 90 meters from the bottom of a cliff. What is the angle of depression?

b) tan θ = 68/90; θ = tan^{-1}(68/90); $\theta \approx$ 37°.

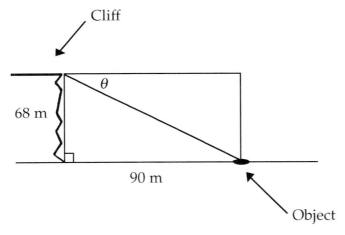

10. **Figure 8.81** shows a truss used in the construction of a building. This particular truss is a 1:4 truss, which means that for every four units of horizontal distance of construction material, the truss rises or falls one unit.

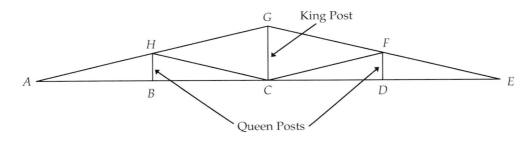

Figure 8.81

a) Assume the truss is 16 feet long, *C* is the midpoint of *AE*, and *B* and *D* are midpoints of *AC* and *CE*, respectively. How tall is the king post? How tall are the queen posts?

10. a) 2 feet; 1 foot

b) At what angle do *AE* and *AG* meet?

b) tan*A* = 1/4; ∠*A* ≈ 14°.

c) Find ∠*HGC*.

c) ∠*HGC* and ∠*GAC* are complementary. So if ∠*GAC* (from (b)) = 14°, then ∠*HGC* = 76°.

11. In **Figure 8.82**, the radius of the circle inside the hex-head bolt is 1/4 inch. What is the length of one side of the bolt?

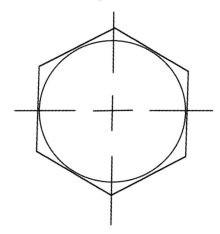

Figure 8.82

11.

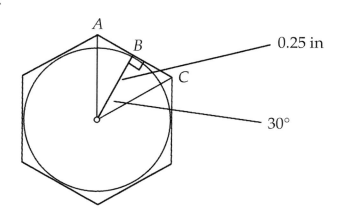

One-half the side of the hexagon (*AC*) can be found using special 30-60-90-degree angle relationships or by using trigonometric ratios.

tan(30°) = *BC*/0.25; *BC* ≈ 0.144 inches. The side of the hexagon is 2(0.144) = 0.29 inches.

12. When a container of sand is dumped on a flat surface such as a table top, it forms a pile such as the one shown in **Figure 8.83.** The angle that the sand makes with the surface is called the **angle of repose** and is denoted by the symbol ϕ. Note that the angle of repose does not change for a given medium. If more of the medium is dumped onto the pile, the pile spreads outward with the angle remaining unchanged.

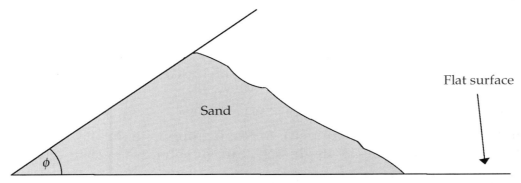

Figure 8.83

a) Find the angle of repose for sand if the pile on the flat surface is 8 inches high and 34 inches in diameter.

12. a) The radius of the pile is 17 in. So tan(angle of repose) = 8/17 and the angle is approximately 25°.

b) The angle of repose for gravel is approximately 30°. If the diameter of a pile of gravel is 16 meters, how high is the pile?

b) tan(30°) = h/8; $h \approx$ 4.6 meters.

13. Agricultural surveying involves measuring angles and distances with simple instruments. For most surveys, the intent is to find the horizontal distance between two points. To do this, the "slope distance" is measured using a surveyor's tape, the percent slope (the slope expressed as a percent) is measured using a hand level or a surveying level, and then the horizontal distance is calculated.

a) What is the angle formed by the horizontal and the ground if the ground slope is 11%?

13. a) $\tan \theta$ = 11/100; $x \approx$ 6.3°.

b) What is the true horizontal distance (THD) if the slope distance (SD) is 375.24 feet?

b)

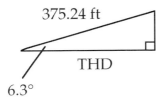

375.24 ft

THD

6.3°

cos(6.3°) = THD/375.24; THD ≈ 373.0 ft.

14. A machinist used a sine bar to set an angle. If the gage block height was 5.764 inches on a 10-inch sine bar, what angle was set?

14. $\sin\theta$ = 5.764/10; $\theta \approx$ 35.2°.

For 15–16, solve the right triangles with the following properties.

15. Isosceles with a hypotenuse of 6 cm

15. Angles: 45°, 45°, 90° (given); Sides: Hypotenuse = 6 cm (given); legs = 4.2 cm.

To find the two equal legs, the Pythagorean theorem can be used or the relationships in 45°-45°-90° triangles, in which case, leg $= \dfrac{6}{\sqrt{2}} \approx$ 4.2 cm.

16. An acute angle of 35° and hypotenuse of 21 inches

16. Angles: 55°, 35° (given), 90° (given); Sides: leg opposite the 35° angle = 12 in, leg adjacent to the 35° angle = 17.2 in; hypotenuse = 21 in (given).

The legs can be found using the sine and cosine of the 35° angle: sin(35°) = opp/21, opp $\approx$ 12.0 in; cos(35) = adj/21, adj $\approx$ 17.2 in.

17. Find $\angle x$ in **Figure 8.84.**

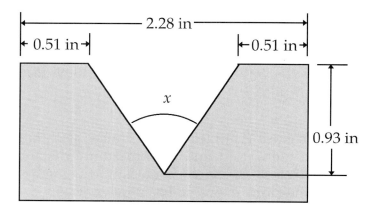

Figure 8.84

17.

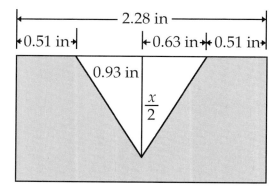

Using the figure, we know that tan(x/2) = 0.63/0.93; x/2 $\approx$ 34.1° and $\angle x \approx$ 68.2°.

For 18–19, use **Figure 8.85** and the information given to solve the triangle.

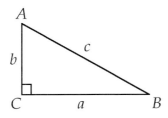

Figure 8.85

18. Given: $b = 7$, $\angle B = 48°$; find $\angle A$, a, and c.

18. $\angle A = 180 - (90 + 48) = 42°$.

   To find c: $\sin(48°) = 7/c$; $c \approx 9.4$.

   To find a: $\tan(48°) = 7/a$; $a \approx 6.3$.

19. Given: $a = 4.8$, $b = 7.2$; find $\angle A$, $\angle B$, and c.

19. To find $\angle A$: $\tan A = 4.8/7.2$, $\angle A \approx 33.7°$.

   To find $\angle B$: $\tan B = 7.2/4.8$, $\angle B \approx 56.3°$.

   To find c: $c^2 = 7.2^2 + 4.8^2$, $c \approx 8.7$.

20. To determine the width of a canyon, a surveyor stood at a location on one side of the ravine where the angle of elevation of the top of a tree located on the opposite side of the ravine was 28°. Moving away from the ravine 100 feet directly behind the first location, the angle of elevation of the top of the tree was 18°. Find the width of the ravine. (Hint: Find two equations with two unknowns, and solve the system of equations. See **Figure 8.86.**)

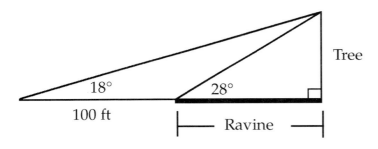

Figure 8.86

20. Let h = height of the tree and w = width of the ravine.

   $\tan(28°) = h/w$; $\tan(18°) = h/(w + 100)$.

   Solving each equation for h: $h = 0.5317w$ and $h = 0.3249(w + 100)$.

   Substituting: $0.5317w = 0.3249(w + 100)$

   $w \approx 157$ feet.

21. From point *A* in **Figure 8.87,** the angle of elevation of the top of a nearby hill is 12°, and from a point *B* directly behind *A*, the angle of elevation is 9.5°. If the distance between points *A* and *B* is 500 feet, find the height of the hill.

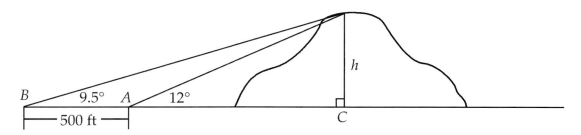

Figure 8.87

21. **Let *h* = height of the hill and *d* = distance from *A* to *C*.**

    $\tan(12°) = h/d$; $\tan(9.5°) = h/(d + 500)$.

    Solving each equation for *d*: $d = h/\tan(12°)$; $d = (h/\tan(9.5°)) - 500$.

    Substituting: $h/\tan(12°) = (h/\tan(9.5°)) - 500$

    $h/0.1673 - h/0.2126 = 500$.

    Multiplying by $(0.1673)(0.2126)$: $0.2126h - 0.1673h = (500)(0.1673)(0.2126)$

    $0.0453h = 17.78$; $h \approx 393$ feet.

Section 8.5 Trigonometric Ratios for Non-Acute Angles

What You Need to Know

- How to find the sine, cosine, and tangent of a given acute angle

What You Will Learn

- To determine the sign of the sine, cosine, and tangent of a non-acute angle

- To solve real-world problems that can be modeled with oblique triangles

Materials

- None

Because trigonometry can be used to solve non-right triangles where angles could be obtuse, it becomes necessary to find trigonometric ratios for angles greater than 90° (see Exercise 5). As you will soon discover, the signs of the trigonometric ratios for these angles are not always positive as they are for acute angles. By examining angles on a coordinate plane, the signs of the ratios for any angle can be found.

When referring to an angle, we often name one of the two sides of the angle the **initial side** and the other side the **terminal side.** If the rotation from the initial side to the terminal side is counterclockwise, the angle is positive (see **Figure 8.88a**). If the rotation is clockwise, the angle is negative (see **Figure 8.88b**).

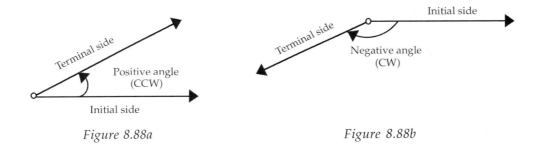

Figure 8.88a   *Figure 8.88b*

An angle is said to be in **standard position** when its vertex is at the origin and its initial side coincides with the positive x-axis (see **Figure 8.89**).

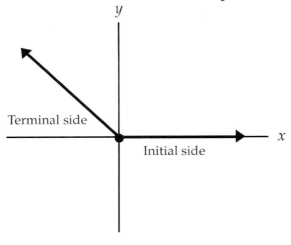

Figure 8.89

Figures 8.90a and **8.90b** show angles in standard position. Each figure shows a point with coordinates (x, y) on the terminal side of the angle. The distance from the origin to (x, y) is labeled r. To find r, we can use the Pythagorean theorem. And because $r = \sqrt{x^2 + y^2}$, it is always a non-negative value.

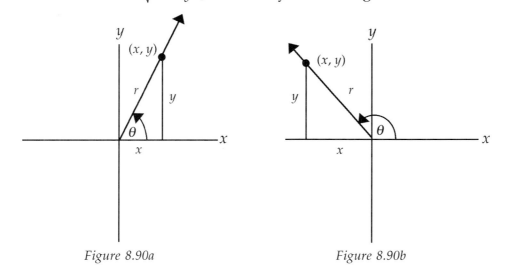

Figure 8.90a   *Figure 8.90b*

Referring to Figures 8.90a and 8.90b, the sine, cosine, and tangent ratios can be defined as follows:

$$\sin\theta = \frac{y}{r}; \quad \cos\theta = \frac{x}{r}; \quad \tan\theta = \frac{y}{x} \quad \text{when the denominators do not equal 0.}$$

In Figure 8.90a, the terminal side of the angle is in Quadrant I, where x is a positive number and r is a positive number. So $\cos\theta$ is a positive number. But in Figure 8.90b, x is a negative number and r is a positive number, so $\cos\theta$ is a negative number.

Example 10

Find the sine, cosine, and tangent of a positive angle θ if the point $(-6, -8)$ is on its terminal side.

Solution:

Figure 8.91 shows the point $(-6, -8)$ on the terminal side of angle θ.

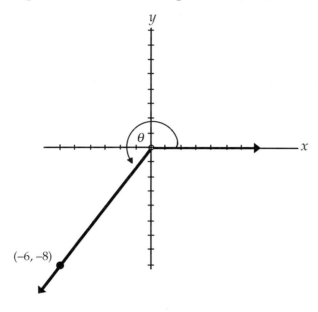

Figure 8.91

Because $(x, y) = (-6, -8), r = \sqrt{(-6)^2 + (-8)^2} = 10.$

$$\sin\theta = \frac{y}{r} = \frac{-8}{10} = -\frac{4}{5}; \quad \cos\theta = \frac{x}{r} = \frac{-6}{10} = -\frac{3}{5}; \quad \tan\theta = \frac{y}{x} = \frac{-8}{-6} = \frac{4}{3}$$

Exercises 8.5

I. Investigations *(SE pages 365–369)*

1. a) Find the sine, cosine, and tangent of a positive angle θ if the point (5, –12) is on its terminal side.

1. a)

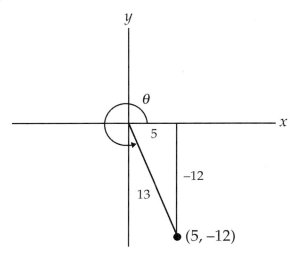

$$\sin\theta = y/r = -12/13; \cos\theta = x/r = 5/13; \tan\theta = y/x = -12/5.$$

   b) Find the cosecant, secant, and cotangent of a positive angle θ if the point (–3, 4) is on its terminal side.

b)

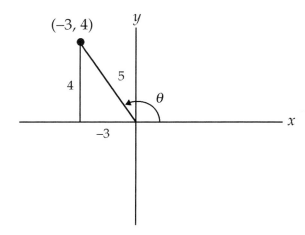

$$\csc\theta = r/y = 5/4; \sec\theta = r/x = 5/-3; \cot\theta = x/y = -3/4.$$

c) Find the sine, cosine, and tangent of a negative angle θ if the point $(-1, -2)$ is on its terminal side.

c)

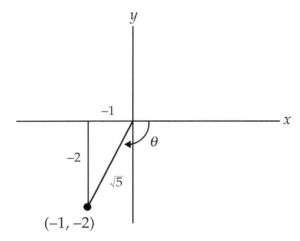

$\sin\theta = y/r = \dfrac{-2}{\sqrt{5}}$ or $\dfrac{-2\sqrt{5}}{5}$; $\cos\theta = x/r = \dfrac{-1}{\sqrt{5}}$ or $\dfrac{-\sqrt{5}}{5}$; $\tan\theta = y/x = -2/-1 = 2$.

2. a) Consider an angle with the point $(-3, 4)$ on its terminal side. In which quadrant does the terminal side lie?

2. a) Quadrant II

b) Is the cosine of this angle positive or negative? Explain why.

b) The cosine of the angle is negative because

$$\cos\theta = \frac{x}{r} = \frac{\text{negative}}{\text{positive}} = \text{negative}.$$

c) Consider other angles with terminal sides in the same quadrant. What can you say about the sign of the cosine of these angles?

c) The cosine of any angle in Quadrant II is negative.

d) Using reasoning similar to that used in (a)–(c), complete **Table 8.6,** showing whether the indicated trigonometric ratios for an angle in standard position are positive or negative.

Terminal Side Quadrant	Sine	Cosine	Tangent
I			
II			
III			
IV			

Table 8.6

d)

Terminal Side Quadrant	Sine	Cosine	Tangent
I	Positive	Positive	Positive
II	Positive	Negative	Negative
III	Negative	Negative	Positive
IV	Negative	Positive	Negative

3. Complete **Table 8.7,** showing whether the indicated trigonometric ratios for an angle in standard position are positive or negative.

Terminal Side Quadrant	Cosecant	Secant	Cotangent
I			
II			
III			
IV			

Table 8.7

3.

Terminal Side Quadrant	Cosecant	Secant	Cotangent
I	Positive	Positive	Positive
II	Positive	Negative	Negative
III	Negative	Negative	Positive
IV	Negative	Positive	Negative

4. **Quadrantal angles** are angles whose terminal side lies on either the *x*-axis or the *y*-axis. Complete **Table 8.8** by finding the values of the sine, cosine, and tangent for the indicated quadrantal angles.

Angle	Sine	Cosine	Tangent
0°			
90°			
180°			
270°			

Table 8.8

4.

Angle	Sine	Cosine	Tangent
0°	0	1	0
90°	1	0	Undefined
180°	0	–1	0
270°	–1	0	Undefined

5. A triangle that is not a right triangle is called an **oblique triangle.** An oblique triangle can have three acute angles or two acute angles and one obtuse angle. To solve an oblique triangle, the measures of three parts must be known and at least one part must be a side.

   If three sides or two sides and an included angle of an oblique triangle are given, the **Law of Cosines** can be used to solve the triangle.

   Law of Cosines

   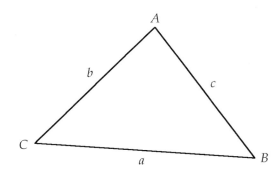

   Figure 8.92

   For a triangle with labeling as in **Figure 8.92,**

   $a^2 = b^2 + c^2 - 2bc \cos A$

   $b^2 = a^2 + c^2 - 2ac \cos B$

   $c^2 = a^2 + b^2 - 2ab \cos C$

a) Use the Law of Cosines to find the distance BC in Figure **8.93**.

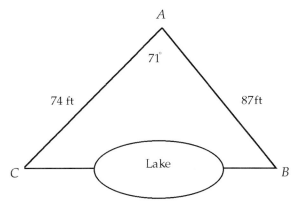

Figure 8.93

5. a) $(BC)^2 = 74^2 + 87^2 - (2)(74)(87)(\cos(71°))$; $(BC)^2 \approx 8853$; $BC \approx 94$ feet.

b) Find angles A and B in **Figure 8.94**. (Hint: Recall that angles with measures greater than 90° and less than 180° lie in Quadrant II and their cosines are negative.)

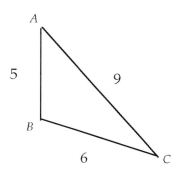

Figure 8.94

b) To find $\angle A$:

$$6^2 = 5^2 + 9^2 - 2 \cdot 5 \cdot 9 \cdot \cos A$$

$$\cos A = \frac{25 + 81 - 36}{2 \cdot 5 \cdot 9}$$

$$\cos A = 0.7778$$

$$A \approx 39°$$

To find $\angle B$:

$$9^2 = 5^2 + 6^2 - 2 \cdot 5 \cdot 6 \cdot \cos B$$

$$\cos B = \frac{25 + 36 - 81}{2 \cdot 5 \cdot 6}$$

$$\cos B = -0.333$$

$$B \approx 109.5°$$

6. A simple proportion exists between the sides and angles of an oblique triangle. This proportion is known as the **Law of Sines.**

Law of Sines

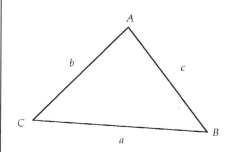

Figure 8.95

Take Note:

The Law of Sines is deceptively simple to use. However, when it is used to find missing angles, it can present problems. These problems arise because the sine ratio is incapable of distinguishing between acute and obtuse angles.

For a triangle with labeling as in **Figure 8.95,**

$$\frac{a}{\sin A} = \frac{b}{\sin B} = \frac{c}{\sin C}.$$

Examine the triangle in **Figure 8.96.** Because only one side of the triangle is given, the Law of Cosines cannot be used to solve the triangle. But two angles and one side are known, so the Law of Sines can be used to solve the triangle.

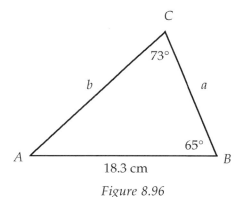

Figure 8.96

For example, to find side b, we know:

$$\frac{b}{\sin(65°)} = \frac{18.3}{\sin(73°)}; b = \frac{18.3(\sin(65°))}{\sin(73°)} \approx 17.3 \text{ cm.}$$

a) Use the Law of Sines to find side a of the triangle in Figure 8.96.

6. **a)** $\angle A = 180° - (73° + 65°) = 42°$. **So,**

$$\frac{a}{\sin(42°)} = \frac{18.3}{\sin(73°)}; a = \frac{18.3(\sin(42°))}{\sin(73°)} \approx 12.8 \text{ cm.}$$

b) Suppose $\triangle RST$ has $s = 14.3$ ft, $\angle T = 86°$, and $\angle S = 48°$. Use the Law of Sines to find side t.

b) $\dfrac{14.3}{\sin(48°)} = \dfrac{t}{\sin(86°)}$; $t = \dfrac{14.3(\sin(86°))}{\sin(48°)} \approx$ **19.2 ft.**

7. If two sides and an angle opposite one of the sides of an oblique triangle are given, the Law of Sines can be used to solve the triangle. This situation is often called the **ambiguous case** because there may be two, one, or no triangles fitting the description. To help understand this situation, consider triangle ABC in which $a = 10$ cm, $b = 15$ cm, and $\angle A = 26°$.

a) Use a ruler and protractor to draw this triangle accurately. Explain why there are two locations possible for vertex B. Make a sketch of each of the two triangles.

7. a) These sample sketches are reduced in size due to space considerations.

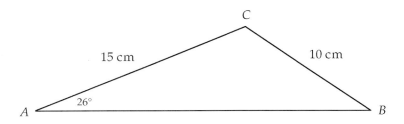

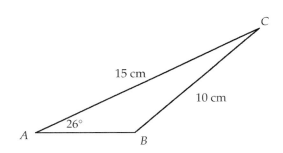

In each case, a, b, and $\angle A$ have the measures given, so two triangles are possible.

b) Use a ruler and protractor to measure the remaining parts of each of your sketches. To help keep track of the triangles, label the larger triangle "1" and the smaller triangle "2."

b) Sample answer: In triangle 1, $\angle C$ is approximately 114°, $\angle B$ is approximately 40°, and side $c \approx 21$ cm. In triangle 2, $\angle C$ is about 15°, $\angle B$ is about 137°, and side $c \approx 6$ cm.

c) Use the Law of Sines to find $\angle B$. How could you use the result to find the other possible $\angle B$?

c) $\dfrac{\sin B}{15} = \dfrac{\sin(26°)}{10}$**, so** $B = \sin^{-1}\left(15\dfrac{\sin(26°)}{10}\right) \approx 41°$**. The other possible value of** $\angle B$ **is** $180° - 41° = 139°$**.**

d) Draw a line from point C perpendicular to side AB. This line is called an altitude of the triangle. Use right triangle trigonometry to find the length of the altitude.

d) The altitude's length is $15\sin(26°) \approx 6.6$ **cm.**

e) In (a)–(c), you've seen a situation where there are two possible solutions. Assume $\angle A$ and side b remain unchanged. For what value(s) of side a will there be exactly one solution? For what value(s) of side a will there be no solutions? (Hint: Use the information from (d) to help formulate your answers.)

e) If a**'s length is the same as the altitude's length, there will be only one triangle possible. There is only one triangle if** $a \geq 15$ **cm because triangle 2 would not be possible. If the length of** a **is less than the altitude, there are no possible solutions because there would be no triangle.**

II. Projects and Group Activities *(SE pages 369–370)*

8. Global Positioning Systems (GPS) are satellite radio positioning systems that provide information to GPS users anywhere on or near the surface of Earth. Civilians can now purchase handheld receivers, some of which cost less than $100. These receivers can relay to the user his or her position coordinates in terms of longitude (position east or west of the prime meridian) and latitude (position north or south of the equator), along with velocity and time information.

One formula that can be used to find the distance between two points on a sphere when the latitude and longitude of the two points are known is as follows:

Distance $= R * a$, where R is the radius of Earth (≈ 4000 miles) and

$a = \cos^{-1}(\cos(\text{lat}A) * \cos(\text{lat}B) * \cos(\text{lon}B - \text{lon}A) + \sin(\text{lat}A) * \sin(\text{lat}B))$.

The value of a must be expressed in radians.

a) Katy Trail State Park in Missouri is a 200-mile biking and hiking trail that parallels the Missouri River as it runs east and west across the middle of the state. Bikers and hikers often find a GPS receiver helpful and convenient in locating their position on the trail. Suppose you begin your hike at Frontier Park in St. Charles (N38°48.403′W90°29.075′) and end it at Jungs Station Rd. (N38°43.451′W90°32.822′). Use the above formula to find out how far you are from your starting point.

8. a) In the solution below, the latitudes and longitudes have been expressed in radians before finding the measure of angle a. If you choose to work in degree mode, then once a is found, it must be expressed in radians prior to finding the distance, $R*a$.

N38°48.403′ = N.677305 radians; W90°29.075′ = W1.57925 radians; N38°43.451′ = N.675864 radians; W90° 32.822′ = 1.58034 radians.

Let A be the starting position and lat$A$ = N.677305 and lon$A$ = W1.57925. Let B be the ending position and lat$B$ = N.675864 and lon$B$ = W1.58034. Hence,

$a = \cos^{-1}(\cos(0.677305)*\cos(0.675864)*\cos(1.58034 - 1.57925) +$

$\sin(0.677305)*\sin(0.675864) = 0.001673$ radians.

The distance can then be calculated. $D = 4000*0.001673$, which is approximately 6.7 miles.

b) Use a GPS receiver to find two locations in your area. Then use the given formula to find the distance between the two points.

b) **Answers will vary.**

III. Additional Practice *(SE pages 370–373)*

9. Determine whether the sine of 160° is positive or negative.

9. A 160° angle is in Quadrant II. Its sine is positive.

10. Determine whether the cosine of 230° is positive or negative.

10. A 230° angle is in Quadrant III. Its cosine is negative.

11. Determine whether the tangent of 190° is positive or negative.

11. A 190° angle is in Quadrant II. Its tangent is positive.

12. Determine whether the sine of –30° is positive or negative.

12. A –30° angle is in Quadrant IV. Its sine is negative.

13. The sine of a given angle x is positive. If $0° \leq x < 360°$, what are the possible values of x?

13. $0° < x < 180°$

For 14–18, use the Law of Cosines.

14. A ship sets sail with a bearing of 45° traveling at 22 miles per hour. A second ship sets sail from the same location at the same time. The second ship has a bearing of 168° and is traveling at 15 miles per hour. Find the distance between the two ships after two hours (see **Figure 8.97**).

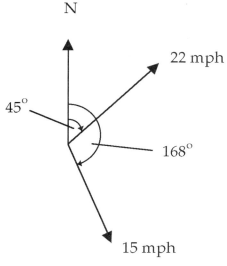

Figure 8.97

14. After two hours, the first ship will be 44 miles from port and the second ship will be 30 miles from port.

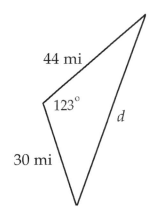

Using the Law of Cosines: $d^2 = 44^2 + 30^2 - 2(44)(30)(\cos(123°)) = 1936 + 900 - (2)(44)(30)(-0.5446) = 4273.74; d \approx 65.4$ miles.

15. Two planes leave Dallas at the same time. The first plane is traveling at 600 miles per hour with a bearing of 128°. The second plane is traveling at 550 miles per hour with a bearing of 234°. After 30 minutes, how far apart are the two planes?

15. **After 1/2 hour, the first plane is 300 miles from the airport, while the second plane is 275 miles from the airport.**

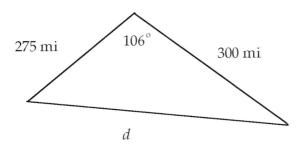

Using the Law of Cosines to find the distance between the two planes:

$d^2 = 300^2 + 275^2 - 2(300)(275)(\cos(106°)) =$
$90{,}000 + 75{,}625 - (2)(300)(275)(-2756) = 211{,}105.2; \; d \approx 459.5 \text{ miles.}$

16. **Figure 8.98** shows three holes drilled in a plate. If the radius of hole A is 4 cm, the radius of hole B is 8 cm, and the radius of hole C is 2.5 cm, find the distance from the center of hole A to the center of hole C.

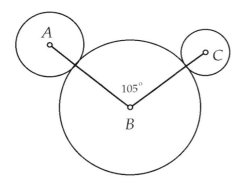

Figure 8.98

16. **Because $AB = 12$ cm, $BC = 10.5$ cm, and $\angle B = 105°$, the Law of Cosines can be used to find AC:**

$(AC)^2 = 12^2 + 10.5^2 - 2(12)(10.5)(\cos(105°)) =$
$144 + 110.25 - (2)(12)(10.5)(-0.2588) = 319.472; \; AC \approx 17.9 \text{ cm.}$

17. Solve for side x in **Figure 8.99.**

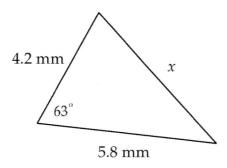

Figure 8.99

17. $x^2 = 4.2^2 + 5.8^2 - (2)(4.2)(5.8)(\cos(63°)) = 29.16$; $x \approx 5.4$ mm.

18. Solve for $\angle x$ in **Figure 8.100.**

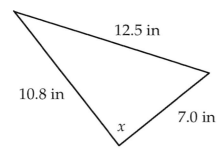

Figure 8.100

18. $12.5^2 = 10.8^2 + 7.0^2 - 2(10.8)(7.0)(\cos(x))$; $\cos x = \dfrac{10.8^2 + 7.0^2 - 12.5^2}{2(10.8)(7.0)}$;

$\cos x = 0.06210$; $x \approx 86.4°$.

For 19–23, use the Law of Sines.

19. A hot air balloon is directly above an extension of a line passing between two ground observers located at points A and B in **Figure 8.101.** The angle of elevation from observer A to the balloon is 22°, and the angle of elevation from observer B to the balloon is 50°. If points A and B are 300 meters apart, how far is the balloon from observer A?

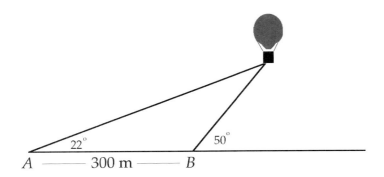

Figure 8.101

19. Assume the point at the balloon is C. Then $\angle ABC = 130°$ and $\angle ACB = 28°$. Using the Law of Sines to find the distance d from the balloon to observer A:

$$\frac{d}{\sin(130°)} = \frac{300}{\sin(28°)}; d = \frac{300(\sin(130°))}{\sin(28°)} \approx 490 \text{ meters.}$$

20. Use the Law of Sines to solve Exercise 20 in Section 8.4.

20.

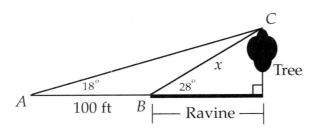

In triangle ABC, angles ABC and ACB are 152° and 10°, respectively.

Using the Law of Sines to find x:
$$\frac{x}{\sin(18°)} = \frac{100}{\sin(10°)}; x = \frac{100(\sin(18°))}{\sin(10°)} \approx 178 \text{ feet.}$$

Using right triangle trigonometry to find the width w of the ravine:
$\cos(28°) = w/178; w \approx 157$ feet.

21. Two tourists located 500 feet apart on opposite sides of the St. Louis Gateway Arch note the angles of elevation to the top of the arch to be 61° and 77°. Find the height of the arch.

21.

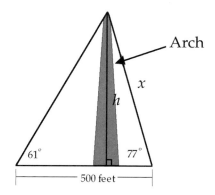

Using the Law of Sines to find x:
$$\frac{x}{\sin(61°)} = \frac{500}{\sin(42°)}; \; x = \frac{500(\sin(61°))}{\sin(42°)} \approx 654 \, \textbf{feet.}$$

Using right triangle trigonometry to find the height: $\sin(77°) = h/654$; $h \approx 637$ **feet.**

22. Solve for $\angle A$ in **Figure 8.102.**

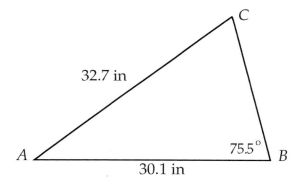

Figure 8.102

22. First find $\angle C$: $\dfrac{32.7}{\sin(75.5°)} = \dfrac{30.1}{\sin C}$; $\sin C = \dfrac{30.1(\sin(75.5°))}{32.7}$; $C \approx 63.0°$. **Then**
$\angle A = 180 - (75.5 + 63.0) = 41.5°$.

23. Solve for the length of side *BC* in **Figure 8.103.**

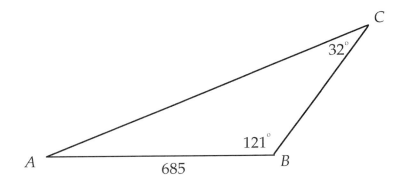

Figure 8.103

23. First find ∠*A*: 180 − (32 + 121) = 27°.

Using the Law of Sines: $\dfrac{685}{\sin(32°)} = \dfrac{BC}{\sin(27°)}; BC = \dfrac{685(\sin(27°))}{\sin(32°)} \approx 587$.

Chapter 8 Summary

Classifying triangles

Pythagorean theorem and its converse

Similar triangles

Trigonometric ratios of acute angles: sine, cosine, tangent

Solving right triangles

Trigonometric ratios of non-acute angles: sine, cosine, tangent

Optional (found in exercises):

      Trigonometric ratios: secant, cosecant, cotangent

      Law of Cosines

      Law of Sines

Chapter 8 Review *(SE pages 375–378)*

1. Write a summary of the important mathematical ideas found in Chapter 8.

1. **Answers will vary. Following are some of the important ideas that should be listed:**

Similar figures have the same shape but not necessarily the same size.

The corresponding angles of similar triangles are equal, and the corresponding sides are proportional.

If the corresponding angles of two triangles are equal, then the triangles are similar. Also, if the corresponding sides of two triangles are in proportion, the triangles are similar.

The properties of similar triangles can be used to solve real-world problems.

Triangles can be classified by their sides: scalene, isosceles, and equilateral. They can also be classified by their angles: acute, obtuse, and right.

Two polygons are similar if their corresponding angles are equal in measure.

Trigonometry is the study of the relationships between the sides and angles of triangles.

Right triangle trigonometry can be used to model real-world situations.

Three of the trigonometry ratios for a right triangle are

$$\text{tangent}\left(\text{which is defined as }\frac{\text{length of the side opposite an acute angle}}{\text{length of the side adjacent to the same angle}}\right);$$

$$\text{sine}\left(\text{which is defined as }\frac{\text{length of the side opposite an acute angle}}{\text{length of the hypotenuse}}\right); \text{ and}$$

$$\text{cosine}\left(\text{which is defined as }\frac{\text{length of the side adjacent to an acute angle}}{\text{length of the hypotenuse}}\right).$$

Trigonometric ratios can be used to solve right triangles. That is, they can be used to calculate all of the sides and angles of the triangle.

The $\sin^{-1}$, $\cos^{-1}$, and $\tan^{-1}$ functions on the calculator can be used to find missing angles when the trigonometry ratio is known.

(Optional) The sign of trigonometric ratios of non-acute angles can be found by placing an angle in standard position and determining where the terminal side is.

(Optional) The Law of Cosines can be used to solve an oblique triangle in which three sides or two sides and an included angle are given.

(Optional) The Law of Sines can be used to solve an oblique triangle in which one side and two angles are given.

2. The sides of $\triangle ABC$ are 45 m, 45 m, and 60 m. The sides of $\triangle MNO$ are 15 ft, 20 ft, and 15 ft. Are the two triangles similar? Explain.

2. Yes, the two triangles are similar because the sides are proportional.

3. Find the distance across the lake in **Figure 8.104.**

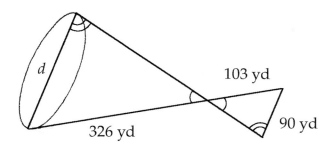

Figure 8.104

3. **Using similar triangles:** $\dfrac{90}{d} = \dfrac{103}{326}$; $d = \dfrac{(90)(326)}{103} \approx 285$ **yards.**

4. During a solar eclipse, the Moon passes between Earth and the Sun. The geometry of a solar eclipse is such that during the period of totality (when the entire Sun is behind the Moon from Earth's perspective), the Moon almost fits over the Sun, that is, they appear to be the same size.

   The average distance from Earth to the Sun is about 1.50×10^{11} m, and the average distance from Earth to the Moon is about 3.85×10^{8} m. If the Moon's diameter is about 3.476×10^{6} m, how large is the Sun?

4. **The Sun is about 1.4×10^{9} m in diameter, or just under 400 times the diameter of the Moon.**

5. If the numbers in (a)–(c) represent the lengths of the sides of a triangle, classify the triangle as right, acute, obtuse, or nonexistent.

   a) 20, 41, 80

5. **a) Not a triangle**

   b) 25, 15, 20

   b) $20^2 + 15^2 = 25^2$, so the triangle is a right triangle.

   c) 11, 9, 8

   c) $11^2 < 9^2 + 8^2$, so the triangle is acute.

6. The length of the hypotenuse of a right triangle is 18 feet, and one of the acute angles is 68°. Solve the right triangle by making a sketch and labeling all the sides and angles.

6. **Angle $A = 180 - (90 + 68) = 22°$; $\sin(22°) = BC/18$; $BC \approx 6.74$ m; $\sin(68°) = AC/18$; $AC \approx 16.7$ m.**

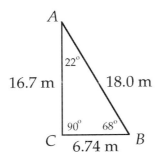

7. From a window 40 feet above ground level, the angle of elevation to the top of a nearby building is 47°. The angle of depression to the ground level of the building is 26°. Find the height of the nearby building (see **Figure 8.106**).

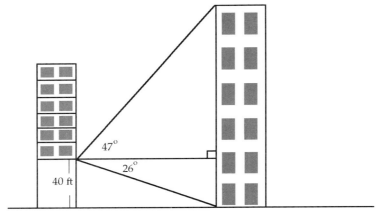

Figure 8.106

7. **Find the distance between the two buildings: $\tan(26°) = 40/x$; $x \approx 82$ ft.**

Find the height of the building above 40 ft: $\tan(47°) = y/82$; $y \approx 87.9$ ft.

Height of the building: 87.9 ft + 40 ft $\approx$ 127.9 feet.

8. To drill holes A, B, C, and D shown in **Figure 8.107,** the distances AD, DC, and DB must be determined. Find them to the nearest 0.01 inch.

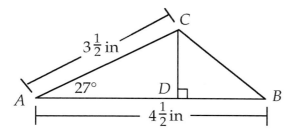

Figure 8.107

8. **To find DC: $\sin(27°) = DC/3.25$; $DC \approx 1.48$ in.**

   To find DB, first find AD: $\cos(27°) = AD/3.25$; $AD \approx 2.90$ in.

   $DB = 4.50 - 2.90 = 1.60$ in.

9. The Ferris wheel at Chicago's Navy Pier is modeled after the one that was built for the 1893 Chicago World Columbian Exposition. The attraction can accommodate up to 240 passengers and offers a fantastic view of Chicago's skyline and lakefront.

   Find the height of the Ferris wheel if the angle of depression from atop the ride to an object 40 feet from the bottom of the ride is 75.1°.

9. **$\tan(75.1°) = x/40$; $x = 150$ feet.**

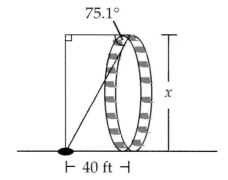

10. In **Figure 8.108,** $\triangle ABC \sim \triangle XYZ.$

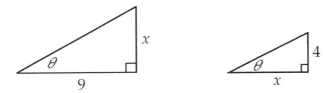

Figure 8.108

Find the six trigonometric ratios of θ. (Hint: First find the value of x.)

10. **Because the two triangles are similar,** $\dfrac{9}{x} = \dfrac{x}{4}; x^2 = 36; x = 6.$ **The hypotenuse**

of the larger triangle is $\sqrt{9^2 + 6^2} = \sqrt{117}$ **or** $3\sqrt{13}$. **The hypotenuse of the**

smaller triangle is $\sqrt{4^2 + 6^2} = \sqrt{52}$ **or** $2\sqrt{13}$. **So,**

$\sin\theta = \dfrac{6}{3\sqrt{13}} = \dfrac{2}{\sqrt{13}}; \cos\theta = \dfrac{9}{3\sqrt{13}} = \dfrac{3}{\sqrt{13}}; \tan\theta = \dfrac{6}{9} = \dfrac{2}{3}; \csc\theta = \dfrac{\sqrt{13}}{2}; \sec\theta = \dfrac{\sqrt{13}}{3};$

$\cot\theta = \dfrac{3}{2}$.

11. Carpenters often use a tool called a carpenter's square to mark off right angles (see **Figure 8.109**). The square (along with some knowledge of trigonometry) can also be used to construct other angles.

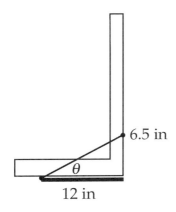

Figure 8.109

a) Suppose a carpenter draws a line along the short side of the square and places a mark at 12 inches. If a second mark is placed at 6.5 inches on the long arm and the two marks are connected, what angle θ is formed?

11. a) $\tan\theta = 6.5/12; \theta \approx 28.4°.$

b) Suppose a carpenter wants to construct a 50° angle. Explain how he or she might do this.

b) **Draw a line along the short side of the square and place a mark at 12 inches. Place a mark at 14.3 inches (tan(50°) = x/12; x = 14.3 in). Then connect the two marks.**

12. A machinist wants to drill three holes in a metal plate as shown in **Figure 8.110**. If the first hole is placed at the origin of a coordinate grid with 1 cm unit distances, the holes are 2 cm apart, and the line forms a 62.3° with the x-axis, find the coordinates of holes A and B.

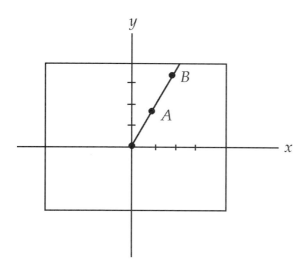

Figure 8.110

12. **Point A: x-coordinate: $\cos(62.3°) = \dfrac{x_A}{2}$; $x_A \approx 0.93$.**

   y-coordinate: $\sin(62.3°) = \dfrac{y_A}{2}$; $y_A \approx 1.8$. The coordinates of point A are

   (0.93, 1.8). Point B: x-coordinate: $\cos(62.3°) = \dfrac{x_B}{4}$; $x_B \approx 1.9$.

   y-coordinate: $\sin(62.3°) = \dfrac{y_B}{4}$; $y_B \approx 3.5$.
   The coordinates of point B are (1.9, 3.5).

13. According to *The Boston Globe* (June 2, 1997), Amtrak plans to install a hydraulic device in its trains of the future so that lateral g forces (the forces that cause a passenger to lean as the train rounds a curve at high speed) can be minimized. To do this, they plan to tilt the train car as much as 10 degrees. Tilting allows trains to pass through curves at a high rate of speed and adds to passenger comfort and safety.

If the hydraulic devices beneath the train car are 4 feet apart, how much longer does one device need to be than the other in order to tilt the car at an angle of 10° (see **Figure 8.111**)?

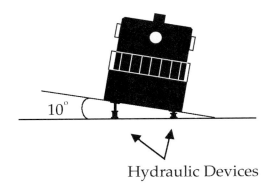

Hydraulic Devices

Figure 8.111

13. $\sin(10°) = h/4$; $h = 0.7$ feet, or about $8\frac{1}{2}$ inches.

14. Determine whether the sine of 275° is positive or negative.

14. A 275° angle is in Quadrant IV. Its sine is negative.

15. Determine whether the tangent of –187° is positive or negative.

15. A –187° angle is in Quadrant II. Its tangent is negative.

16. Solve for $\angle A$ and $\angle B$ in **Figure 8.112**.

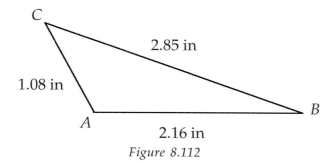

Figure 8.112

16. **Using the Law of Cosines:**

$\angle A: 2.85^2 = 1.08^2 + 2.16^2 - 2(1.08)(2.16)\cos A$

$$\cos A = \frac{2.85^2 - 1.08^2 - 2.16^2}{-2(1.08)(2.16)}; A = \cos^{-1}\left(\frac{2.85^2 - 1.08^2 - 2.16^2}{-2(1.08)(2.16)}\right) \approx 119°.$$

$\angle B: 1.08^2 = 2.85^2 + 2.16^2 - 2(2.85)(2.16)\cos B$

$$\cos B = \frac{1.08^2 - 2.85^2 - 2.16^2}{-2(2.85)(2.16)}; B = \cos^{-1}\left(\frac{1.08^2 - 2.85^2 - 2.16^2}{-2(2.85)(2.16)}\right) \approx 19°.$$

17. Solve for $\angle A$ and side a in **Figure 8.113**.

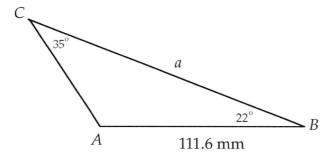

Figure 8.113

17. $\angle A = 180 - (35 + 22) = 123°.$

Using the Law of Sines: $\dfrac{111.6}{\sin(35°)} = \dfrac{a}{\sin(123°)}; a = \dfrac{111.6(\sin(123°))}{\sin(35°)} \approx 163 \text{ mm.}$

Handout 5.1

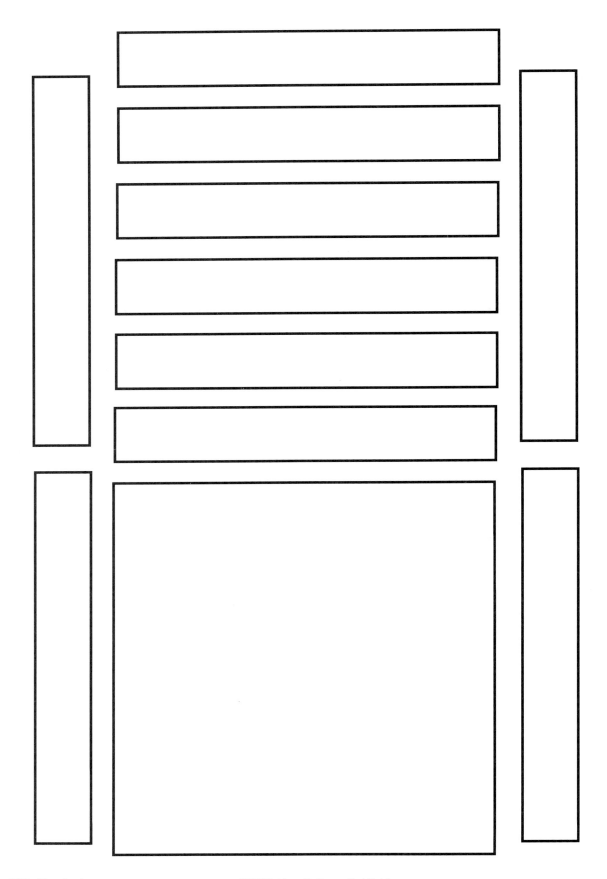

©2003 Key College Publishing

Residuals and Their Role

In many of the modeling situations in this text, you will be asked to make a scatter plot of a given set of data and use the regression features of your calculator to determine an equation that best models the data. After finding the regression equation of the line or curve, you must decide whether your model adequately describes the relationship shown by the scatter plot.

To see how well the line or curve fits the data, plot both the data and the graph of your model on the same set of axes. Does your model fall on each of the data points? If not, then you must examine how close the line or curve is to the data points.

To do this, statisticians usually study the vertical distance between each data point and the curve or line. See **Figure 1.**

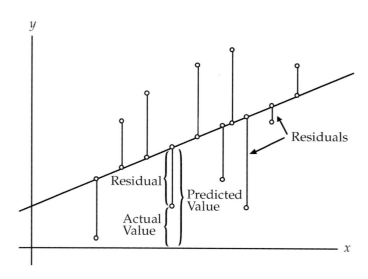

Figure 1

These differences between the *y*-values of the actual data and the *y*-values of the model are called the deviations, residual errors, or simply residuals.

Residuals are the errors between the actual and the predicted values: (error = actual – predicted).

Examination of the residuals helps you determine how well your model fits the data. If the model is a good fit, the residuals should be small relative to the data. If an extremely large residual occurs, further investigation of the data point is needed in order to find the cause.

Your residuals should be randomly distributed. If a pattern or trend is evident, it is possible that your model can be improved.

A scatter plot of the residuals versus the independent variable is known as a residual plot (see **Figure 2**). Valuable information about the adequacy of your model can be obtained from such graphs.

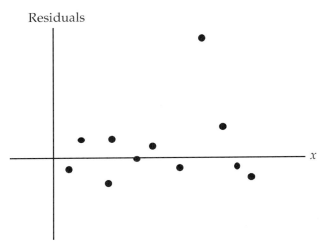

Plot showing an extremely large residual

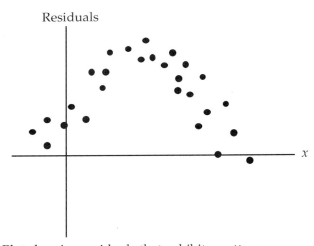

Plot showing residuals that exhibit a pattern

Figure 2

TI-83 Instructions for Finding the Least-Squares Line

Here's how to use the TI-83 to calculate the least-squares line and its residuals.

1. Enter the explanatory variable values into L1 and the response variable values into L2. Then make a scatter plot of the data in an appropriate window.

2. To calculate the equation of the least-squares line, first press STAT, CALC, LinReg($ax + b$).

To complete the command, you need to tell the calculator which list contains the data for the explanatory variable, which list contains the data for the response variable, and where you want the least-squares equation stored. Complete the entry to match the one shown in **Figure 3** and then press ENTER.

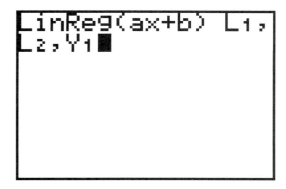

Figure 3

3. Press Y = . Has the regression equation been stored as Y1? Press GRAPH to view the scatter plot with a graph of the least-squares line.

4. In assessing whether a linear regression model is adequate to describe the relationship shown by a scatter plot, statisticians examine the errors, also called residuals. The errors or deviations of the regression model from the actual data can be calculated by subtracting the predicted y-values as determined from your model from the actual y-values as given by your data (error = actual – predicted).

The TI-83 computes the residuals automatically and stores them in a list named RESID whenever you use a built-in regression command.

To see the residuals displayed, place the cursor on the header of list L3, press INS, and select RESID from the LIST menu. Your screen should look like **Figure 4.**

```
L2          ▮▮▮▮▮  L3      12
   13                   ------
   21
   24
   16
   24
   20
   15
Name=RESID
```

Figure 4

Press ENTER again and the residuals will appear. A plot of the residuals versus the independent variable is called a residual plot. This plot gives you information about how well your model describes the data. The points in the residual plot should be randomly scattered and reasonably small.

Handout 7.1

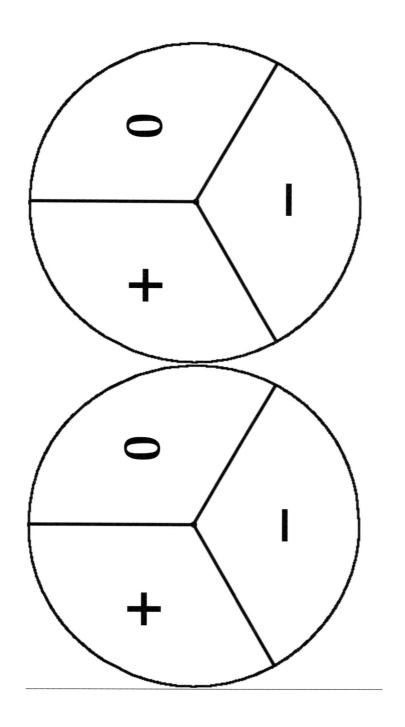

Handout 7.2

©2003 Key College Publishing

Handout 7.3

Student Questionnaire

Circle Yes or No for each question except number 1.

1. What is your present age? _____

2. Are you a full-time student? Yes   No

3. Do you live on campus?   Yes   No

4. Do you drive a car (or ride as a passenger in a car) to get to classes? Yes   No

5. Do you own a car? Yes   No

6. Do you own a home? Yes   No

7. Do you own a cellular phone? Yes   No

8. Do you work at least part time while taking classes? Yes   No

9. Do you work full time while taking classes? Yes   No

10. Is English your first language? Yes   No

11. Do you usually get more than seven hours of sleep at night? Yes   No

12. Do you play organized college sports (either intercolegiate or intermural)?
    Yes   No

13. Did you vote in the last presidential election? Yes   No

14. Do you regularly watch MTV? Yes   No

15. Do you usually eat a breakfast meal in the morning? Yes   No

16. During the past year, have you attended a live concert more than four times?
    Yes   No

17. During the past year, have you been involved in a motor vehicle accident?
    Yes   No

18. Do you drink coffee in the morning? Yes   No

19. If you were given the option of a paid vacation at a beach resort or at a resort
    in the mountains, would you choose the beach? Yes   No

20. When you listen to a song, do you pay more attention to the music than to
    the words? Yes   No

Histograms

There are many kinds of graphs that are used to display data. Some examples are line graphs, pie charts, bar graphs, and dot plots. For presentation of quantitative data in written reports or at formal meetings, a particular kind of bar graph called a **histogram** is often preferred.

Consider the following data on major earthquakes (see **Table 1**). For each of the hundred years from 1900 to 1999, the number of earthquakes of Richter magnitude 7.0 or greater has been recorded, and the yearly totals are:

13	14	8	10	17	26	32	27	18	32
36	34	32	33	32	18	26	21	21	14
8	11	14	23	18	17	19	20	22	19
13	26	13	14	22	24	21	22	26	21
23	24	27	41	31	27	35	26	28	36
39	21	17	22	17	19	15	34	10	15
22	18	15	20	15	22	19	16	30	27
29	23	20	16	21	21	25	16	18	15
18	14	10	15	8	15	6	11	8	7
13	10	23	16	15	25	22	20	16	11

Table 1 (Source: United States Geological Survey)

Of course, the mean value would sum up the data with a single number. But in order to present the data in a way that shows the entire data set in a more easily understood way, we can first group the data into **classes** containing a range of earthquake occurrences like "from 0 to 4 major earthquakes per year," "from 5 to 9 major earthquakes per year," and so on. Such a **grouped frequency table** for these data is shown as **Table 2.**

©2003 Key College Publishing

Number of Earthquakes per Year	Frequency
0–4	0
5–9	6
10–14	16
15–19	27
20–24	24
25–29	13
30–34	9
35–39	4
40–44	1

Table 2

The table by itself reveals some information that cannot be easily seen in the raw data list. For instance, a typical year has around 20 major earthquakes. But also we see that it is relatively rare for either fewer than 10 or more than 29 major quakes to occur in a single year.

A histogram of the grouped frequency data (see **Figure 1**) shows each class as a bar (usually, although not always, vertical). The width of the bar is equal to the (constant) class width (five in this case), and the length or height of the bar is measured on a scale that shows the frequency, or count, of the observations in each class.

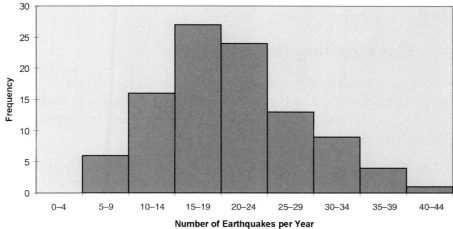

Figure 1

Although it would be extremely difficult to look at the original data for this situation and make any real sense of it, the histogram is a more comprehensible summary of the entire data set. Its one disadvantage is that sorting the data into groups results in a loss of the individual measurements. This is usually far outweighed (for most purposes) by the added clarity and ease of finding patterns in the data.

Much of the information contained in the original data set can be found by reading the histogram. For example:

- The total number of years of data involved can be determined by adding the total of all the bar heights.

- During most years, there have been from 10 to 59 major earthquakes.

- The number of years during which there were fewer than 15 major earthquakes can be found by adding the heights of the first two bars, giving a total of 23 years.

However, it is not possible just from the histogram to determine, say, during how many years there have been more than 32 earthquakes. This is because a single bar includes 30, 31, 32, 33, and 34 earthquakes per year, and it is *not* possible to distinguish the frequencies of each value. It *is* possible to determine that there were 14 years with *at least* 30 major earthquakes by totaling the heights of the last three bars (9 + 4 + 1 = 14 years).

It is also not possible to determine the smallest or the greatest number of major earthquakes that occurred in any single year just by reading the graph. The grouping has eliminated the individual data values.

Because the histogram has a quantitative scale on both axes, the actual area of each bar is proportional to the frequency of occurrence of the values represented by the bar. If the bars touch, as they do in Figure 1, and if the entire shaded portion of the histogram were cut out as a unit, it could physically balance at the mean of the distribution.

Constructing Histograms

The purpose of a histogram is usually to convey a maximum amount of information about the distribution of a set of measurements with a minimum of numerical detail. However, for this to occur, several guidelines for histogram construction should be followed.

- Most histograms should contain from 5 to 20 groups, depending on how large the data set is. Too few groups and any pattern in the data will be lost, whereas too many groups leads to too much random fluctuation in bar height, defeating the purpose of the graph. A rule of thumb that is often used sets the number of groups at approximately the value of the square root of the number of values in the data set.

- The width of each group (that is, the range of values contained in the group) should be the same as for all other groups. Otherwise, the pictorial information will be distorted.

Labeling of the data scale on the graph should make it clear where each bar of the histogram starts and stops. In the example on earthquake data (Figure 1), there is no question of the 30–34 group overlapping the 35–39 group. However, some people might label this histogram differently, as shown in **Figure 2,** in order to simplify the data scale and make it appear less cluttered.

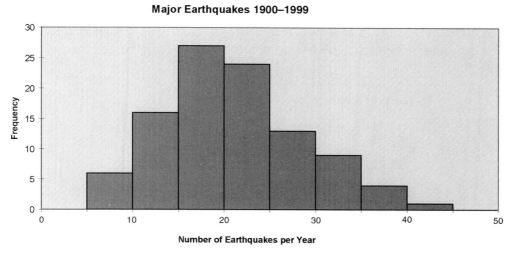

Major Earthquakes 1900–1999

Figure 2

In this case, it is not clear which group would contain a value of exactly 40, because it is at the boundary of two different groups: 40 appears to be the upper limit of one group and the lower limit of another. When histograms are labeled this way, it is a good idea to include a note of explanation, such as "Lower limits are included in each group." The labeling of histograms created with computer spreadsheets or calculators is often done automatically, and it may not be clear exactly how the labeling is related to the group data. Caution should be exercised when interpreting such graphs.

Relative Frequency Histograms

Data suitable for presentation in a histogram is often more easily interpreted in the form of a **relative frequency histogram.** Such a graph would differ from that already discussed only in the treatment of the frequencies on the vertical scale. Each of the group frequencies would be replaced by a relative frequency, or fraction of the total number of observations. Thus, the first nonzero class, containing years when 5–9 major earthquakes occurred, would have a relative frequency of 6/100 = 0.06 because there were 100 total yearly observations included. These numbers would usually be expressed as percentages. The percentage is just the numerator of the fraction that represents the ratio of actual frequencies to the total number of observations, provided that the denominator is written as 100. There were 5–9 major earthquakes recorded in 6% of the years between 1900 and 1999. The relative frequency histogram would look like **Figure 3.**

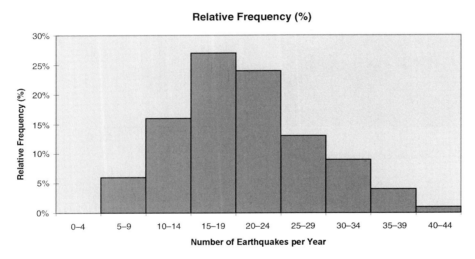

Figure 3

Handout 7.5

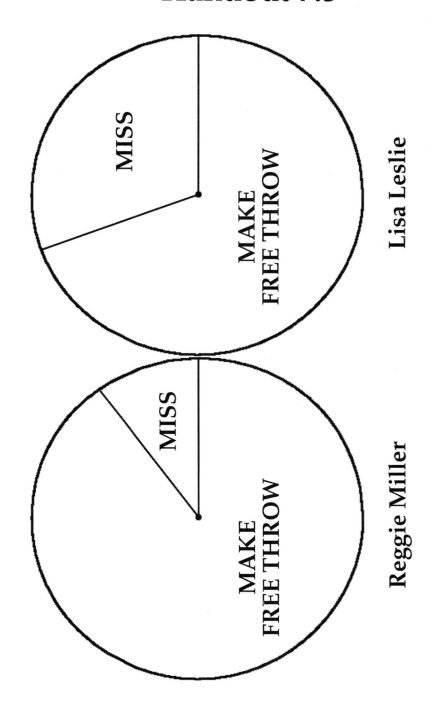

Lisa Leslie

Reggie Miller

Handout 8.1

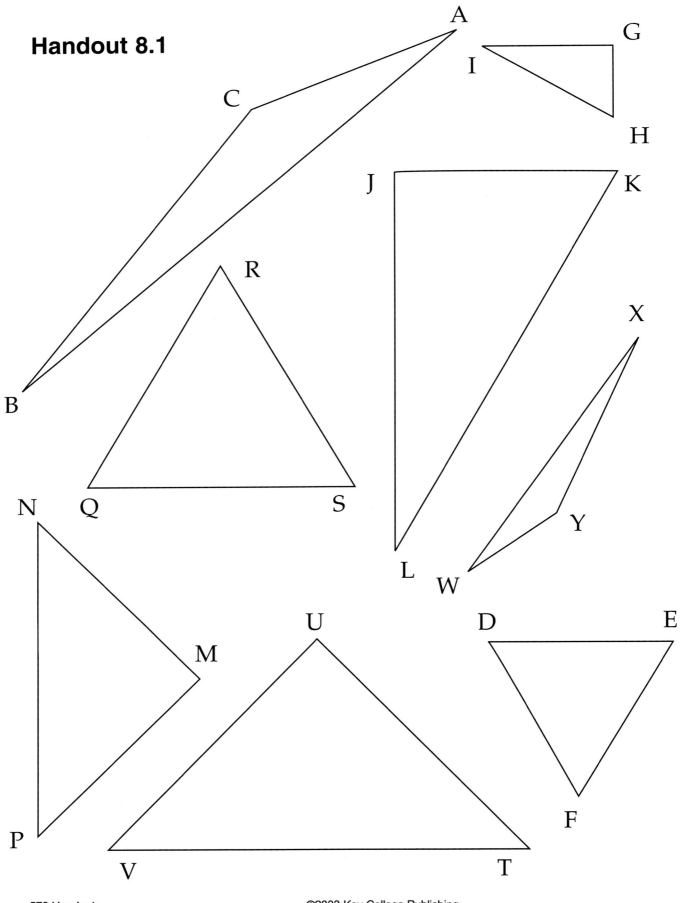

Handout 8.2

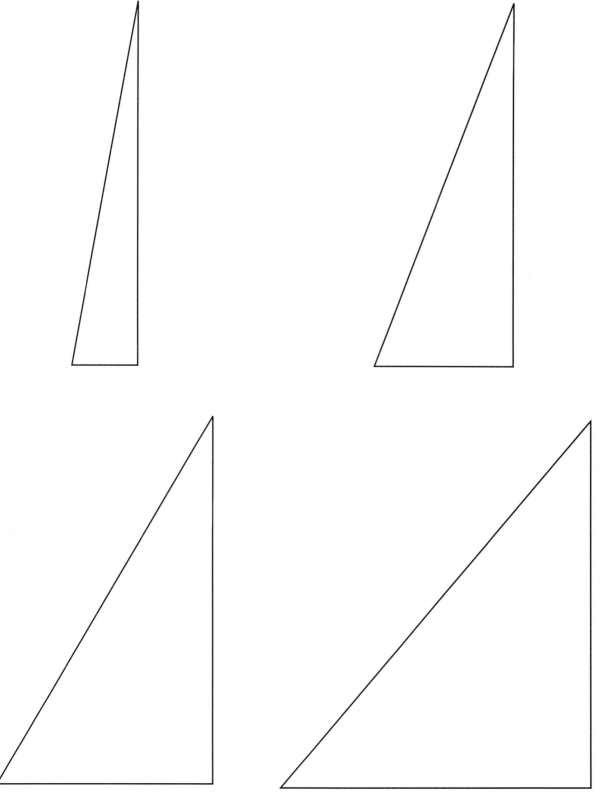

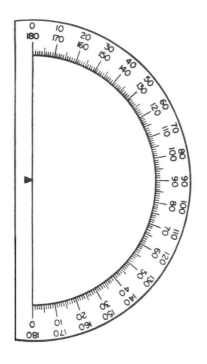

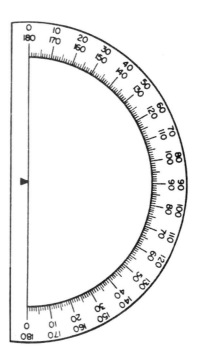

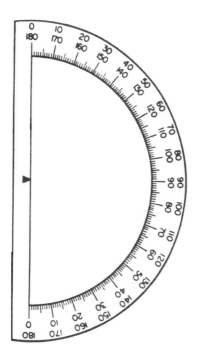

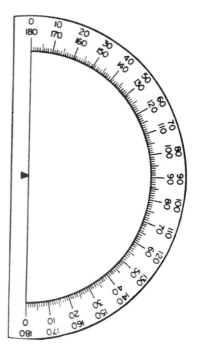

Appendix

Units and Conversions

Quantity	SI	U.S. Customary	Conversions
Length	meter (m)	foot (ft)	1 m = 3.281 ft = 39.37 in
		1 ft = 12 inches (in)	1 in = 2.54 cm
		1 yard (yd) = 3 ft	1 mi = 1.609 km
		1 mile (mi) = 5280 ft	1 km = 0.6214 mi
		1 in = 1000 mils	
Area	square meter (m^2)	square foot (ft^2)	1 hectare = 2.471 acres
	1 hectare = 10,000 m^2	1 acre = 43,560 ft^2	
Volume	cubic meter (m^3)	cubic foot (ft^3)	1 l = 0.03531 ft^3
	1 liter (l) = 1000 cm^3	1 ft^3 = 7.48 US gallons (gal)	
		1 gal = 231 in^3	
Weight	newton (N)	pound (lb)	1 lb = 4.448 N
		1 lb = 16 ounces (oz)	
		1 ton = 2000 lb	
Mass	kilogram (kg)	slug	1 kg = 2.205 lb*
	1 metric ton (tonne) = 1000 kg		1 oz = 454 g*
Time	second (s)	second (s)	

* Note: These are convenience conversions between mass units and weight units that are strictly valid only at sea level on the Earth.

Geometry Formulas

Two-Dimensional Figures

Figure	Area	Perimeter
Circle radius r, diameter d	$A = \pi r^2 = \pi d^2/4$	Circumference $C = 2\pi r = \pi d$
Rectangle length l, width w	$A = lw$	$P = 2l + 2w$
Square side s	$A = s^2$	$P = 4s$
Triangle base b, height h	$A = \dfrac{1}{2}bh$	
Parallelogram base b, height h	$A = bh$	
Trapezoid bases b_1, b_2, height h	$A = \dfrac{1}{2}(b_1 + b_2)h$	

Three-Dimensional Figures

Figure	Volume	Surface Area
Rectangular box length l, width w, height h	$V = lwh$	$SA = 2lw + 2lh + 2wh$
Cube side s	$V = s^3$	$SA = 6s^2$
Sphere radius r	$V = \dfrac{4}{3}\pi r^3$	$SA = 4\pi r^2$
Circular Cylinder radius r, diameter d, height h	$V = \pi r^2 h$	$SA = 2\pi rh = \pi dh$
Cone or Pyramid base area B, height h	$V = \dfrac{1}{3}Bh$	

Metric Prefixes

Prefix	Meaning
tera-	trillions
giga-	billions
mega-	millions
kilo-	thousands
hecto-	hundreds
deca-	tens
deci-	tenths
centi-	hundredths
milli-	thousandths
micro-	millionths
nano-	billionths
pico-	trillionths
femto-	quadrillionths

Index
to the Student Edition

A

acute angles, calculating, 347–348
acute triangles, 311
adding
      probabilities, 235, 237–239
      radicals (roots of numbers), 59–61
      rational expressions (fractions), 158–163
Addition Rule for Probabilities, 237–239
ambiguous case of Law of Sines, 369
angles
      angle of depression, 357
      angle of repose, 359
      complementary angles, 333, 342
      degrees (to measure angles), 311
      initial and terminal sides, 362
      Law of Cosines, 366–367
      Law of Sines, 368–369
      quadratic functions, 366
      radians (to measure angles), 333, 339
      secants, cosecants, cotangents, 337
      sines, cosines, tangents, 330–332, 335
      standard position, 363
annulus, 60
arithmetic mean, 63
asymptotes, 119–125
      horizontal, 123–125
      oblique, 129
      vertical, 119–122
axis of symmetry, parabolas, 6

B

Basic Multiplication Principle, 209–210

C

chance. *See* probability
circle, equation of, 92–96
combining radicals, 56–59
common denominators, 158–163
complementary angles, 333, 342
complementary events, 225–226
completing the square, 28
compound events, 205–211
conditional probabilities, 238
confidence intervals, 271–272
conjugate pairs, 148

continuous variables, 228
converse of Pythagorean theorem, 312
corresponding sides of similar triangles, 305
cosecant of an angle, 337
cosine of an angle, 330–332, 335
      Law of Cosines, 366–367
      obtuse angles, 362–364
cotangent of an angle, 337

D

data
      drawing conclusions from, 268–272
      finding probabilities from, 222–226
degrees (to measure angles), 311
dependent systems of equations, 186
dependent variables, 7
depression, angle of, 357
deterministic models, 211
discrete variables, 228
discriminant of a quadratic equation, 24
distance formula, 91–92
dividing rational expressions, 142–144
divisor methods, 149
drawing conclusions from data, 268–272
      confidence intervals, 271–272
      significance testing, 276
      standard deviation, 275

E

elimination method, 186
empirical probabilities. *See* experimental
      probabilities
equally likely events, 207
equation of a circle, 92–96
equations, systems of, 173, 177–178
      inconsistent, 185
      linear equations, 179–182
equilateral triangles, 310
error, margin of, 272
events, 202
      complementary, 225–226
      compound events, 205–211
      independent, 236
      mutually exclusive, 238
excluded values, 119
expected monetary values, 256
expected values of probability distributions,
      253–255
experimental probabilities, 200, 222–226
      complementary events, 225–226
      conducting probability experiments,
      202–203

drawing conclusions from data, 268–272

modeling through simulations, 280, 283–287

mutually exclusive events, 238

significance testing, 276

sum of probabilities, 225

exponents

fractional, 73–77

positive and negative, 72

properties of, 73

solving radical equations, 77–80

extraneous solutions, 80

F

factoring

quadratic equations, 14

rational expressions, 139–141

first differences, 23

fourth-degree models, 39

fractions

adding and subtracting, 158–163

fractional exponents, 73–77

frequency histogram, 251

functions, definition of, 7

G

geometric mean, 63

golden ratio, golden rectangle, 188

graphing

asymptotes, 119–125, 129

frequency histogram, 251

grouped frequency tables, 270

quadratic equations, solving, 18–20

quadratic functions, 11–13

residual plots, 36

tree diagrams, 212

grouped frequency tables, 270

H

horizontal asymptotes, 123–125

I

inconsistent systems of equations, 185

independent events, 236

independent variables, 7

index of radicals, 54, 60

initial side of an angle, 362

inverse variation, 127

isosceles triangles, 310

J

joint probabilities, 235–239

L

Law of Cosines, 366–367

Law of Sines, 368–369

like denominators, fractions with, 158–159

like radicals, 59

linear equations, systems of, 179–182, 185

linear models, 1–2

lowest terms, reducing rational expressions to, 139–141

M

margin of error, 272

mathematical models

definition of, 33

deterministic vs. probabilistic, 211

right triangles for, 345–350

simulations, 280, 283–287

multiplying

Basic Multiplication Principle, 209–210

Multiplication Rule for Probabilities, 236–237

probabilities, 236–237

Product Property of radicals, 56–59

rational expressions, 142–144

mutually exclusive events, 238

N

negative

exponents, 72

roots of numbers, 54–56

trigonometric ratios, 362–364

non-acute angles, trigonometric ratios of, 362–364

nonlinear behavior, modeling. *See* polynomial functions

O

oblique asymptotes, 129

oblique triangles, 366

obtuse angles, trigonometric ratios of, 362–364

obtuse triangles, 311

odds, definition of, 214. *See also* probability

outcomes, 202

P

p-value, 277
parabolas, 6
   discriminants, 24
   graphing, 11–13
   solving parabolic equations, 13–20
polygons, 302
   similar polygons, 313
polynomial functions, 3
   of degree two. *See* quadratic functions
   distance formula, 91–92
   equation of a circle, 92–96
   parabolas, 6, 11–20, 24
   power functions, 22
   quadratic equations. *See* quadratic
      equations, solving
   quotients of. *See* rational expressions
population (in statistics), 224
power functions, 22
principal roots of numbers, 54–56
probability, 200–204, 222–226
   combining (joint probabilities), 235–239
   complementary events, 225–226
   compound events, 205–211
   computing theoretical probabilities,
      208–211
   conditional probabilities, 238
   drawing conclusions from data, 268–272
   expected values, 253–255
   experimental. *See* experimental
      probabilities
   finding from data, 222–226
   modeling through simulations, 280,
      283–287
   mutually exclusive events, 238
   odds, 214
   probability distributions, 249–255
   significance testing, 276
   subjective probabilities, 200
   sum of probabilities, 225
   theoretical probabilities, 200, 203
   uniform probability distribution, 256
Product Property of radicals, 56–59
properties
   of exponents, 73
   of radicals (roots of numbers), 56–59
proportions (in statistics), 224
Pythagorean theorem, 299, 312
Pythagorean triples, 321

Q

quadrantal angles, 366
quadratic equations, solving, 13. *See also*
      quadratic functions
   discriminants, 24
   factoring, 14
   graphs for, 18–20
   quadratic formula for, 16–17
quadratic formula, 16–17
quadratic functions, 3, 7. *See also* quadratic
      equations, solving
   distance formula, 91–92
   equation of a circle, 92–96
   graphing, 11–13
   modeling data with, 33–36
   parabolas, 6, 11–13, 13–20, 24
   solving radical equations, 77–80
   standard form of, 10
quadratic regression, 34, 37–39
quartic functions, 39
Quotient Property of radicals, 56–59
quotients of polynomial functions. *See* rational
      expressions

R

radians (to measure angles), 333, 339
radical form, 54
radicals (roots of numbers), 54–56
   adding and subtracting, 59–61
   conjugate pairs, 148
   properties of, 56–59
   solving radical equations, 77–80
randInt command, 282
random variables, 249–255
rational expressions, 112, 119–125, 139–141
   adding and subtracting, 158–163
   multiplying and dividing, 142–144
   solving, 156–158
ratios, 112
reducing rational expressions to lowest terms,
      139–141
repose, angle of, 359
residual plots, 36
right triangles, 311–312. *See also* trigonometry
   30°-60°-90° triangles, 321
   45°-45°-90° triangles, 320
   modeling with, 345–350
   Pythagorean triples, 321
   secants, cosecants, cotangents, 337
   sines, cosines, tangents, 330–332, 335
   solving, 348–350
risk factors, 238

roots of numbers (radicals), 54–56
    adding and subtracting, 59–61
    conjugate pairs, 148
    properties of, 56–59
    solving radical equations, 77–80
roots of quadratic equations, 14

S

sampling, 268–272
    sample variability, 272
    standard deviation, 275
scalene triangles, 310
secant of an angle, 337
second differences, 23
semiperimeter, 62
sign of trigonometric ratios, 362–364
significance testing, 276
similar polygons, 313–314
    corresponding sides, 305
    similar triangles, 303–308
simplified rational expressions, 139–141
simplifying radicals, 56–59
simulations, 280, 283–287
sine of an angle, 330–332, 335
    Law of Sines, 368–369
    obtuse angles, 362–364
solving
    quadratic equations, 13–20, 24
    radical equations, 77–80
    rational equations, 156–158
    right triangles, 348–350
    systems of equations, 173, 177–178
    systems of linear equations, 179–182
standard deviation, 275
standard form
    equation of a circle, 94
    quadratic equation, 13
    quadratic functions, 10
standard position of an angle, 363
statistically significant differences, 276
statistics, 201. *See also* probability
    confidence intervals, 271–272
    margin of error, 272
    population and proportion, 224
    standard deviation, 275
subjective probabilities, 200
substitution method, 176–177
    systems of linear equations, 179–182
subtracting
    radicals (roots of numbers), 59–61
    rational expressions (fractions), 158–163
surface area, comparing to volume, 137–139
systems of equations, 173, 177–178

dependent, 186
inconsistent, 185
linear equations, 179–182

T

tangent of an angle, 330–332, 335
    obtuse angles, 362–364
terminal side of an angle, 362
theoretical probabilities, 200, 203
    computing, 208–211
    theoretical probabilities, 208–211
tick marks, 310
tree diagrams, 212
trials, 202–203
    experiments. *See* experimental probabilities
    sums of probabilities, 225
triangles, 302–303, 309–314. *See also*
    trigonometry
    30°-60°-90° triangles, 321
    45°-45°-90° triangles, 320
    Law of Cosines, 366–367
    Law of Sines, 368–369
    modeling with right triangles, 345–350
    oblique triangles, 366
    Pythagorean theorem, 299, 312
    Pythagorean triples, 321
    right triangles, 311–312, 345–350
    secants, cosecants, cotangents, 337
    similar triangles, 303–308
    sines, cosines, tangents, 330–332, 335
    solving right triangles, 348–350
    types of, 310–314
trigonometric ratios, 330–335
    obtuse angles, 362–364
trigonometry, 299, 328–335
    Law of Cosines, 366–367
    Law of Sines, 368–369
    secants, cosecants, cotangents, 337
    sines, cosines, tangents, 330–332, 335
    solving right triangles, 348–350

U

uniform distribution, 256
unlike denominators, fractions with, 160–163
unlike radicals, 59

V

variability of samples, 272
variables
    continuous and discrete, 228
    dependent and independent, 7
    random, 249–255
variation, inverse, 127
vertex of a parabola, 6
vertical asymptotes, 119–122
volume, comparing to surface area, 137–139

Z

Zero Product Property, 14
zeros of quadratic functions, 14